AQA GCSE (9-1) Design and Technology

All Material Categories and Systems

**Bryan Williams • Louise Attwood • Pauline Treuherz
Ian Fawcett • Dan Hughes • Dave Larby**

CONTRIBUTORS: Andrea Bennett • Stuart Douglas • Saul Taylor

Approval message from AQA

This textbook has been approved by AQA for use with our qualification. This means that we have checked that it broadly covers the specification and we are satisfied with the overall quality. Full details of our approval process can be found on our website.

We approve textbooks because we know how important it is for teachers and students to have the right resources to support their teaching and learning. However, the publisher is ultimately responsible for the editorial control and quality of this book.

Please note that when teaching the *AQA GCSE Design and Technology* course, you must refer to AQA's specification as your definitive source of information. While this book has been written to match the specification, it cannot provide complete coverage of every aspect of the course.

A wide range of other useful resources can be found on the relevant subject pages of our website: www.aqa.org.uk.

HODDER EDUCATION
AN HACHETTE UK COMPANY

Although every effort has been made to ensure that website addresses are correct at time of going to press, Hodder Education cannot be held responsible for the content of any website mentioned in this book. It is sometimes possible to find a relocated web page by typing in the address of the home page for a website in the URL window of your browser.

Hachette UK's policy is to use papers that are natural, renewable and recyclable products and made from wood grown in well-managed forests and other controlled sources. The logging and manufacturing processes are expected to conform to the environmental regulations of the country of origin.

Orders: please contact Hachette UK Distribution, Hely Hutchinson Centre, Milton Road, Didcot, Oxfordshire, OX11 7HH. Telephone: +44 (0)1235 827827. Email education@hachette.co.uk Lines are open from 9 a.m. to 5 p.m., Monday to Friday. You can also order through our website: www.hoddereducation.co.uk

ISBN: 9781510401082

First published in 2017 by
Hodder Education,
An Hachette UK Company
Carmelite House
50 Victoria Embankment
London EC4Y 0DZ

www.hoddereducation.co.uk

Impression number   10 9

Year   2024

Cover photo © Joerg Metzer \ Restless photography \ Alamy Stock Photo

Illustrations by Integra

Typeset in India

Printed and bound by CPI Group (UK) Ltd, Croydon, CR0 4YY

A catalogue record for this title is available from the British Library.

MIX
Paper | Supporting
responsible forestry
FSC™ C104740

CONTENTS

4 Non-exam assessment

5 The written paper

Introduction to AQA GCSE Design and Technology

This book has been written to help you fully understand the wide range of technical content and approaches to designing that you need for GCSE Design and Technology.

The AQA GCSE requires you to develop skills and understanding in both designing and making.

When designing, you will need to understand what the user or client requires and then provide them with a functional prototype which can be tested, so that you and your user or client can judge how successful it is.

Making involves working with a range of materials, some of which may be new to you, but others you will have already used before starting the course. Only by gaining experience working with materials first-hand can you start to understand their properties, and the ways in which those materials can be used for a variety of purposes.

The course is divided into three main parts:
1   Core technical principles
2   Specialist technical principles
3   Designing and making principles.

All of these parts will be assessed in the final examination, which is a two-hour written paper. This is worth 50 per cent of your final marks.

The other 50 per cent is for the non-examined assessment (NEA) project you will complete in the final year of your course.

The written paper is covered in Section 4 of this book and the NEA project in Section 5.

How to use this book

Section 1 Core technical principles

This section covers all the basic information you need across a wide range of materials and technologies. You need to understand all of this content, even if you are working in a single material area in the specialist technical principles section which follows. However, you will not be expected to have an in-depth knowledge of particular materials as this is the purpose of Section 2.

The content in this section provides a sound preparation for dealing with technical issues and understanding which materials should be selected when designing products or systems.

There are a number of exercises and activities throughout this section; these will help ensure that you fully understand what you have read.

Section 2 Specialist technical principles

In this section greater depth in your technical understanding is expected. However, you are not expected to know everything about everything. Instead you will gain your understanding by learning about a restricted range of materials selected from a choice of timber, metals, polymers, textiles and paper/board. You may learn about some aspects of the technical content using a range of materials, but with other aspects only use a single material. If you are working with mechanical and electronic systems you will still need to study at least one other material area as systems components are constructed from other materials.

Section 3 Designing and making principles

You will have two opportunities to demonstrate your understanding of this section: in the examination paper, and through your non-examined assessment (NEA) project. The NEA project, which is carried out in the last year of the course, will be based on a selected theme set by the examination board. You will be expected to produce a prototype and a design portfolio. Section 3 is intended to provide you with the design ability to select tools and equipment for manufacture, which you will need to produce a successful outcome for your project. You will also need to be able to discuss issues that have environmental, social or economic implications, or describe the work of established designers or companies with a reputation for good design.

Sections 4 and 5: preparing for assessment

These sections of the textbook deal with the NEA project and the written examination in more detail. They provide a useful guide and helpful advice on how to set about preparing and completing your project, and revising and answering the written paper in an organised and efficient manner.

Finally, the book includes a glossary of words or phrases and abbreviations that you are not familiar with. When you find a term with which you are unfamiliar, use the glossary to help you learn and improve your understanding.

Features to help you

A range of different features appear throughout the book to help you learn and improve your knowledge and understanding of designing and making.

What will I learn?
Clear learning objectives for each topic explain what you need to know and understand.

Science links
Where this icon appears, the content applies science skills and knowledge to design and technology.

Key words
All important terms are defined.

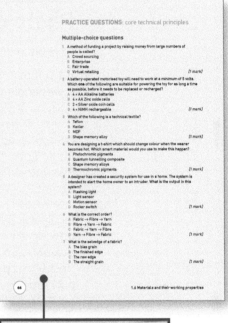

Key points
Summaries of key points appear at the end of each topic to help you remember the most important aspects of a topic, and to help you with revision.

Stretch and challenge
Stretch and challenge activities will help you to develop your understanding further. They may ask you to complete further research on a topic, or to consider more complex or challenging topics.

Practice questions
These questions appear at the end of each section and are designed to help you prepare for the written exam.

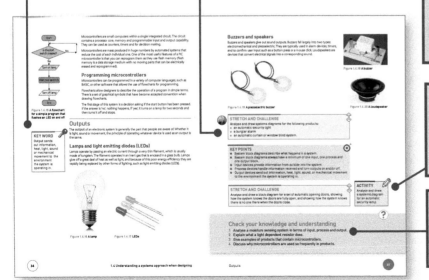

Activity
Short activities are included to help you to understand what you have read. Your teacher may ask you to complete these.

Check your knowledge and understanding
These short questions test your knowledge and understanding of each topic.

Acknowledgements

The authors and publishers would like to thank the following schools and students for the examples of their student work:

Abingdon School in Abingdon; Ripley St Thomas CE Academy in Lancaster; West Island School in Hong Kong; Oasis Academy, St Joseph's; Bedford Modern School in Bedford; Highgate School in London; Nonsuch High School for Girls in Surrey and Cameron Farquar at St George's School, Harpenden.

AQA material is reproduced by permission of AQA.

Photo credits

All photos not listed below are copyright of the authors.

p.1 © Tomislav – Fotolia; **p.2** © Zoe – Fotolia; **p.3** top © A.P.S. (UK) / Alamy Stock Photo, middle © Westend61 GmbH / Alamy Stock Photo, bottom © imageBROKER / Alamy Stock Photo; **p.4** © Simon Rawles / Alamy Stock Photo; **p.5** top © Julie g Woodhouse / Alamy Stock Photo, bottom © Alex Hinds / Alamy Stock Photo; **p.6** top © imageBROKER / Alamy Stock Photo, bottom © artpartner-images.com / Alamy Stock Photo; **p.7** left © Trinity Mirror / Mirrorpix / Alamy Stock Photo, right © Kumar Sriskandan / Alamy Stock Photo; **p.8** top © Alex Segre / Alamy Stock Photo, bottom © STANCA SANDA / Alamy Stock Photo; **p.9** left © Dario Sabljak / Alamy Stock Vector, right © Art Directors & TRIP / Alamy Stock Photo; **p.10** top © Cultura Creative (RF) / Alamy Stock Photo, bottom © Jeff Gilbert / Alamy Stock Photo; **p.12** © marina zagulyaeva culligan / Alamy Stock Photo; **p.13** © martin33 – Fotolia; **p.15** © Anna Vaczi/123RF; **p.16** top left © Alonso Aguilar Als/123RF, top right © Nigel Hicks/Alamy Stock Photo, bottom © Saskia Massink – Fotolia **p.17** © Vladislav Gajic - Fotolia.com ; **p.18** © Doug Houghton/Alamy Stock Photo; **p.20** top © Stuart Aylmer/Alamy Stock Photo, bottom © Donatas1205/123 RF; **p.21** top © Bryan Fisher - Fotolia.com, bottom © Scanrail/123 RF, ©Ensuper/Shutterstock; **p.23** top © ANUCHA RUENIN/123RF, bottom © luchschen/123RF; **p.24** top © The Lighthouse / Universal Images Group/REX/Shutterstock, middle © leonello calvetti /123RF, bottom © pooja sarda/Shutterstock.com; **p.25** top © Serg_v /12RF, middle © Pablo Hidalgo/123RF, bottom © Tim Gainey / Alamy Stock Photo; **p.26** top © Will Thomass/Shutterstock.com, bottom © MARTYN F. CHILLMAID/SCIENCE PHOTO LIBRARY; **p. 27** top left © Michael Grant Travel / Alamy Stock Photo, top right © RKTPHOTO:Rachel K. Turner / Alamy Stock Photo, middle © Robert Wilkinson / Alamy Stock Photo, bottom © Miguel Aguirre Sánchez / Alamy Stock Photo; **p.28** © Nejron Photo/Shutterstock.com, bottom © diter – Fotolia; **p.29** © Carolyn Jenkins / Alamy Stock Photo; **p.31** top © Arthur S. Aubry/Photodisc/Getty Images/ Science, Technology & Medicine 2 54**; p.32** top © Awe Inspiring Images – Fotolia, bottom © MARTYN F. CHILLMAID/SCIENCE PHOTO LIBRARY; **p.33** top left © Jirawat Jerdjamrat / Alamy Stock Photo, top right © David J. Green - electrical / Alamy Stock Photo, middle left © Dmitrij Skorobogatov/Shutterstock.com, middle right © Ultra Secure Direct (www.ultrasecuredirect.com) (Reference: 004-0310), bottom © jayfish / Alamy Stock Photo; **p.34** left © Roman Samokhin – Fotolia, right © Schneider Sebastian / Alamy Stock Photo; **p.35** left © NNL_STUDIO/Shutterstock.com, top right © ANDREW LAMBERT PHOTOGRAPHY/SCIENCE PHOTO LIBRARY, bottom right © David J. Green - electrical / Alamy Stock Photo; **p.36** top © Rob Bouwman - Fotolia.com (1.5.0), left © Hero Images Inc. / Alamy Stock Photo, right © Stephane106 / Alamy Stock Photo; **p.37** top © Finnbarr Webster / Alamy Stock Photo, bottom © LJSphotography / Alamy Stock Photo; **p.38** left © Hugh Threlfall / Alamy Stock Photo, right © OJO Images Ltd / Alamy Stock Photo; **p.39** top © Ted Foxx / Alamy Stock Photo, bottom © Julia Hofer/123RF; **p.42** © Stephen Dorey /

Photo credits

1 Core technical principles

This section of the book will help you understand that designers need to use up-to-date information and be aware of developments that are happening around the world.

New materials and technologies are being developed all the time, and sometimes designers find new uses for materials that were originally intended for something else. Designers often need to deal with complex technical systems, and using systems thinking can help us to simplify our approach. This section also looks at mechanical devices that have moving parts and a range of materials, some of which you will use when designing and making.

This section includes the following topics:

1.1   New and emerging technologies

1.2   Energy generation and storage

1.3   Developments in new materials

1.4   Systems approach to designing

1.5   Mechanical devices

1.6   Materials and their working properties

At the end of this section you will find practice questions relating to core technical principles.

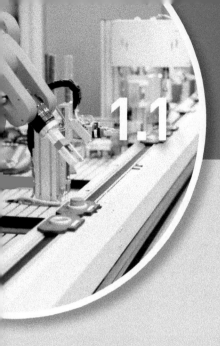

New and emerging technologies

What will I learn?

In this topic you will learn about:

→ the impact of new and emerging technologies on the design and organisation of the workplace, buildings and the place of work, tools and equipment
→ enterprise based on the development of an effective business innovation
→ the impact of resource consumption on the planet
→ how technology push/market pull affects choice
→ changing job roles due to the emergence of new ways of working driven by technological change
→ changes in fashion and trends in relation to new and emergent technologies
→ respecting people of different faiths and beliefs
→ how products are designed and made to avoid having a negative impact on others
→ the positive and negative impacts new products can have on the environment
→ the contemporary and potential future use of production techniques and systems
→ how the critical evaluation of new and emerging technologies informs design decisions.

Technology is transforming the way we live in many ways, from the way we communicate with one another, to the way we manufacture and use the objects we produce. As new technologies continue to emerge we need to consider how this will affect our world.

Technologies such as self-driving cars, robots that schedule our meetings, solar panel phone screens and 3D printers that can produce all sorts of parts are all being developed, and will impact upon people's lives in a variety of ways.

Industry

Industry and the way we manufacture is constantly changing and evolving. The Industrial Revolution in the late 1700s was due to the development of the emerging technology of the time: steam power. It brought significant changes to the way things were made and saw momentous and influential changes and innovation in manufacturing equipment. Prior to the Industrial Revolution, products were generally made in small workshops or at home, but emerging technology enabled products to be made faster and more economically in factories.

A factory usually consists of buildings housing machinery where workers make goods or operate machines turning materials into products. The first factories only contained a small number of machines and workers. Modern factories tend to be located with access to transportation; they are located close to good road, rail or seaport links in order to get the raw materials in and then transport the goods produced out. The buildings are usually large warehouses, containing manufacturing machinery used for assembly-line production. In some industries the factories are split into areas known as 'shops'. In car manufacturing, for example, the 'shops' generally include:

- a pressing shop
- an axle shop
- a body shop
- a paint shop
- a plastics shop
- a casting shop.

The tools and equipment needed in the workplace varies depending upon what is being manufactured. Manufacturers that mass produce items such as drinks cans will use dedicated automated machinery that produces only those items. Certain industries that make products in batches require greater flexibility. To enable them to vary batches they will use more adaptable equipment. Press formers are adaptable because the die that does the cutting can be changed to press different products, such as wristwatch cases, stainless steel pans for cooking, knives and spoons, door handles and coins. Machines that can be controlled by computers, computer numerically controlled (CNC) machines, among others can offer this capability. CNC machines include lathes, milling machines, plasma cutters and water jet cutters. Robot arms can also be programmed to carry out different tasks.

Figure 1.1.1 The Nissan car plant factory at Washington near Sunderland, UK

Automation

Repetitive tasks in mechanised assembly lines are now frequently carried out by robots, relieving human operators of these monotonous and often tedious tasks. This has been enabled by the development of computers and processors to control the robots and mechanised areas of factories.

Automated systems produce products of a consistent high quality at a low product cost because of the numbers of products they can produce quickly. However, they are very expensive to set up, they require a specialist workforce and if the system breaks down it can be extremely costly. The modern automated manufacturing workplace is designed and organised to make sure that people get the products they want, in the correct numbers and when they want them.

Figure 1.1.2 An automated production line producing newspapers

The use of robotics

Robot arms are now extensively used in many industries. They can do many jobs on the production line by being provided with different tools. The tasks that they can perform include welding, spray painting, assembly tasks, pick-and-place tasks, packaging and labelling, as well as product inspection. They perform these tasks with high speed and precision that make the initial costs worthwhile for many industries.

Figure 1.1.3 Industrial robots putting together car bodies on an assembly line

STRETCH AND CHALLENGE

Explain the benefits and drawbacks of automated systems in factories.

ACTIVITY

Make a list of the advantages and disadvantages of robots carrying out repetitive tasks in factories.

Enterprise

Enterprise is a skill where individuals or organisations are prepared to show initiative and take risks in order to make things happen. Businesses and entrepreneurs look for business opportunities and gaps in markets in order to undertake new ventures. **Innovation** involves bringing these new ideas to the market, inventing and developing ideas into products. Entrepreneurs look to come up with new ideas and approaches by thinking creatively. New ideas often result in new products or processes, or improving existing products. Innovation is often made possible by new and emerging technologies that allow products or processes to be developed in a completely new or improved way.

Crowdfunding is a method of funding a project by raising money from large numbers of people. Instead of getting institutions or a few very wealthy individuals to contribute large amounts of money, it turns this around by using the internet to get lots of people to contribute small amounts of money. The project that requires finance is described on a website that is run by a crowdfunding organisation which uses different communication methods including social media to raise the money required. There are different types of crowdfunding: donations – where people give money because they believe in the cause, equity – where money buys shares in the business, and debt – where the money is loaned and then paid back with interest added. This method of funding is made possible by the development of the internet and social media.

Looking to the future, **virtual marketing and retailing** are areas where enterprising businesses are showing interest. Virtual marketing uses marketing techniques in which social networks and websites are used to increase brand awareness. It does this by getting websites or users to pass on marketing messages to other websites and users; this hugely increases the messages' visibility and effect. The retail sector is facing the challenge of adapting to trends such as showrooming, where shoppers visit shops to examine a product before buying it online at a lower price, which means consumers go elsewhere to make a purchase. Retailers are therefore looking to give shoppers a whole retail world to explore in virtual reality by using developing technologies. Recently, Tesco used the walls of a subway station in Seoul in Korea to display more than 500 of its most popular products with barcodes which customers could scan using an app on their smartphone; the products were then delivered directly to customers' homes.

ACTIVITY
Make a list of products you think were innovative when they were first sold.

KEY WORDS
Co-operative a business owned, governed and self-managed by its workers.

Fair trade a movement that aims to achieve fair and better trading conditions and opportunities that promote sustainability for developing countries.

Figure 1.1.4 Workers at a Fair trade banana plantation wash and package banana bunches.

A **co-operative** is a business owned and self-managed by its workers. Worker-owners work in the business, govern it and manage it; they set production schedules and determine working conditions. Generally, worker co-operatives have higher productivity than conventional companies although the difference is usually quite small. In most worker co-operatives, all shares are held by the workforce with no outside or consumer owners. Tech co-operatives have grown out of new technologies. These worker-owned co-operatives provide technical support and consulting to other companies with communications and computer technology goods and services.

Fair trade is a movement that was created to help producers in developing countries (relatively poor countries with low levels of industrial production). Its aim is to achieve fair and better trading conditions and opportunities that promote sustainability. It supports the payment of higher prices to exporters in developing countries to improve social and environmental standards. It focuses primarily on products which are typically exported from developing countries to developed countries. The use of social media and its accessibility has enabled organisations to connect with people directly, and has encouraged like-minded people to support the Fair trade movement.

STRETCH AND CHALLENGE
Explain the benefits and drawbacks of worker co-operatives.

1.1 New and emerging technologies

Sustainability

We are using the Earth's resources all the time when we manufacture products. We are manufacturing more and more products using resources such as oil, metal ores and timber at an increasingly high rate. Collecting and processing raw materials, converting them into products and then using them consumes huge amounts of energy. Using these resources has an impact on the planet, which we need to minimise for future generations. **Sustainability** is about meeting our own present-day needs without compromising the needs of future generations. New technologies can be used to help us manufacture products more sustainably.

A **finite resource** is a resource that does not renew itself quickly enough to meet the needs of future generations. Examples are coal, oil and natural gas. Plants and organic materials, with the aid of heat and pressure over millions of years, become coal, oil or natural gas. Minerals and metal ores, for example cassiterite (tin) and chromite (chromium), are also considered finite resources, as once they are completely exhausted there is no natural way to renew them. Hydraulic fracturing, or fracking, is a new technology designed to extract gas and oil from shale rock, but is proving to be a controversial method of extraction because of concerns regarding environmental impact.

Non-finite resources are resources that can replenish quickly enough to meet our needs. Examples include water and plant life, such as trees. There are also common renewable energy resources that we are now using, such as solar, geothermal and wind power. Technological advancements have allowed us to use these renewable resources more effectively and to generate more energy from them, reducing our reliance on finite resources.

Designers need to think about the life cycle of a product. They need to consider the environmental impact of the product from the raw materials required, how long the product will last before it wears out, and its inevitable disposal at the end of its life. There are different ways designers can do this, including:

- using low-impact materials
- conserving resources by using recycled materials
- reducing material usage
- designing products that use less or no energy when the product is in use
- ensuring a prolonged lifetime
- making sure materials and components can be easily recycled and recovered at the end of their lifetime.

These factors all contribute to reducing the **ecological footprint** of the product.

The effect that a company or organisation has on people and communities is often referred to as its **social footprint**. Companies have a responsibility to consider human rights and the working conditions of their workforce. Companies with a good social footprint take care of their workers, in terms of health and safety, workforce equality, child labour, and wider social issues that affect communities in their supply chains.

Disposal of waste

At the end of a product's useful life its disposal has to be taken into consideration. There are a number of ways waste can be dealt with.

Landfill has been the most common method of organised waste disposal, and continues to be so in many places across the world. It simply involves putting waste into the ground

Figure 1.1.5 An open cast coal mine. Coal that has been formed over millions of years is a finite resource on a human timescale.

Figure 1.1.6 Solar panels fitted to the tiled roof of a UK home are an example of a non-finite energy source.

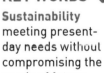

KEY WORDS

Sustainability meeting present-day needs without compromising the needs of future generations.

Ecological footprint the impact of a person or community on the environment; the amount of land needed to supply the natural resources they use.

Social footprint the impact a company or organisation has on people and communities.

Figure 1.1.7 Landfill site in Bogotá, Colombia, South America

and burying it. This poses many problems as it can cause pollution of the local environment, such as contamination of groundwater or soil. Decaying organic waste also generates methane gas – a greenhouse gas which is flammable and can be explosive. Many other problems occur, including the smell coming from sites, effects on local wildlife and noise pollution, which can lead to reduced property prices.

Alternatives to landfill are being developed across the globe.

Resource recovery is a process that recovers certain disposed materials for specific use. Some materials can be recycled, some composted and some can even be used in energy generation. Waste recycling reduces the use of newly created materials, while composting and energy generation reduces landfill and gives further use to materials. Energy generation is often referred to as **energy recovery**.

Incineration is a waste treatment process that involves the burning of waste materials. It can reduce the mass of waste by up to 85 per cent, so significantly reducing the volume of landfill. However, there are concerns about pollutants in the gas emissions from incinerators, including large amounts of CO_2 (carbon dioxide – the main greenhouse gas) being discharged into the atmosphere.

ACTIVITY

Make a list of finite and non-finite resources.

STRETCH AND CHALLENGE

Write an article for a design magazine that explains why designers should consider what happens to the products they design at the end of the product's useful life.

KEY WORDS

Technology push where new technology or materials are developed and designers take the opportunity presented by this to design new products.

Market pull where users want a product to be improved or redeveloped to meet their needs.

People

Designers create new products because of customer need or because of developments in technology. When designing these new products, consideration must be taken into how they work, as well as any social, moral, cultural, environmental and market factors.

Technology push is where new technology or materials are developed and designers take this a an opportunity to design new products. The iPad developed by Apple is an example of a product developed through technology push. The iPad was the first tablet computer to reach the market. What is often noticeable about products developed as a result of technology push is that consumers didn't know they wanted a product until after it has been launched.

Market pull is where users want a product to be improved or redeveloped to meet their needs. Market research is carried out and analysed in detail to identify what needs exist and how existing products can be improved or reinvented to meet those needs. The car industry has seen the redevelopment and relaunch of some iconic models from the 1950s and 1960s, such as the Volkswagen Beetle, the BMC's Mini (redeveloped by BMW) and the Fiat 500. All these models have been developed to meet the desires of the modern car user.

Figure 1.1.8 New Volkswagen Beetle at the 2011 International Motor Show in Frankfurt, Germany

Changing job roles

As technology develops, it impacts on the workplace and the jobs within it. In certain industries the jobs remain the same but new technologies are used to perform the tasks. Designers now use CAD software to perform tasks that would have required pencil and paper before; office workers use computers rather than typewriters; more bank clerks use telephones and computers with telephone and internet banking rather than sitting behind a counter in a bank.

Information technology and the use of computers in the workplace has caused some jobs to disappear and some to change. Robots controlled by computers are now common on production lines replacing factory workers; but these robots need to be designed, manufactured, programmed, and maintained, which creates new jobs. Large warehouses now only need a handful of staff to control stock due to computers, but as a result of online ordering of goods, we have seen a growth in logistics – the selection, postage and delivery of items to customers. As some jobs are lost, new ones are created, such as programmers, system analysts and computer technicians.

Culture

Culture is the values, beliefs, customs and behaviours used by groups and societies to interact with each other and the world. We are born into or raised in some of these groups (such as ethnic groups) while others we may choose to join (such as political and religious groups). Some people adjust their beliefs as they grow and learn. It is important to learn and respect the views and beliefs of your own culture as well as that of others. Accepting and learning about the views, beliefs, experiences, arts and crafts of different cultures can enrich our lives and can be of great benefit to designers.

Figure 1.1.9 People buying food at a Chinese street food stall: an example of multicultural Britain.

Fashions, trends and the challenge to keep up with the latest technology have a say in the changing nature of many products.

The airbag for cyclists is an example of new technology and fashion combining. It is a neck-worn system that incorporates an air bag that fully inflates to absorb the shock of any impact. Fashion drives styles, colour and materials. Fashion styles can be dependent on particular groups of people. This is often seen when musicians or celebrities in popular culture wear a specific style of clothing. As celebrities change their tastes or new celebrities become popular, the style and fashion changes with them.

Figure 1.1.10 Fashion for winter 1969–70 from Paris

Fashion can also change with world events. During the Second World War people were only allowed a certain amount of fabric to make clothes. As a result, fashion had to change in order to fit the limited resources of the time. Fashion is also affected by different cultures and subcultures. Punk is a subculture that includes music and fashion worn by bands such as the Sex Pistols in the late 1970s; it grew out of a rebellious attitude towards society which was reflected in the clothing.

ACTIVITY
List as many subcultures as you can that have created fashion trends.

STRETCH AND CHALLENGE
Write an essay that describes how a multicultural Britain has had an impact on the design and development of products.

Figure 1.1.11 **A punk**

Society

Designers have a responsibility to design products that meet the needs of everyone in society. Different groups within society also have different needs, which designers must take into account. Think about how you might open a jar with only one hand or with only partial use of your hands; or how you might get into the bath or step into a shower without full movement of your knee joint. It is extremely important that designers take into account the needs of all groups of people, including those with a disability. For example, motor neurone disease is a degenerative disease that leads to muscle weakness and wasting. Eye movement is more resistant to degenerative disease, and technology has been developed that enables people with motor neurone disease to control computers with just their eyes.

ACTIVITY
List what settings could be changed on a tablet to help the elderly.

Figure 1.1.12 **Older people find tablets easier to use than smartphones.**

Figure 1.1.14 **Islamic traditions do not approve of images of people: geometric patterns are favoured.**

As we get older we all go through some fundamental changes. Ageing makes many things more difficult. As we age our vision and hearing deteriorate and our motor skills decline. Young people use their smartphones frequently, but watch a 75 year old use a mobile phone and you will notice that changes in vision and motor control can make small screens difficult for them. Older people aren't afraid to try new technology if they can see its usefulness; in fact, older people are one of the largest groups to use tablets. This is because the main difference between a tablet and a phone is screen size, and therefore the elderly find it easier to use.

It is also important that designers understand different cultures so that the products they design do not cause offence. For instance, colour can have different meanings in different cultures – in China red symbolises good fortune while in South Africa it is the colour of mourning. In many Islamic traditions the use of images of people is not approved of and geometric patterns are favoured.

STRETCH AND CHALLENGE
Use notes and sketches to show how you think this device works.

Figure 1.1.13 **An elderly person using a device to help open a jar**

Environment

Designers need to understand the impact the products they design have on the environment. As the Earth's resources are consumed (such as oil, metal ores and timber), the process of converting these raw materials into products uses huge amount of energy, which creates pollution and carbon dioxide, affecting the environment.

The continuous improvement of a product by designers leads to the product becoming more efficient and having a better level of performance. If the product is electrically powered, this increased efficiency and performance level will lead to a reduced environmental impact caused by the product when it is in use, because of the reduced amount of energy required in operation.

Improved efficiency and performance in the workplace are also beneficial to the environment and can be brought about by staff training, and creation and investment in new technology. An example of how continuous improvement in the workplace can be encouraged is by getting production line workers to suggest improvements to the product and the way it is made.

Manufacturing products uses energy, of which a huge proportion is produced by burning fossil fuels. Burning fossil fuels is a major cause of pollution (with pollutants such as smoke, sulphur dioxide, carbon monoxide, and carbon dioxide). Smoke causes soot deposits and can cause breathing difficulties. Carbon monoxide is a poisonous gas and carbon dioxide is a greenhouse gas that contributes to global warming.

Global warming is the gradual increase in the average temperature of the Earth's atmosphere and oceans that is affecting the Earth's climate. The Earth's atmosphere behaves like a greenhouse – the Sun's rays pass through it and warm the Earth. The heat produced radiates out towards space but some of it does not pass back out through the atmosphere and is reflected back down trapping some of the heat. When greenhouse gases such as carbon dioxide and methane accumulate high in the atmosphere, more heat gets trapped and the planet warms up further. The burning of fossil fuels gives off carbon dioxide, which accumulates and this increases the greenhouse effect and the planet warms up more. Technology is being developed that uses alternative energy sources instead of fossil fuels. Ways of generating energy from the sea, the wind, the sun, rivers, and the heat stored underneath the earth's surface all continue to be developed.

ACTIVITY
Make a list of the Earth's resources used by designers and manufacturers that cannot be replaced.

STRETCH AND CHALLENGE
Explain how global warming is caused and the effects it is having on the planet.

Figure 1.1.15 Engineer watching CNC lathe progress on screen

Figure 1.1.16 The Nissan factory in Sunderland operates a JIT production method.

Production techniques and systems

The development of computers and processors has led to the automation of a lot of the areas within manufacturing factories. **Computer-aided design (CAD)** allows designers to draw, design and model on screen, and if linked to a compatible machine allows **computer-aided manufacturing (CAM)**. In other words, it allows CAD drawings to be made by machine. CAD even allows products to be designed in one location and made at a factory or location in another part of the country, or even another part of the world. CAM can create a faster production process and generally only uses the necessary amount of raw materials.

Opportunities to make parts for all sorts of equipment have been created by the advent of 3D printing. Small-scale car production is now being developed where all the exterior body parts are 3D printed and assembled.

A **flexible manufacturing system (FMS)** is a system where production is organised into cells of machines performing different tasks. A range of **computer numerically controlled (CNC)** machines are put in each cell, such as a CNC miller and a CNC lathe. The parts that are manufactured in the cell are generally handled by a material handling system that could be a robot arm. Advantages of flexible manufacturing systems are their high flexibility; they can produce different products simultaneously, and they can be set up to produce new products quickly and easily, saving time and effort.

Just in time (JIT) production is a method of organising a factory so that materials and components are ordered to arrive at the product assembly plant just in time for production. Many companies operate JIT systems, such as Nissan, Toyota and Dell. It helps to create **lean manufacturing**, which means it focuses on giving customers value for money by reducing waste. The advantages of JIT production and lean manufacturing are:
- a reduced need to keep large stockpiles of components and materials
- less space needed to keep stocks of components and materials
- smaller numbers of finished products to be stored and put into stock
- less waste.

KEY WORDS

Computer-aided design (CAD) using computer software to draw, design and model on screen.

Computer-aided manufacturing (CAM) manufacturing products designed by CAD.

Flexible manufacturing system (FMS) a system in which production is organised into cells of machines performing different tasks.

Computer numerically controlled (CNC) machine tools that are controlled by a computer.

Just in time (JIT) a production method that means materials and components are ordered to arrive and the product assembly point just in time for production.

Lean manufacturing focusing on reduction of waste when manufacturing.

How the critical evaluation of new and emerging technologies informs design decisions

Products do not last forever; they wear out in time and are either recycled, reused or thrown away. It is important that designers think about how they can make a product last as long as possible and what can be done with the product at the end of its useful life. It is better for the environment if, rather than replacing a product at relatively short intervals, a product has a longer lifespan. This uses fewer materials. However, a longer lifespan also means that manufacturers sell fewer products.

Planned obsolescence

Some manufacturing companies plan or design products to have a short useful life. They do this so they will become obsolete; this means that they will become unfashionable, or they will no longer function after a certain period of time. The companies then produce new and improved products at short intervals. They do this for a variety of reasons; it could be because of technology improving or just to keep their sales at a steady level. This is of course very wasteful. It impacts on the environment and creates disposal issues. Apple is an example of a business that uses **planned obsolescence**. With each new model of their iPhone, they change the connections, meaning that people have to buy new leads, chargers and headphones alongside their new phone.

Design for maintenance

Maintenance means performing functions on a product that will help to keep it functioning correctly throughout its life. Designing for maintenance can mean a variety of things, from the facility to change batteries to allowing access into the product to repair or replace worn out components. Some products, such as personal computers (PCs), are made up of different modules, which allows a module to be repaired or replaced in the event of a fault rather than the entire product. This system of using modules also means that some modules can be replaced in order to upgrade as technology improves.

Ethics

People like the things they buy to be of a good quality and low cost. To keep prices low, manufacturing companies can cut costs in a variety of ways, but in doing so there are other costs. For example, automating factories can cut workforce costs by employing fewer people, but this means some people lose their jobs. Other methods of reducing costs include manufacturing in a country where labour costs are lower than the UK. In these cases, workers are paid less and work in poor conditions, and often the pollution created in manufacturing may be much higher. Also, environmental costs may come from transporting the products between countries.

The environment

Manufacturing products requires the use of energy and materials which impacts on the environment. Designers need to consider how they can reduce the impact on the environment. The main considerations should centre on the life of the product and the material used, and what will happen to the product at the end of its useful life.

Figure 1.1.17 **Old computers for disposal**

End of life disposal

How we dispose of products can have a significant impact on the environment. It is better if the materials can be recycled than if they are buried in underground rubbish dumps as landfill. Recycling means reprocessing a material so that it can be used again. It reduces the need for new materials and therefore causes less environmental impact. For example, plastics can be sorted according to their reuse. Every plastic bottle or container has a recycling symbol and number that helps in the sorting process and can lead to it being reshaped into new products. Aluminium drinks cans are recycled by going through a re-melt process; they are turned into ingots that are then used to make new cans. Glass is often reused – for example, milk in some areas is still delivered to houses in milk bottles. The bottles are returned to the dairy, sterilised and reused for further milk deliveries. Jars can become containers for other items, but at the end of their useful life they are crushed, melted and moulded into new bottles and jars or other glass items. Glass does not degrade so it can be recycled again and again.

If a product eventually has to go into a landfill then ideally it should be made of biodegradable materials. Biodegradable means that the materials naturally break down quickly when in landfill to naturally occurring substances. Non-biodegradable materials (such as many plastics) can take hundreds of years to break down. Supermarkets started charging for plastic bags to encourage their reuse as, like most packaging materials, they quickly end up in landfill or polluting land and sea. This is now a legal requirement.

ACTIVITY

List as many products as you can think of where maintenance can prolong their lifespan.

STRETCH AND CHALLENGE

Write an article that discourages manufacturers from producing products that have planned obsolescence.

KEY POINTS
- Automation is the use of computers to control machinery in factories with minimal human involvement.
- Enterprise is a skill where people take risks to bring new products to the market.
- Sustainability is about meeting our own present-day needs without compromising the needs of future generations.
- Culture is the values, beliefs, customs and behaviours displayed by different groups of people.
- Just in time (JIT) production is a method of organising a factory so that materials and components are ordered to arrive at the workplace just in time for production.
- Planned obsolescence is when a product is deliberately designed to have a short life span or to go out of fashion.

Check your knowledge and understanding

1 Explain what is meant by a finite resource.
2 Give an example of a non-finite resource.
3 Explain why robots are used so extensively in many modern industries.
4 Give an example of a product developed as a result of technology push.
5 Discuss the advantages and disadvantages of just in time and lean manufacturing systems.

1.2 Energy generation and storage

What will I learn?

In this topic you will learn about:
- → using fossil fuels for energy generation
- → alternative energy sources: their increasing use and different types
- → how energy is stored
- → batteries and their advantages and disadvantages.

Without energy most of the things we do would be impossible. Over the last one hundred years we have become increasingly dependent on electricity and the energy sources that we rely on to provide it. Electricity can be produced in a number of ways. The majority of electricity in the United Kingdom is produced by burning fossil fuels, although an increasing amount is produced by using alternative technologies, which rely on the use of renewable sources of energy. Non-renewable sources, such as fossil fuels, are consumed and will eventually run out, whereas renewable sources are naturally replenished.

Fossil fuels

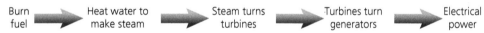

Burn fuel → Heat water to make steam → Steam turns turbines → Turbines turn generators → Electrical power

Figure 1.2.1 **Energy conversion**

Traditionally, Britain has relied on **fossil fuels** such as coal, gas and oil to provide its energy. Fossil fuels were formed over millions of years from dead organisms – coal from trees, and oil and gas from marine organisms.

All fuels have to be burnt to produce heat.
- In electricity generation the heat is used to convert water into steam at very high pressure and temperature, which is used to drive turbines connected to generators which produce electricity.

Burning any fossil fuel produces carbon dioxide, which adds to the greenhouse effect and possible **global warming**.

Fossil fuel power stations can be built almost anywhere provided you can get the fuel to them, however, a water supply is needed for cooling so they are normally found near rivers or the sea.

Coal

Although most deep coalmining in the UK has stopped, and the use of open pit mining has reduced, we still obtain 23 per cent of our electricity from coal-fired power plants. This means we have to import coal from abroad.

Mining and burning coal produces waste and atmospheric pollution, which poses environmental problems. Waste tips, stockpiles and open pits look unsightly and hazardous, and sulphur dioxide fumes from coal power stations add to atmospheric pollution and cause acid rain, which damages trees and lakes.

An advantage of coal is that it doesn't require any processing before burning, although it is usually crushed. There are still sufficient reserves of coal to last hundreds of years.

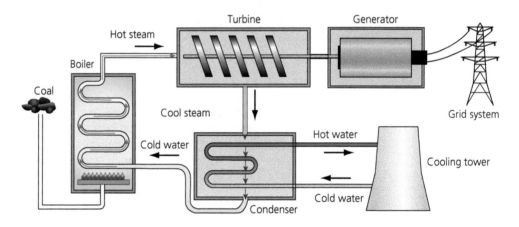

Figure 1.2.2 **A coal-fired power station**

Natural gas

Natural gas is currently the main source of power for electricity production in the UK. Natural gas is used for heating and cooking as well as for industrial uses. It can be burnt directly and does not require crushing like coal. It is easy to transport through pipelines. Mainly consisting of methane, gas is found deep underground with coal and oil deposits. It has to be processed before it can be used to remove water and other impurities.

As gas-fired power stations can be brought into service quickly, they will remain important as a replacement for less reliable sources, such as wind power on days when there is insufficient wind available for the wind turbines.

Gas from the seas around Britain accounts for some of the gas we use, but the majority comes from pipelines connected to Europe and in liquefied form (LNG) shipped from around the world in tankers. There are deposits of shale gas under Britain, but there is considerable discussion about using these, as there might be pollution of water supplies and a risk of small earthquakes called seismic tremors.

Oil

Oil is hardly used for electricity production (in the UK). Oil is used for some heating systems (typically in rural areas away from a main gas supply) and is the main fuel used in road and sea transportation. It does need to be processed by refining from crude oil before it can be used. Stocks of oil will run out before coal. Although there is no immediate shortage, the amount of oil in the North Sea has reduced, which reflects a worldwide trend.

Nuclear power

Twenty-two per cent of the United Kingdom's electricity comes from nuclear reactors, in which uranium atoms are split to produce heat. This process is known as **fission**. A vast amount of energy can be produced by this process from a relatively small amount of uranium. The energy produced as heat is used to convert water to superheated steam, which in the same way as other power stations, drives turbines connected to generators to generate electricity.

The current set of nuclear power stations are expected to close by 2025 because they are getting old and reaching the end of their serviceable life. Several replacement nuclear power stations are planned. The cost of safely disposing of unused nuclear power stations is high, and there have been several well-publicised incidents at nuclear plants in Japan, Ukraine and the United States which have resulted in nuclear material leaking. Nuclear waste is highly hazardous and can have long-lasting effects on the health of humans and animals for thousands of years if not dealt with carefully.

Figure 1.2.3 **Nuclear reactors use uranium to produce heat, which creates steam to drive turbines connected to generators that produce electricity.**

Renewable energy

Recently, due to concerns over pollution and the possibility that some sources of fuel might eventually run out or become uneconomic to obtain, there has been much greater support for **renewable** sources of power, such as wind or solar energy.

Renewable energy sources provide 25 per cent of the electricity we use. Unlike fossil fuels they tend not produce any waste or significantly add to global warming by producing gases.

Source	% electricity production in UK
Coal	23%
Oil	<1%
Natural gas	30%
Nuclear	22%
Renewable	25%

Table 1.2.1 **UK electricity production, 2015**

Wind

Wind energy has been used for thousands of years, and before the advent of steam power, windmills were a very common sight scattered across the countryside. Today we use tall towers with propeller-like blades driving a generator. Although sometimes found as single towers, more often they are grouped together to form a 'wind farm'. You may also see much smaller versions attached to caravans or boats to charge their batteries.

The best places to put wind turbines are on the coast, offshore, on a hilltop, or between hills or mountains, so that the wind supply is reliable.

As the blades are so long the tower needs to be very high, but the land underneath can still be used.

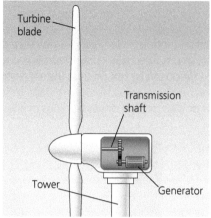

Figure 1.2.4 The UK government has set a target to generate 15 per cent of all energy from renewable sources by 2020. Building new wind turbines will help towards meeting this target.

Solar

Solar technologies are either passive (for example, positioning a building to gain heat from the sun, by placing most windows to the south side of the building) or active (photovoltaic cells), depending on how they catch solar energy and convert it into power.

The amount of solar energy that reaches the Earth is vast, easily outstripping all of the combined fossil fuel and uranium deposits in a single year. The largest difficulty in gaining solar energy is position on Earth: being closest to the equator is the most efficient, and near a polar cap is the least. The difference between day and night is also an issue. The placement of photovoltaic panels is another issue as they take up valuable land. This has been solved to a degree by placing cells on roofs and the sides of tall buildings thereby lessening their environmental impact.

Figure 1.2.5 Solar panels are often placed on roofs and the sides of tall buildings.

Tidal (marine)

Tidal energy relies on the gravitational pull of the Moon, which causes the change in water levels that we call tides. It has been estimated that Britain could provide around 20 per cent of its energy needs from tidal power.

The most common form of tidal power is the tidal barrage. A tidal barrage is a long dam that is built across the mouth of a river where it meets the sea. These are known as 'estuary barrages'.

They can be constructed so that the incoming tide passes through turbines generating electricity. The water can then be held in the high tide state upstream of the barrier, until it is needed. It is then released at low tide, flows out through the turbines, again turning them and generating electricity.

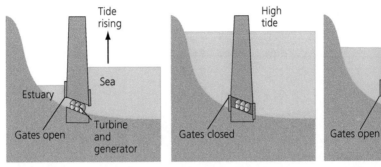

Figure 1.2.6 **Tidal energy**

Disadvantages are that the river would never completely empty, flooding mudflats, which provide a habitat for thousands of birds. There may also be problems of waste disposal as many towns discharge sewage into rivers.

Water or hydroelectricity

Like wind power we have been harnessing the power of moving water for many years. Hydroelectrical schemes use a dam to block a river. Once sufficient water has built up behind the dam, it is stored in a reservoir and it can be channelled through turbines that are used to turn generators for producing electricity. The main cost is the building of the dam, but it has the advantage that as the water is held, electricity can be produced very quickly by opening valves that control the flow to the turbines. The environmental effects are from the flooding of a valley and reducing the water flow below the dam that can affect the growing of crops and river ecosystems.

Figure 1.2.7
Hydroelectricity channels water through turbines to turn generators producing electricity.

Renewable energy

Wave

Although the power contained in wave movements is very obvious, it has proved difficult to harness. Most ideas seem to rely on the up and down movement of the waves either being converted into mechanical energy and moving rams or pistons, or to compress air so that it is forced through a turbine. Like wind power the waves can vary a lot in size and power, so positioning is critical.

Figure 1.2.8 Pelamis offshore wave generator

Biomass

Biomass involves growing plants so that they can be burnt, or using decaying plant or animal materials to produce heat.

To reduce the amount of fossil fuel being used, some vegetable oils are treated after being used for cooking. After processing, they are suitable for diesel engines used to power large trucks used for deliveries and for small-scale electricity generation.

Plants such as oilseed rape or willow are grown as biomass crops, and are harvested so they can be burnt in furnaces. This burning to produce heat causes some atmospheric pollution so measures need to be taken to reduce this, although it is not as severe as burning fossil fuels. The big advantage for biomass is that replacement crops can be grown very quickly to ensure a constant supply. However, the disadvantage is this land could have been used for growing crops for food.

KEY WORDS

Renewable energy energy from a source that is not depleted when used, such as wind or solar power.

Hydroelectricity the process which uses a dam to block a river in a valley and channels water through turbines that are used to turn generators for producing electricity

Biomass growing plants so that they can be burnt, or using decaying plant or animal materials to produce heat.

Energy storage systems, including batteries

Sometimes we do not want to use the energy we generate immediately, but instead want to capture it for use at a later date. Pumped storage systems, flywheels, clockwork, capacitors and rechargeable batteries are examples of devices used to store energy for use at a later time.

Kinetic pumped storage systems

A similar system to hydroelectric power generation is used in pumped storage systems. Here a hydroelectric dam system is used with two reservoirs one at a low level, and one up a mountain.

They are a good way of dealing with sudden high demand for electricity. The classic example of this is a commercial break in the middle of a popular TV programme when everybody pops into the kitchen to put the kettle on for a hot drink. At moments like that the electrical power supply system might not be able to cope with demand,

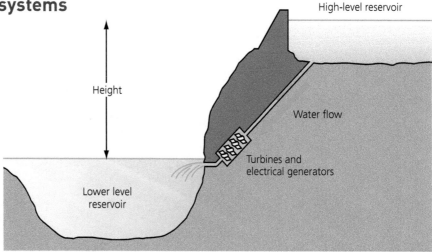

Figure 1.2.9 Kinetic energy conversion

so an additional fast-acting top-up is needed. Nuclear and coal-fired stations are slow to respond so something else is needed. Gas-fired stations are often working at full capacity in winter months.

The pumped storage system instantly releases electricity into the system by opening valves to allow water to flow from the high reservoir to the lower one through turbines. This cannot be sustained for long, but it is usually long enough for the other power stations to catch up with demand and avoid the need for power cuts. During the night, demand is usually low, but the nuclear and coal-fired stations cannot reduce their output significantly, so the excess electricity is available at low cost. This cheap excess electricity is used to pump the water back up to the top reservoir, ready for sudden peaks in demand the next day.

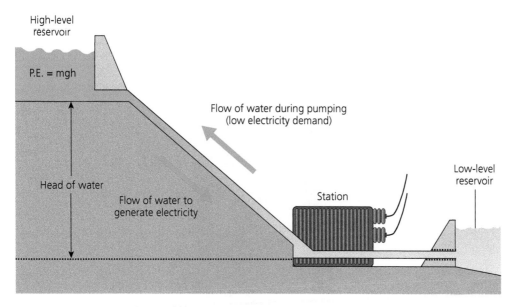

Figure 1.2.10 A pumped energy storage system

Mechanical energy storage

Energy can be stored in other forms. An example of this is the use of a flywheel. We have used flywheels in machines for centuries; they are a heavy spinning mass that continues to spin when the driving force stops.

A recent development used in some buses in Belgium and Switzerland is a flywheel which is accelerated to a very high speed. When spinning at speed it stores the energy (rotational energy) and slows as it is released. Electricity is used to accelerate and slow the flywheel, so an energy conversion takes place. High-efficiency bearings are an important part of this technology and frequently magnetic bearings are used to reduce friction, which in turn increases the efficiency of the system. In this system, energy is also stored when the buses are braking.

Clockwork is still used for storing energy in toys and mechanical devices. Normally a key is used to wind up a spiral spring by forcing it into a smaller space, where it is stored as potential energy with a system of gears being used to release the energy slowly.

Figure 1.2.11
A traditional flywheel

Figure 1.2.12 Clockwork is used for storing energy in toys and mechanical devices.

Electrical energy storage

Capacitors

Capacitors are the most popular non-chemical method of storing electricity and their invention predates the battery. They consist of two plates of opposite polarity; when the capacitor is charged, the positive charges migrate to one plate and the negative to the other.

Supercapacitors are used in electric vehicles because they can be recharged many more times than batteries – millions as opposed to thousands of times – which makes them suitable for use in regenerative braking systems where the energy used to slow the vehicle is stored and used to drive the motor.

Batteries

Although there are many types of devices consisting of one or more electrochemical cells, we tend to refer to them all as batteries.

There are two main types of battery: primary or 'single-use', which we use and discard; and secondary or 'rechargeable'. Both are extremely useful when we need electrical power in locations where mains electricity would be difficult or even impossible to access, such as moving cars.

Examples of primary types include alkaline batteries, such as those used in clocks, and zinc-carbon, which although cheaper than alkaline batteries do not last as long or store as much electrical energy in the same space. Typically, they produce about 1.5 volts per cell.

Rechargeable batteries are more expensive to purchase than alkaline batteries, but are cheaper to use as they can be recharged many times. There is, however, a limit to how many times they can be recharged. Nickel cadmium rechargeable batteries last longer if they are completely discharged before recharging. Lithium ion ones are more adaptable.

Rechargeable batteries typically have a cell voltage of 1.2 volts, so in a 12-volt device you would need ten rechargeable batteries but only eight single-use batteries.

Batteries come in a variety of shapes and sizes, with large lead–acid batteries for cars, trucks and even submarines at one end of the scale; and miniature batteries such as those used in hearing aids at the other. For all batteries there are issues related to safe disposal, as they contain harmful chemicals and metals that must not be allowed to contaminate groundwater supplies.

Figure 1.2.13 Capacitors

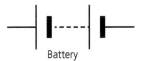

Battery

Figure 1.2.14 Battery circuit symbol

ACTIVITY

Most modern toys now use batteries rather than clockwork. Find out when batteries became the most popular energy storage method for toys. Explain the advantages of this development. What sort of problems did the change cause?

Figure 1.2.15 Batteries are useful when we need electrical power where mains is unavailable.

Check your knowledge and understanding

1 What is the best time of day to store energy in a pumped storage system?
2 What advantages do fossil fuel have that make them so popular?
3 Briefly explain why growing crops for biofuels might not solve the world's energy problems.
4 How can a flywheel store energy?
5 Give two disadvantages of secondary batteries.

Developments in new materials

What will I learn?

In this topic you will learn about:

→ modern materials
→ smart materials
→ composite materials
→ technical textiles.

Modern materials

Modern materials are new materials developed to have properties that are useful when designing and making products. They are produced through the invention of new or improved processes.

Graphene

Graphene is a very thin two-dimensional material layer of carbon that was discovered and extracted from graphite in 2004. It is a very strong and very light material. It is harder than diamond, about 300 times stronger than steel and is currently the lightest known material. Graphene is transparent and conducts electricity and heat even better than copper. It is extremely flexible, which is unusual for such a tough, strong material. Due to this strength, graphene is being developed for use in protective clothing, vehicles and even buildings.

Graphene is in the early days of development, but you can already buy some products that contain it. Conductive ink is made by mixing tiny graphene flakes with ink, enabling you to print onto paper and use this print to conduct electricity. This was previously possible using semi-conductive ink, but the use of graphene has greatly improved how well electricity is conducted.

Graphene is also used in solar cells. These cells need conductive materials that allow light to get through them. Graphene is a suitable material due to its high conductivity and transparency. Unfortunately, it is not very good at collecting the electrical current produced inside the solar cell, so further developments need to be made. Researchers are looking at modifications, such as the use of graphene oxide – this material is less conductive but more transparent and a better collector of charge. Graphene has also been used in paint to act as a barrier between materials and the corroding effects of oxygen and water. In the future, vehicles could be made corrosion resistant with such paint.

Electronics companies are investigating the use of graphene in touchscreens as it is able to transfer electrons at much faster speeds than silicon. This, paired with its flexibility, could lead to foldable televisions or computers, and due to its transparency, could lead to virtual curtains or intelligent windows.

KEY WORD

Modern material a material that has recently been developed for specific applications.

Figure 1.3.1 **Sheet graphene**

Metal foam

Metals such as aluminium (and sometimes steel or titanium) can be made into a foam by injecting gas into the metal when it is a liquid state. This creates a foam that is very lightweight and has high compressive strength – a porous material that can absorb energy well.

Metal foam can be used as soundproofing or for crash protection in vehicles as it is light enough to be carried in cars without reducing their speed. It is also being developed in the use of body armour, again because it is lightweight and strong. Metal foams are also used in filtration due to their porosity, and are currently being developed for use in prosthetics (artificial body parts) for animals. The hope is to create prosthetics that are lighter and more comfortable for the user.

Figure 1.3.2 Aluminium foam

Titanium

Titanium is a fairly new metal in comparison to others (such as steel, copper and aluminium), and it is the fourth most abundant metal, making up about 0.62 per cent of the Earth's crust. It is particularly useful due to its high corrosion resistance, even to salt water and chlorine. It also has a high strength-to-density ratio, making it suitable for use in applications such as knee replacements.

Figure 1.3.3 Concorde

The supersonic Concorde planes were coated in a heat-proof titanium skin that could stretch as much as 250 millimetres during flight. Its strength and low density made it a good choice of material for aircrafts, and when exposed to the elements it doesn't go rusty (it can resist the effects of the rain and moisture in the air by reacting with oxygen to produce titanium oxide, which keeps out water and air).

Titanium can be alloyed with other metals such as aluminium and vanadium. It is also frequently used as titanium oxide. In this form, it is used as a white pigment in plastic and paint and as a sunscreen in cosmetic products.

Coated metals

Coated metals include anodised aluminium, nickel-plated steels and polymer-coated aluminium.

Anodised aluminium is aluminium with a thick oxide layer created by passing a current through an electrolytic solution. This process increases resistance to corrosion and wear, and makes it easier to paint and glue.

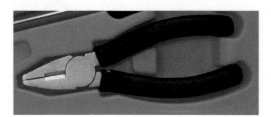

Figure 1.3.4 Dip coated tool

Nickel-plated steel uses the process of electrolysis to add the metal nickel to a steel object. This allows us to use a strong and relatively cheap material such as steel and coat it in a thin layer of more expensive nickel, which is corrosion resistant and can add an attractive finish to a product.

Polymer-coated aluminium can be created by dip coating or powder coating the aluminium. Polythene is often used in powder form for dip coating, whereby the metal is dipped into a liquid polymer. This process is used for products such as dishwasher racks and tool handles. Powder coating is similar but the powder is sprayed on to the metal and then cured in an oven. These products tend to be more expensive than standard ones as they are particularly hardwearing. Both processes create a layer of plastic around the metal which reduces the impact of exposure to air and moisture, and provides an attractive and potentially colourful finish to the product.

Liquid crystal display

A liquid crystal display (LCD) is a laminated material of two layers of glass with a liquid crystal core. It is a thin flat panel that lets light go through when a voltage is applied, or blocks the light when the voltage switched off.

LCDs are often used in items such as digital watches and flat-screen televisions. Many LCDs work well by themselves when there is other light around but others need a back light (such as for smartphones, computer monitors, and televisions).

Nanomaterials

Nanomaterials have tiny parts less than 100 nanometres in size. A nanometre is one billionth of a metre. Nanomaterials are used as surface coatings or thin films, such as on computer chips. They are also used in sports equipment, such as tennis rackets and golf shoes, where nanoparticles are added to materials to make them stronger while not substantially increasing the weight. Nanoparticles are also used in clothing such as socks due to their antibacterial properties, to reduce the absorption of sweat.

Figure 1.3.5 **Teflon-coated pan**

Teflon

Teflon is the trade name for a polymer called polytetrafluoroethene or PTFE. It is best known for being used on the surface of pans to ensure that food being cooked does not stick. Teflon, like many things, was invented by accident, but is now used in a variety of different products. It is a very slippery material, so is used in clothing to make it difficult for dirt to stick to the fabric. PTFE is unreactive, so it is also used to make pipes and containers for chemicals.

Corn starch polymers

Corn starch polymers were developed to replace oil-based thermoplastics. They are often used in the manufacture of products which are intended to have a short lifespan, such as disposable cutlery. This is because they are degradable, which means they will break down over time. This is a real advantage environmentally, as oil-based polymers can cause huge environmental concerns, partly due to the period of time they remain in the environment. Oil-based polymers are often washed into the oceans where they exist for many years and pollute the environment or are ingested by marine life.

Figure 1.3.6 **Plastic waste**

Corn starch polymers are made from polylactic acid found in high starch vegetables such as potatoes, corn and maize. These are all renewable sources, which is beneficial, and they don't take as much energy to produce as the extraction of oil does. Corn starch polymers decompose within a shorter time than oil-based polymers so they don't litter our environment; some have even been developed to become a food for marine life when they are discarded.

Corn starch polymers are also food safe, so are used in the production of food packaging in order to restrict the effect that packaging has on the environment.

Figure 1.3.7 **Biodegradable plastic products**

Modern materials

> **KEY WORD**
> **Smart material** a material that changes its properties in response to changes in its environment.

Smart materials

Smart materials are materials that have one or more properties that can be significantly changed in a controlled fashion by external stimuli. These influences could be, for example, stress, temperature, moisture or pH. These changes are often reversible when the environment changes again.

Thermochromic pigments

Thermochromic materials change colour at specific temperatures. They are available as plastic, ink, and dyes for textiles and paint. They have many uses, for example:

- test strips on batteries – when the strip is pressed at each end and if the battery is in good condition, current flows through a printed resistor under the thermochromic film; this heats the resistor, producing a colour change
- plastic strips used as thermometers that are applied to children's foreheads
- colour indicators on drinks cans to show whether the contents are cold enough
- mugs that change colour when hot water is added
- baby spoons that change colour when the food is too hot; these are a safety feature to ensure the child's food is a safe temperature
- t-shirts that change colour.

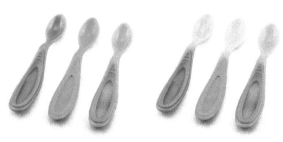

Figure 1.3.8 A thermochromic spoons. The spoons on the right have been heated.

Shape memory alloys (SMAs)

If materials made from **shape memory alloys (SMAs)** are bent or deformed, they will return to their original shape when heated. This is a useful property when:

- a response to a change in temperature is needed (for example, in fire alarm systems or controllers for hot water valves in showers)
- movement is needed from an electrical current (for example, in electric door locks, rotary movement and artificial muscles in robot arms); the temperature change can be achieved by passing an electrical current through the thin product or wire
- a damaged product needs to be repaired (for example, if someone bends a glasses frame, it can be returned to its original shape by being heated).

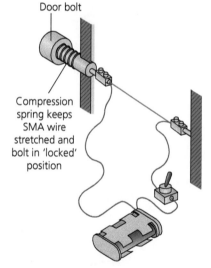

Figure 1.3.9 An electric door lock is an example of an SMA.

Photochromic pigments

Photochromic materials change colour if the level of light changes. Examples include:

- lenses in sunglasses that become lighter or darker depending on the light conditions
- security markers that can only be seen in ultraviolet light
- photochromic windows that change their transparency according to how much light there is; these reduce glare and help prevent cooling systems in buildings from overloading
- bracelets that change colour to tell you if you are getting too much sun; these can be an indicator to tell you to reapply sun cream.

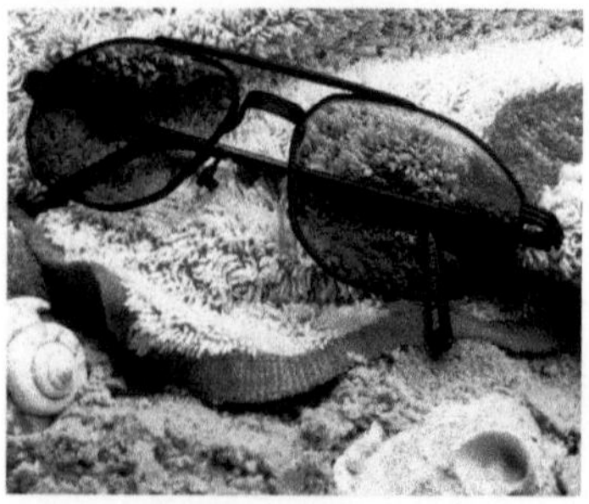

Figure 1.3.10 Photochromic sunglasses change colour according to the amount of light.

Composites

Composites are materials made from two or more different materials, and combine the properties of the materials they are made from. The earliest known composites were bricks, made by combining straw and mud. The mud when dried is good at withstanding compressive forces but is prone to cracking, but when straw is added to the mix it provides tensile strength which prevents cracking when the brick dries.

Figure 1.3.11 **Bricks made from straw and mud**

Figure 1.3.12 **A reinforced concrete bridge**

Concrete is an example of what we call a particle composite. It is a mixture of cement, sand and stones; water is added during the manufacturing process. Combining these materials creates a composite with very good compressive strength, however, if concrete is to be used somewhere where it needs tensile strength, steel reinforcing is added. The steel provides strands to add further tensile strength to the material.

Glass-fibre reinforced polymer (GRP), combines strands of glass fibres which are strong but brittle, with a flexible polymer. This makes a composite material that is tough but not brittle. GRP is used to make hulls for yachts and, increasingly, in car bodies.

Figure 1.3.13 **A yacht with a GRP hull**

Using combinations of properties from different materials allows for the development of new products and materials. Fibre-reinforced composites are becoming popular for use in products which need to be lightweight but incredibly strong, even though the composites can be extremely expensive. Carbon-fibre reinforced plastic (CRP) has a very high strength-to-weight ratio so is popular in products which take very harsh loading and need to be lightweight, such as those used in the aerospace industry and motor racing. It gives a high-quality finish and a better strength-to-weight ratio, but is more expensive than glass-fibre reinforced plastic. Composite materials are also used in medical applications and in orthopaedic surgery.

Figure 1.3.14 **Carbon-fibre-based composites are used in modern F1 racing cars.**

Technical textiles

Technical textiles are textile materials and products that are manufactured for their technical and performance properties rather than their aesthetic characteristics.

Figure 1.3.15 Fencing suits

Conductive fabrics

Conductive fabrics are fabrics that have either conductive fibres woven into them or conductive powders impregnated into them. These are often called electronic textiles or **e-textiles**. Material such as conductive thread is useful for use in circuits that power LEDs, etc. However, the material is not a brilliant conductor so it has limitations.

Conductive materials have been built in to competitors' jackets for fencing contests to help with scoring systems. During a game, opponents score by hitting a scoring area on the front of the suit with their foil (sword). When the metal of the foil makes contact with this panel a 'strike' is recorded. This is possible due to the use of conductive fibres on the panel of the suit.

KEY WORD

Technical textiles textile materials and products that are manufactured for their technical and performance properties.

Fire-resistant fabrics

Fire-resistant fabrics have multiple uses. Not only are they are used for items that are often exposed to flames, such as fire fighters' suits, but also for items such as children's nightwear and cotton/viscose furnishings. Such items must be given a flame resistant finish by law.

Nomex is a brand name for a fire-resistant fabric made from a type of polymer called a meta-aramid. It is used in the production of fire fighters' suits, etc. to ensure the best protection from fires. Nomex thickens when heated, therefore increases protection, while still staying supple and flexible so that it does not impair movement. It is a lightweight material that is flame resistant and protects the wearer from heat. It is also breathable, which is important for those wearing it in active environments, and is durable and abrasion resistant.

Kevlar

Kevlar is a material formed by combining terephthaloyl chloride and para-phenylenediamine. These threads can then be woven to create an incredibly strong material. When layers of woven Kevlar are combined with layers of resin, the result is a very light material that has a tensile strength (stretching or pulling strength) over eight times greater than that of steel wire. Unlike most plastics, Kevlar does not melt, which makes it perfect for use in extreme temperatures as it can withstand up to 450°C (850°F). It can also withstand very low temperatures of up to −196°C (−320°F). Kevlar can also resist attacks from many different chemicals.

Kevlar is often used in the production of personal armour such as bullet proof vests, helmets, face masks, motorcycle safety clothing and so on. It can also be used in sports equipment, such as inner linings for bicycle tyres and table tennis bats, due to its high strength-to-weight ratio.

Figure 1.3.16 Kevlar bulletproof vest

Gore-Tex

Gore-Tex has been designed to be a waterproof yet breathable textile. It is used in clothing to provide a waterproof product that also releases perspiration vapour, and is therefore more comfortable to wear than traditional waterproof materials. Gore-Tex contains a layer of plastic based on PTFE (Teflon) which contains lots of tiny pores. Each hole is too small for water droplets but big enough for sweat to pass through. There are approximately 14 million pores per square millimetre. This modern material was developed in 1969 and is now commonly used in outdoor wear. Waterproof jackets, walking boots and trainers are common products for its use. The construction makes a 'breathable' fabric which can also be combined with insulation to make outdoor clothing that blocks the wind and keeps you dry and warm.

Figure 1.3.17 Gore-Tex walking boots

Microfibres

A microfibre is a very thin synthetic fibre. Microfibres are often used for outdoor clothing and sportswear because they are breathable, durable, crease resistant and easy to care for. Very fine polyamide and polyester microfibres are used for sportswear and lingerie garments, and Tencel microfibre is used for shirts.

Some microfibres incorporate **microencapsulation**, this means the very thin fibres hold chemicals in tiny capsules. These capsules gradually break, releasing chemicals like perfumes, insecticides or antiseptics. This means that the wearer can benefit from these functions throughout their time of use. This technology has been used in the manufacture of outdoor clothing that repels mosquitos and other biting and stinging insects. It was originally developed for the military and is now sold worldwide for holiday clothing.

Microencapsulation has also been used to add scent to fabrics. Uses range from clothing which holds a perfume released over time to soft furnishings. Soft furnishings include more novelty-based products such as curtains for children's bedrooms that smell of sweets to bedsheets that smell of lavender to encourage a good night's sleep.

ACTIVITIES

1 Choose one type of modern or smart material. Create a list of different applications that it could be used for.
2 What is the difference between a modern and a smart material? Create a mood board that includes examples of the differences between modern and smart materials.

KEY WORD

Microencapsulation very thin fibres hold chemicals in tiny capsules, which break open releasing the chemicals

STRETCH AND CHALLENGE

Write an article for a design magazine explaining how the development of smart materials will affect the work of designers. This should include at least two case studies of products that could be changed to include smart materials.

Technical textiles

KEY POINTS

- Graphene is a very new material that is very light and flexible.
- Polymer coated metals can be dip or powder coated depending on the required finish.
- Teflon is used on non-stick pans.
- Corn starch polymers degrade much quicker than oil-based polymers in the right conditions.
- Smart materials change their properties in response to changes in the environment.
- Thermochromic materials change colour at specific temperatures.
- Shape memory alloys (SMAs), if deformed, will return to their original shape when heated.
- Photochromic materials change colour if the level of light changes.
- Composite materials combine the properties of the materials that were used to make them.
- Kevlar is a very light and strong material.
- Gore-Tex is a material used in the manufacture of outdoor clothing as it is breathable and waterproof.

Check your knowledge and understanding

1 Give an example of a product in which Kevlar may be used.
2 Give an example of a product in which a glass-fibre reinforced polymer (GRP) might be used.
3 Explain what composite materials are and why they are developed.
4 Explain what is meant by the term 'smart material'.
5 Give an example of a product in which photochromic pigments may be used.
6 Discuss the advantages and disadvantages of corn starch polymers.
7 Name a composite material.
8 Explain how microencapsulation works.
9 Explain the difference between GRP and carbon-fibre reinforced plastic (CRP).

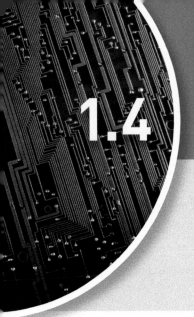

1.4

Understanding a systems approach when designing

What will I learn?

In this topic you will learn about:
- → input devices
- → process devices
- → output devices.

A systems approach explained

To work out how any product containing electronics works, you need to ask the following questions:

1 What does the product do? (What is the **output?**)
2 How does it do it? (What is the **process**?)
3 What happens to allow this to work? (What is the **input**?)

Figure 1.4.1 **Simple systems diagram**

These three elements of output, process and input are what we call a systems approach to designing. We use systems diagrams like Figure 1.4.1 to show this information.

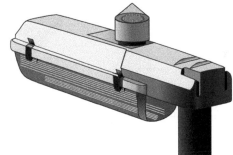

Figure 1.4.2 **A street light sensor**

If you look at the street light shown in Figure 1.4.2 you will probably be able to work out that the light will come on when it gets dark; this describes the output of the product. You will also probably be able to work out that something is switching on the light; this describes the process. You may be able to work out that a light sensor is sensing that it has got dark; this describes the input. The system diagram for the light sensor would look like Figure 1.4.3.

Figure 1.4.3 **Systems diagram for the light sensor**

Designing an ice alarm system

If we wanted to design an alarm that tells us when the temperature is below zero and therefore that we might expect ice or frost, we would start with a simple systems diagram. You may decide that you want to hear an audible warning when it gets cold, so start thinking in terms of input, process and output.

A **system diagram** is very useful to describe what you want to happen in the system and to help you understand how the operations might be achieved.

| INPUT Temperature | → | PROCESS Transistor switching | → | OUTPUT Buzzer sounds |

Figure 1.4.4 Systems diagram for an ice alarm system

Input devices

Input devices are electrical or mechanical sensors that use signals from the environment, such as light levels, temperature and pressure, and convert them into signals that can be passed into processing devices and components.

Light dependent resistor (LDR)

Figure 1.4.5 Light dependent resistor (LDR)

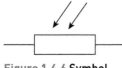

Figure 1.4.6 Symbol for an LDR

A **light dependent resistor (LDR)** is a component in which the resistance to the flow of electrical current through it changes as the light intensity that falls upon it alters. LDRs are used as input devices in light sensing circuits.

LDRs can be found in many products such as street lamps, clock radios and nightlights. In mobile phones LDRs are often used to detect light levels when using the camera – if the level of light is low, the flash is automatically turned on when taking a photo.

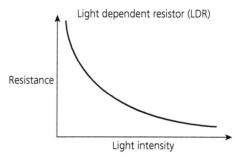

Figure 1.4.7 Resistance against light intensity

Thermistor (temperature sensor)

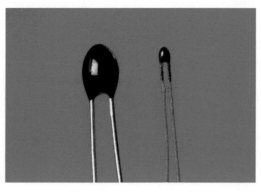

Figure 1.4.8 Thermistor

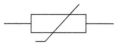

Figure 1.4.9 Symbol for a thermistor

A **thermistor** is a component in which the resistance to the flow of electrical current through it changes with temperature. There are two types of thermistor: those with a resistance that increases with temperature increase (positive temperature coefficient – PTC) and those with a resistance that falls with temperature increase (negative temperature coefficient – NTC).

Thermistors can be found in many products to regulate temperature, such as toasters, coffee makers, refrigerators, freezers, hair dryers and many more.

Switches and pressure sensors

There are a variety of switches that can be used to input signals to processing devices. They can be used to sense when objects apply pressure to them. This is useful when:

- you want to turn something on
- an object has reached as far as you want it to travel
- you want to start something by pressing with your finger
- you want to know if something has closed
- you want to stop a machine if a guard is opened.

Figure 1.4.10 **Rocker switches**

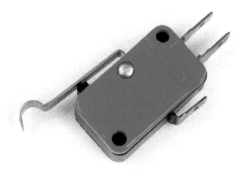

Figure 1.4.11 **Microswitch**

Figure 1.4.12 **Push-to-make switch**

Figure 1.4.13 **Pressure-pad switch**

KEY WORDS

Process devices handle information received and turn outputs on and/or off

Microcontroller a small computer within a single integrated circuit

Processes

Electronic **processes** can be carried out by many components, but in recent years they are more frequently performed by **microcontrollers**. Microcontrollers can be found in many products such as car engine control systems, remote controls, office machines, medical devices, and toys; most electronic products contain one.

A PIC chip (or peripheral interface controller) is a common form of microcontroller that you may have used at school.

Figure 1.4.14 **A PIC microcontroller**

Processes

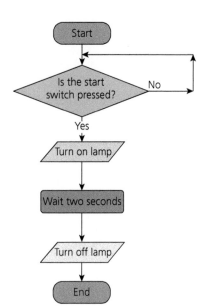

```
Start
  │
  ▼
Is the start ──No──┐
switch pressed?    │
  │                │
  Yes             │
  ▼                │
Turn on lamp      │
  │                │
  ▼                │
Wait two seconds  │
  │                │
  ▼                │
Turn off lamp     │
  │                │
  ▼                │
End
```

Figure 1.4.15 A flowchart for a simple program that flashes an LED on and off

Microcontrollers are small computers within a single integrated circuit. The circuit contains a processor core, memory and programmable input and output capability. They can be used as counters, timers and for decision making.

Microcontrollers are mass produced in huge numbers by automated systems that reduce the cost of each individual one. One of the most useful features of a PIC microcontroller is that you can reprogram them as they use flash memory (flash memory is a data storage medium with no moving parts that can be electrically erased and reprogrammed).

Programming microcontrollers

Microcontrollers can be programmed in a variety of computer languages, such as BASIC, or other software that allows the use of flowcharts for programming.

Flowcharts allow designers to describe the operation of a program in simple terms. There is a set of graphical symbols that have become accepted convention when drawing flowcharts.

The first stage of this system is a decision asking if the start button has been pressed. If the answer is 'no', nothing happens, if 'yes', it turns on a lamp for two seconds and then turns it off and stops.

Outputs

The **output** of an electronic system is generally the part that people are aware of. Whether it is light, sound or movement, the principle of operating whatever device is used as an output is the same.

Lamps and light emitting diodes (LEDs)

Lamps operate by passing an electric current through a very thin filament, which is usually made of tungsten. The filament operates in an inert gas that is encased in a glass bulb. Lamps give off a great deal of heat as well as light, and because of this poor energy efficiency they are rapidly being replaced by other forms of lighting, such as light emitting diodes (LEDs).

> **KEY WORD**
>
> **Output** sends out information, heat, light, sound or mechanical movement to the environment the system is operating in.

Figure 1.4.16 **A lamp**

Figure 1.4.17 **LEDs**

Buzzers and speakers

Buzzers and speakers give out sound outputs. Buzzers fall largely into two types: electromechanical and piezoelectric. They are typically used in alarm devices, timers, and to confirm user input such as a button press or a mouse click. Loudspeakers are devices that convert electrical signals into a corresponding sound.

Figure 1.4.18 A buzzer

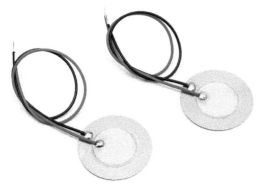

Figure 1.4.19 A piezoelectric buzzer

Figure 1.4.20 A loudspeaker

STRETCH AND CHALLENGE

Analyse and draw systems diagrams for the following products:
- an automatic security light
- a burglar alarm
- an automatic curtain or window blind system.

KEY POINTS

- System block diagrams describe what happens in a system.
- System block diagrams always have a minimum of one input, one process and one output block.
- Input devices provide information from outside into the system.
- Process devices handle information received and turn outputs on and/or off.
- Output devices send out information, heat, light, sound, or mechanical movement to the environment the system is operating in.

STRETCH AND CHALLENGE

Analyse and draw a block diagram for a set of automatic opening doors, showing how the system knows the doors are fully open, and showing how the system knows there is no one there when the doors close.

ACTIVITY

Analyse and draw a systems diagram for an automatic security lamp.

Check your knowledge and understanding

1 Analyse a moisture sensing system in terms of input, process and output.
2 Explain what a light dependent resistor does.
3 Give examples of products that contain microcontrollers.
4 Discuss why microcontrollers are used so frequently in products.

1.5 | Mechanical devices

What will I learn?

In this topic you will learn about:
- → different types of movement
- → what levers and linkages are and what they do
- → the different orders of lever
- → how to convert one type of motion to another
- → how diagrams and symbols are used to represent mechanisms
- → changing the magnitude and direction of forces in rotary systems.

KEY WORDS

Mechanism a device that changes an input motion into a different output motion.

Lever a mechanism that moves around a fixed point (a pivot).

A **mechanism** is simply a device that changes an input motion into a different output motion. It normally changes the amount of movement or the direction of movement in some way to make a job easier to do.

Different types of movement

There are four basic types of motion in mechanical systems:
- **Linear motion** is movement in a straight line, such as on a paper trimmer.
- **Reciprocating motion** is movement backwards and forwards in a straight line, such as the movement of the needle on a sewing machine.
- **Rotary motion** is movement round in a circle, such as a wheel turning.
- **Oscillating motion** is movement swinging from side to side, such as a pendulum in a clock.

Changing magnitude and direction of force
Levers

Levers are possibly the oldest type of mechanism. A lever changes an input movement and force (effort) into an output movement and force (load). A lever moves around a fixed point called a pivot. Scissors and pliers consist of two levers that pivot around one point.

Figure 1.5.1 **A woman cutting hair with a pair of scissors**

Figure 1.5.2 **A pair of pliers**

There are three basic types or orders of lever. The order of lever depends upon the position of the load (L), effort (E) and fulcrum.

- The **load** is the object to be moved
- The **effort** is the force applied to move the load
- The **fulcrum** is the point where the load is pivoted.

A first-order lever has the fulcrum between the effort and the load. If the fulcrum is moved closer to the load less effort is needed to move it (although the load does not move as far).

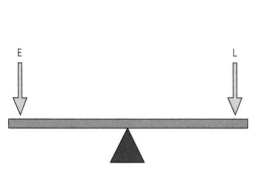

Figure 1.5.3 A first order lever

Figure 1.5.4 A seesaw is an example of a first-order lever.

A second-order lever has the load and effort on the same side of the fulcrum. The load is nearer to the fulcrum, so again less effort is required to move it.

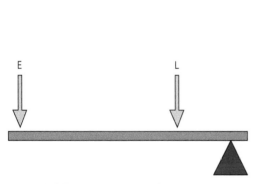

Figure 1.5.5 A second-order lever

Figure 1.5.6 A wheelbarrow is an example of a second-order lever.

In both first- and second-order levers with the fulcrum nearer to the load less effort is required to move the load. To make the effort smaller, the distance the lever moves is greater.

A third-order lever also has the load and effort on the same side as the fulcrum. Unlike first- and second-order levers the load is further from the fulcrum, so the effort required is greater than the load. These levers are used for items such as barbeque tongs and tweezers where the article being picked up is small or awkward, or when handling something that could be squashed or is fragile.

Different types of movement

KEY WORD

Linkages mechanisms that transfer force and can change the direction of movement.

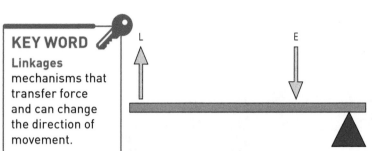

Figure 1.5.7 **A third-order lever**

Figure 1.5.8 **Barbeque tongs are an example of third-order levers.**

Figure 1.5.9 **A mirror on a linkage extension arm**

Linkages

Linkages are widely used in mechanisms to transfer force and can change the direction of movement. They are simply an assembly of levers to transmit motion and force.

Different linkages can be designed to transmit motions and force. The number and shapes of the linkages can change the direction of the force, and the position of the pivots can change the size or magnitude of the force. The closer the pivot is to the output end of the lever the larger the force is at the output.

Here are some examples.

In Figure 1.5.10 the direction of motion is reversed but the magnitudes of the forces are equal as the distances from the pivot are equal.

In Figure 1.5.11 the direction of motion is reversed but the magnitude of the output force is greater than that of the input force when the fixed pivot is closer to the output lever.

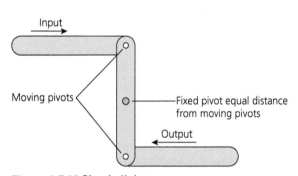

Figure 1.5.10 **Simple linkage**

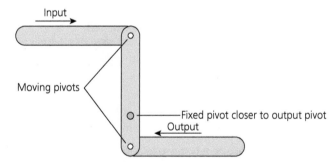

Figure 1.5.11 **Simple linkage with the pivot closer to the output end of the lever**

 and  are small bits, already covered.

In a bell-crank mechanism the direction of motion is turned through 90 degrees, but the magnitude of the output force is greater than that of the input force when the fixed pivot is closer to the output lever.

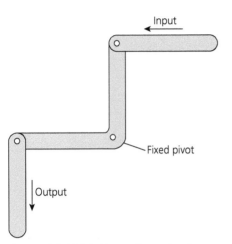

Figure 1.5.12 A bell-crank mechanism

Figure 1.5.13 A bicycle braking system uses a bell-crank mechanism

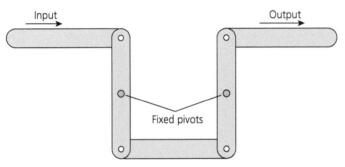

Figure 1.5.14 Parallel motion linkage

In a parallel motion linkage, often referred to as a push-pull linkage, the direction of motion and magnitude of the forces are the same but are used when the force cannot be transferred directly in a straight line.

Figure 1.5.15 This sewing box uses a parallel motion linkage.

ACTIVITY

Look around the school workshop and list as many machines as possible that use linkages.

STRETCH AND CHALLENGE

Using any modelling materials available to you, see if you can create a mechanism that turns rotary motion into oscillating motion (such as in car windscreen washers). When you have created your model, draw a diagram to explain the operation of this mechanism.

Different types of movement

Rotary systems

Cams and followers

A **cam** mechanism has three parts: a **cam**, a **slide** and a **follower**. When the cam rotates, the follower moves up and down in a reciprocating motion. The pattern the follower moves up and down in is governed by the shape of the cam; it can do three things:

- go up (rise)
- go down (fall)
- stay still (dwell).

Cams come in many different shapes to create different combinations of rise, fall and dwell.

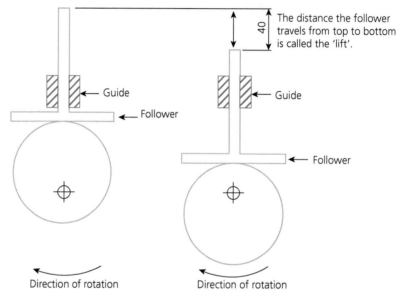

The distance the follower travels from top to bottom is called the 'lift'.

Figure 1.5.16 An eccentric cam mechanism

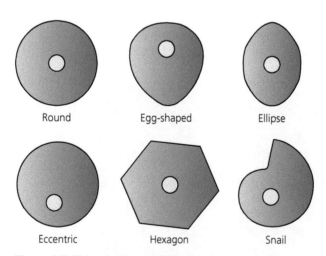

Round Egg-shaped Ellipse

Eccentric Hexagon Snail

Figure 1.5.17 A selection of different shaped cams

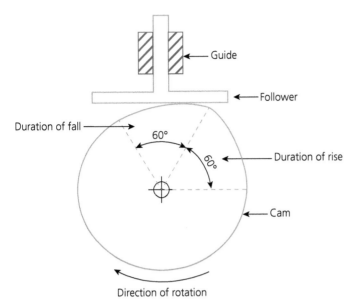

Figure 1.5.18 **The angle of rise and fall of a cam**

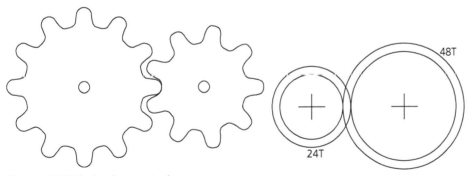

ACTIVITY
Use the internet to find machines that use cam mechanisms, crank and slider mechanisms and rack and pinion mechanisms. Make a chart listing the different machines in the three categories.

As seen in Figure 1.5.17, cams come in a variety of shapes and sizes depending on the purpose they serve. Figure 1.5.18 shows the profile of a cam with the duration of rise and fall of a cam as an angle. In the remaining 240° of the profile the follower stays still and does not move; this is the dwell.

Simple gear trains

A **gear train** is a mechanism for transmitting rotary motion and **torque**. The gears have teeth, and interlock or mesh with one another to transmit the rotary motion. Different-sized gears connected together either increase or decrease the speed of rotation and increase or decrease the torque transmitted.

KEY WORDS

Gear train a mechanism for transmitting rotary motion and torque.

Torque the turning force that causes rotation.

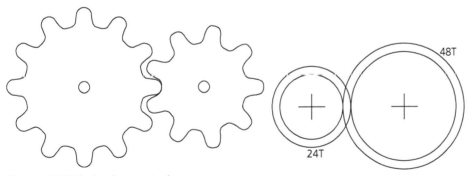

Figure 1.5.19 **A simple gear train**

Figure 1.5.20 **The symbol for a simple gear train**

The velocity ratio for a simple gear train is worked out using the formula:

$$\text{Velocity ratio} = \frac{\text{Number of teeth on the driven gear}}{\text{Number of teeth on the driver gear}}$$

In the gear train in Figure 1.5.20, where the driver gear has 24 teeth and the driven gear has 48 teeth:

$$\text{Velocity ratio} = \frac{48}{24} \qquad \frac{2}{1} = 2:1$$

So the velocity ratio is 1:2, meaning that the driven gear will rotate two times slower, or half the speed than the driver gear.

Different types of movement

Friction is the resistance to motion when one object rubs against another for example, when you rub your fingers on a desk they do not run smoothly across the desk because of friction. Friction slows the motion because it acts in the opposite direction. The **amount of friction** depends on the materials from which the two objects are made. Materials that offer little resistance to a substance moving over them are said to have a low **co-efficient of friction** and a material that offers a high resistance to a substance moving over it is said to have a high co-efficient of friction. In most mechanisms friction is a problem and needs to be reduced. It can be reduced by choosing materials that have a low co-efficient of friction, or it can be reduced by lubrication. Some mechanisms need friction in order to work, such as a belt and pulley systems and braking systems.

Many gear mechanisms use brass for the gears. It is only suitable for applications where forces are small, but it has a low co-efficient of friction so does not need lubrication. Many clocks have brass gears since the forces involved are very small. For the gears in cars steel is generally the favoured material as it is strong and hardwearing. This is important as a lot of torque is transmitted through these gears. For marine applications, stainless steel is the preferred material to make pulleys, winches and an assortment of riggings since it will not corrode or rust. Nylon gears offer advantages in some systems because of their ability to operate with minimum or no lubrication, as they have a relatively low co-efficient of friction, they are corrosion-resistant, and are quiet in operation. Below are some examples of gear trains, but there are others you can study.

A **rack and pinion mechanism** can turn rotary motion to linear motion or vice versa. You can see this mechanism in a pillar drill and it is used in most cars in the steering system. A rack is, in simple terms, a flat or straight gear whose teeth mesh with the teeth on a pinion gear. If the pinion is rotated about its centre, the rack will move in a straight line in one direction. If the rack is moved, the linear motion created will be converted into the rotary motion of the pinion. Most rack and pinion systems use steel because it is hardwearing.

Figure 1.5.21 A rack and pinion mechanism

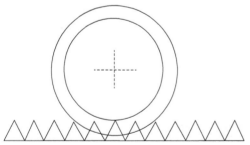

Figure 1.5.22 The symbol for a rack and pinion mechanism

A **crank and slider mechanism** can also turn linear motion to rotary motion or vice versa. By rotating a crank, a slider is made to move back and forwards along a cylinder or track. Alternatively, if the slider produces the input, the crank is forced to rotate. This is what happens with a piston in a car engine. A variety of materials are used in car engines that use crank and slider mechanisms. The sliders, or pistons as they are called in engines, are made of aluminium alloy because of its very high strength-to-weight ratio and because it conducts heat very well. The crankshafts in a car are made from steel because of its strength, toughness, and hardness.

Figure 1.5.23 A section of a car engine with a crank and slider mechanism

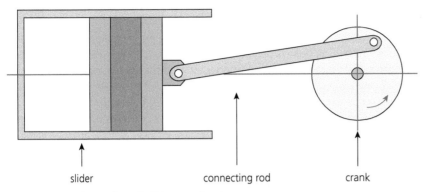

slider   connecting rod   crank

Figure 1.5.24 **A crank and slider mechanism**

A **chain and sprocket mechanism** transfers rotary motion to rotary motion elsewhere in a machine, such as in bicycles and motorcycles. A sprocket is a toothed wheel and a chain is made up of links that can pivot and go over the teeth on a sprocket. Rotary motion is transmitted between the shafts that the sprockets are attached to by the grip that the chain links have on the teeth of the sprocket as it rotates. When the sprocket on the motor is larger than the sprocket it is driving, the powered sprocket will spin faster and the **torque** will increase.

Figure 1.5.25 **A chain and sprocket mechanism**

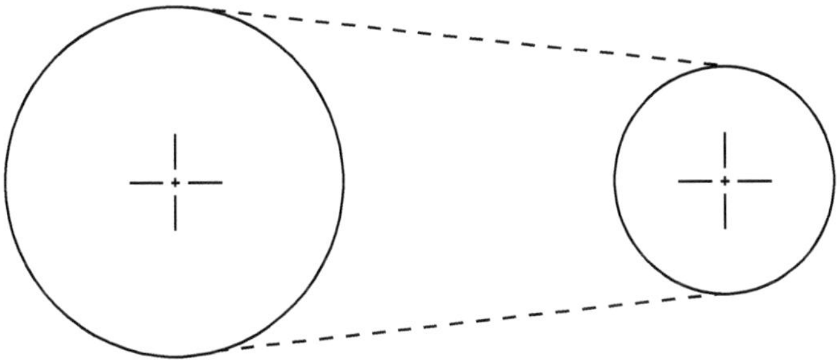

Figure 1.5.26 **Symbol for a chain and sprocket mechanism**

The velocity ratio for a sprocket and chain system is worked out using the formula:

$$\text{Velocity ratio} = \frac{\text{Number of teeth on the driven sprocket}}{\text{Number of teeth on the driver sprocket}}$$

If a bicycle had 60 teeth on the sprocket attached to the pedals and 12 teeth on the sprocket attached to the back wheels:

$$\text{Velocity ratio} = \frac{12}{60} \qquad \frac{1}{5} = 1{:}5$$

So the velocity ratio is 1:5, meaning that the driven sprocket will rotate five times faster than the driver sprocket.

Different types of movement

STRETCH AND CHALLENGE

Explain the advantages and disadvantages of using plastic gears in modern electronic devices such as printers and photocopiers.

Pulleys and belts

Pulley systems transmit rotary motion to rotary motion in machines. A pulley is just a simple wheel with a groove in its rim. Two pulleys connected together by a flexible belt will transmit rotary motion and torque. Different sized pulleys connected together either increase or decrease the speed of rotation and increase or decrease the torque transmitted. It is important that the belt does not slip on the pulley if all the movement is to be transferred. Pulleys are made of a variety materials, including aluminium, while the majority of belts used in pulley systems are rubber. Aluminium is used for the pulleys in machines using large turning forces because of its high tensile strength and relatively low weight. Rubber reinforced with high tensile materials like Kevlar (to prevent stretching) is used for belts so that it gets a good grip on the pulleys. For marine applications, stainless steel is the preferred material to make pulleys, since it will not corrode or rust.

Figure 1.5.27 **An electric motor driving a simple pulley system**

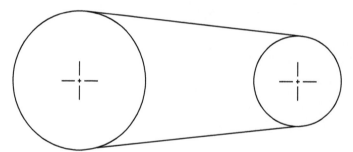

Figure 1.5.28 **The symbol for a simple pulley system**

When the pulley on the motor is larger than the pulley it is driving, the powered pulley will spin faster but the torque will be less. Similarly, when the pulley on the motor is smaller than the pulley it is driving, the powered pulley will spin slower but the torque will be increased.

The speed that two pulleys turn in relation to each other is called the **velocity ratio**. Speed changes are achieved by varying the diameters of the pulleys; for example, a small diameter pulley of 50 millimetres driving a larger diameter pulley of 200 millimetres reduces the speed of rotation. The velocity ratio can be calculated by using a formula:

$$\text{Velocity ratio} = \frac{\text{Diameter of the driven pulley}}{\text{Diameter of the driver pulley}}$$

$$\text{Velocity ratio} \ \frac{200}{50} = \frac{20}{5} = \frac{4}{1} = 4{:}1$$

So the velocity ratio is 4:1, meaning that the driven pulley will rotate four times slower than the driver pulley.

Check your knowledge and understanding

1 Explain what is meant by the term 'torque'.
2 Give examples of machines that use a rack and pinion mechanism.
3 Explain what is meant by the terms 'rise', 'fall', and 'dwell' in a cam mechanism.
4 Give examples of tools that use levers.
5 Discuss the advantages and disadvantages of using stainless steel gears and pulleys in the marine industry.

Materials and their working properties

1.6

What will I learn?

In this topic you will learn about:
→ the different classifications of papers and boards, their properties and common uses
→ the different classifications of natural and manufactured timbers, their properties and common uses
→ the different classifications of metals, their properties and common uses
→ the different classifications of polymers, their properties and common uses
→ the different types and properties of fibres and fabrics, their properties and common uses
→ the physical and mechanical working characteristics of materials.

Materials can be split into different groups according to their origins. These materials have their own working properties. As designers and manufacturers, it is important for us to know which materials are lightest, most durable, absorb water well or are flexible and so on. Understanding these properties allows us to choose the most suitable material for each specific use. We can also be better at designing as we are able to bear in mind the working characteristics of materials available to us when developing our ideas.

Paper and boards

Paper and boards are used for a variety of purposes, from writing and drawing to packaging and model making. Paper and board are made from cellulose fibres found in wood, rags or grasses, which are all renewable materials. Often paper and board is at least part-recycled, making it a more environmentally friendly material than polymer.

Paper and board has been developed to suit a number of different purposes. It can be given texture or watermarks and can be laminated with other materials such as plastic, which can give board waterproof properties.

Paper

Paper is measured in sizes from A0–A6 and in grams per square metre (gsm). Anything less than 200 gsm is considered a type of paper. For example, photocopier paper is usually 80 gsm. Paper was invented by the Chinese and is now used in huge quantities for drawing, writing and printing.

Paper	Properties	Common uses
Bleed proof	Smooth paper, often used with water and spirit based markers Prevents marker bleed (when ink runs and seeps through the paper)	Used for presentation drawings
Cartridge paper	Good quality white paper often with a slight texture Available in different weights	Due to the good-quality surface, it can be used for paints and markers as well as drawing
Grid	Paper printed with different grids as guidelines (these can be isometric or differently sized grids)	Quick model-making and working drawings
Layout paper	Thin translucent lightweight paper Can be drawn on with markers and takes colours well	Initial quick sketching and tracing
Tracing paper	Thin, transparent paper	Tracing copies of drawings

Table 1.6.1 **Paper properties and uses**

Isometric grid paper makes drawing in 3D easier by providing grid lines at 30° angles (as triangles). This allows us to accurately create isometric drawings.

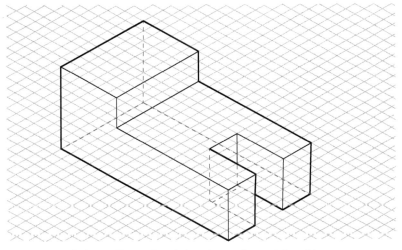

Figure 1.6.1 **A three-dimensional drawing on isometric grid paper**

Boards

Boards (card or cardboard) are always greater than 200 gsm. An example of this is corrugated cardboard which is 250+ gsm.

Corrugated card is a relatively light material and is often used to protect products during transport. It is normally over 2.5 millimetres thick and consists of a fluted or ridged layer between two thin layers of cardboard.

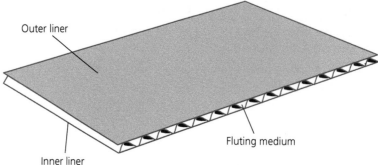

Outer liner

Fluting medium

Inner liner

Figure 1.6.2 Corrugated cardboard

Corrugated cardboard is a good modelling material, but has been more recently used to make load-bearing furniture due to its strength-to-weight ratio. It is a relatively environmentally friendly product since it is possible to manufacture it from recycled materials and to recycle it after use. It is made from a renewable material and has become fashionable in a throwaway society used to more disposable products.

Figure 1.6.3 Corrugated cardboard seat

ACTIVITY

Collect a number of products made from paper and board. These could include cereal boxes, book covers, gift cards, etc. Can you identify the type of paper or board they are made from? Use Tables 1.6.1 and 1.6.2 to help you.

Boards	Properties	Common uses
Corrugated card	Strong, lightweight material Made up of two or more layers and a fluted middle section leading to good insulating properties Available in different thicknesses	Packaging such as pizza boxes, and large boxes used for heavy items that need protecting
Duplex board	Thin board which often has one side that is suitable for printing	Food packaging
Foil-lined board	Board covered on one side with aluminium foil, making it a good insulator of heat	Takeaway or ready-meal packaging
Foam core board	Two pieces of board with a core of foam to increase the thickness Thick board that is very lightweight	Model making, such as architectural models
Inkjet card	Treated so it can be used in all inkjet printers	Printing in inkjet printers
Solid white board	Top quality cardboard, smooth and white Good for printing on	Book covers

Table 1.6.2 Card and board properties and uses

Natural and manufactured timbers

Timber is a common material that you will come across in the school or college workshop which can be used for many popular products. It is readily available and has the benefit of being recyclable, reusable and renewable.

Timber is a natural material; its original source is trees. It can be categorised into two groups: **hardwoods** and **softwoods**.

Hardwoods

Hardwoods such as oak and beech come from deciduous trees – those that generally lose their leaves in autumn. Deciduous trees take a long time to mature before being able to be felled and turned into useable timber; as a result they tend to be more expensive than softwoods. They can be identified by their broad leaves and branches grouped at the top of the tree.

There are a few exceptions to the rule, such as holly, which is a hardwood but keeps it leaves all year around.

Softwood:
- mostly evergreen
- retain leaves all year round
- needle- or scale-like leaves
- bear cones.

Hardwood:
- mostly deciduous
- shed leaves each autumn
- typically flat leaves.

Figure 1.6.4 Hardwood vs softwood

Hardwood	Properties	Common Uses
Ash	Tough and flexible, wide grained. Finishes well	Sports equipment, ladders
Beech	Hard, strong, close grain. Prone to warping and splitting	Furniture, children's toys, workshop tool handles and bench tops
Mahogany	Strong and durable. Available in wide planks. Fairly easy to work but can have interlocking grain	Good quality furniture, panelling and veneers
Oak	Hard, tough, durable, open grain. Can be finished to a high standard.	Timber framed buildings, high quality furniture, flooring,
Balsa	Strong and durable, lightweight, easy to work	Model making, floats and rafts

Table 1.6.3 Common hardwoods

Figure 1.6.5 Cricket stumps made from ash

Ash is commonly used for the manufacture of cricket stumps and bails due the fact that it is a tough and flexible material, making it durable enough to withstand the impact from a cricket ball travelling at speeds of up to 160 kilometres per hour.

Softwoods

Softwoods such as pine and spruce come from coniferous trees, also known as evergreens. Coniferous trees are quick growing and take around ten years to reach maturity before felling. This makes them an extremely sustainable group of materials as they are renewable. They can be identified by their cones and needle-shaped leaves, their branches being located the whole length of the trunk and their triangular shape.

They are most commonly found in products designed to be used indoors and for joinery as most have poor resistance to decay and require the addition of preservatives before being used outside. Remember, most softwoods are less dense than hardwoods, but not always. Balsa is a hardwood as it is comes from a deciduous tree, but it is soft, lightweight and can be easily cut with a craft knife.

Softwood	Properties	Common Uses
Larch	Reddish in colour and has a striking grain pattern Tough but easy to work, although quite resinous and prone to splitting Naturally resistant to rot	Fencing, fence posts, cladding and decking
Pine	Straight grained, light yellow in colour Soft and easy to work Can be quite knotty	Interior joinery and furniture, window frames
Spruce	Creamy white in colour Easy to work with small knots Lightweight with good resonant properties	Bedroom furniture, stringed musical instruments

Table 1.6.4 **Common softwoods**

Figure 1.6.6 Larch shed and fence

Larch is one of the few types of softwood that is used outside. It is used in the manufacture of fencing and cladding for sheds, due to its toughness and natural resistance to rot from moisture. It is so durable in a moist environment that the piles and poles on which Venice is held above the water are built almost exclusively from larch.

Manufactured boards

Manufactured boards were developed as an alternative to natural timbers. They proved so popular and versatile that in many applications they have almost completely replaced the use of natural timber. Kitchen manufacturing and developments in self-assembly furniture are two areas where manufactured boards are almost exclusively used.

Manufactured boards fall into two categories: **laminated boards** are produced by gluing large sheets or veneers together and **compressed boards**, as the name suggests, are manufactured by gluing particles, chips or flakes together under pressure.

Advantages of manufactured boards:
- They are available in much larger sheets than solid timber.
- They have consistent properties throughout the board.
- They are more stable than natural timbers, meaning they are less likely to warp, shrink or twist.
- They can make use of lower grade timber, so can have environmental benefits.
- They can be faced with a veneer or a laminate to improve their aesthetic appearance.
- Due to their consistent quality, they are well suited to CNC machining and volume production.

Medium-density fibreboard (MDF)

Medium-density fibreboard (MDF) is a compressed board that is manufactured from fine fibres of wood combined with a synthetic adhesive (usually formaldehyde resin). The MDF pulp is compressed between two heated plates where the adhesive bonds the fibres together. MDF makes use of low-grade softwood and hardwood timber, along with the waste created from other wood manufacture processing. Care should be taken to limit the dust produced when working with MDF as it can cause respiratory issues due to the size of the fine particles.

The surface of MDF boards is smooth, which makes it easy to apply a high-quality paint finish. The edges of the board are fibrous and so need additional sealing before painting.

The MDF pulp is compressed to 0.25 per cent of its original thickness, which is why MDF is denser than other manufactured boards.

In addition to standard board, MDF is also available in a range of specialist versions, including moisture-resistant board, fire-resistant board and flexible MDF. Flexible MDF has a series of small grooves or kerfs cut into one side of the material that allows the board to bend around a radius. This particular form of MDF is often used in shop fitting applications.

Figure 1.6.7 Medium-density fibreboard

MDF is also commonly faced with a veneer to improve its aesthetics. Common veneered or faced boards include oak-faced, ash-faced and beech-faced. Faced MDF can be single or double sided, and adhesive veneer edging tape of the corresponding material can be applied with an iron or edge banding machine to improve the aesthetic of the exposed MDF edge.

Figure 1.6.8 The application of veneered MDF edging strip

Common uses of MDF include flat-pack furniture, decorative mouldings and shop interiors.

Plywood

Plywood is a laminated board made up several veneers of wood glued on top of each other. Each layer is laid at a 90° angle to the last, so that the grain alternates in direction. The number of layer is always an odd number and the two outside surfaces always have the grain running in the same direction. This arrangement of layers gives plywood a consistent strength across the whole board. The adhesive used to produce plywood is a formaldehyde resin.

Figure 1.6.9 Plywood structure

Natural and manufactured timbers

Where the plywood is going to be visible in use, it is common to find that the outside veneers are made from a more expensive wood, such as birch or oak.

Due to the very nature of the material, veneer-faced boards will differ in appearance from one to the next. While impossible to specify its appearance, timber merchants will use a plywood grading system to help with appropriate selection. 'A grade' plywood will be blemish free and of high quality, whereas 'D grade' plywood will have knots and repair patches visible.

Figure 1.6.10 A grade plywood   **Figure 1.6.11 D grade plywood**

Figure 1.6.12 Maple ply skateboard

One of the advantages of plywood over other manufactured boards is its stiffness. This means that it is hard to bend into other shapes. You will however often see plywood in bent forms. In these cases, the glued veneers will have been compressed in a shaped mould as they dried.

You can clearly see the alternating plywood layers in the laminated skateboard in Figure 1.6.12, where the alternate grain provides a stiff, strong board.

Most plywood is manufactured with interior applications in mind, although it is also available in weatherproof and marine versions, where the adhesive and types of wood veneers are selected to be more suitable for exterior use. Common uses of plywood include laminated flooring, roofing and furniture.

Chipboard

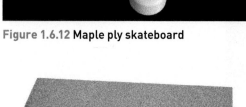

Figure 1.6.13 Chipboard

Chipboard, sometimes referred to as 'particle board', is a manufactured board that is made up of large flakes or chips of timber glued together under pressure. Chipboard is cheaper to produce than MDF and plywood, but is not as strong or durable. It is commonly used in applications where cost is a more important factor than strength or aesthetics. It can be faced with a laminate or plastic film as the unfinished surface is usually rough. Many popular chipboard applications have been superseded by MDF.

Common uses of chipboard include kitchen work surfaces, kitchen cupboards and flooring.

ACTIVITY

Have a look around your house and identify five wooden products. Can you determine what timber they are manufactured from? Think about their function, the environment that they are found in and their appearance. Are they made from hardwood or softwood?

1.6 Materials and their working properties

Metal and alloys

Metals have been used by society since the Bronze Age and Iron Age periods (around 9000BC). Since this time, huge developments have been made in the field of metallurgy (the study of metals). Metals in their many forms and varieties are vital materials in today's society. They play an integral role in the manufacture of buildings, vehicles and household products, ranging from the Shard in London through to the keys to your house.

Metal is a naturally occurring material and is mined from the ground in the form of ore. The raw metal is then extracted from the ore through a combination of crushing, smelting or heating with the addition of chemicals and huge amounts of electrical energy. Most metals can be recycled, saving natural resources and limiting the amount of materials imported from abroad.

Metals can be categorised into two groups: **ferrous** metals and **non-ferrous** metals.

Ferrous metals

Ferrous metals are those that contain iron. Most are magnetic, which is a useful property when it comes to sorting out metals when recycling. Their carbon content means that most are prone to corrosion in the form of rust when exposed to moisture and oxygen.

Their properties, such as hardness and malleability, are directly related to their carbon content. For example, the more carbon that is found in the metal, the harder and less malleable the metal becomes.

A good way of remembering that ferrous metals contain iron is to remember the periodic table symbol for iron. **FE** = iron = **fe**rrous.

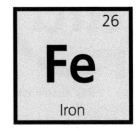

Figure 1.6.14 Iron in the periodic table

Ferrous metal	Composition	Properties	Common uses
Cast iron	Iron and 3.5% carbon	Hard surface but has a brittle soft core Strong compressive strength Cheap	Vices, car brake discs, cylinder blocks, manhole covers
Low carbon steel (mild steel)	Iron and 0.15–0.35% carbon	Good tensile strength, tough, malleable Poor resistance to corrosion	Car bodies, nuts, bolts, and screws, RSJ's and girders
High carbon steel (tool steel)	Iron and 0.70–1.4% carbon	Hard but also brittle Less tough, malleable or ductile than medium carbon steel	Screwdrivers, chisels, taps and dies

Table 1.6.5 Common ferrous metals

Low carbon, or mild steel, is one of the most widely used of the ferrous metals. It has excellent tensile strength and when fabricated into an I-beam cross section, it can be used to produce rolled steel joists (RSJs). These are widely used in the construction of buildings.

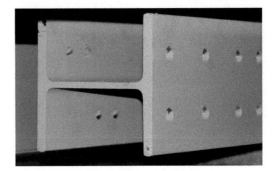

Figure 1.6.15 RSJs used in buildings

Non-ferrous metals

Non-ferrous metals are those that don't contain iron. The absence of iron makes non-ferrous metals desirable for their malleable properties and for their resistance to corrosion. The majority of them are also not magnetic, which means that they can be used in electronic devices and wiring.

After steel, aluminium is the most widely used metal. Aluminium is produced from alumina, which is extracted from an ore known as bauxite. In addition to heating, several chemicals are used to help the separation process, including caustic soda and lime. The alumina then goes through an electrolysis process from which liquid aluminium is then obtained. All of this processing takes a huge amount of energy, which is why aluminium is so regularly recycled. It takes around 95 per cent less energy to recycle aluminium than to produce the raw material from bauxite.

Figure 1.6.16 Aluminium recycling

Non-ferrous metal	Composition	Properties	Common uses
Aluminium	Pure metal (often alloyed with copper and manganese depending on application)	Lightweight, soft, ductile and malleable A good conductor of heat and electricity Corrosion resistant	Aircraft bodies, high-end car chassis, cans, cooking pans, bike frames
Copper	Pure metal	Extremely ductile and malleable An excellent conductor of heat and electricity Easily soldered and corrosion resistant	Plumbing fittings, hot water tanks, electrical wire
Zinc	Pure metal	Weak in its pure state High level of corrosion resistance Low melting point Easily worked	As a galvanised coating in crash barriers, corrugated roofing, intricate die cast products
Tin	Pure metal	Soft ductile and malleable Low melting point Excellent corrosion resistance	Coatings on food and drinks cans, solders

Table 1.6.6 Common non-ferrous metals

Figure 1.6.17 Copper fixings and fittings

Copper is a versatile material due to its wide range of desirable properties. It is commonly found in plumbing fixings and heating systems because it is easy to solder and will not corrode when in contact with moisture.

Alloys

Metals in their pure form can be useful for many purposes, but it is often desirable to adjust their mechanical and physical properties in order to produce a more suitable material for a particular application. An **alloy** is a material that is produced by combining two or more elements together to produce a new material with refined properties. Alloys too can be categorised as ferrous alloys or non-ferrous alloys, depending on the main pure metal that they contain.

Alloy	Composition	Properties	Common uses
Brass – non-ferrous alloy	Alloy of copper (65%) and zinc (35%)	Strong and ductile Casts well Corrosion resistant Conductor of heat and electricity	Castings, forgings, taps, wood screws
Stainless steel – ferrous alloy	Alloy of steel including chromium (18%) and nickel (8%)	Hard and tough Excellent resistance to corrosion	Sinks, cutlery, surgical equipment, homewares
Duralumin	Alloy of aluminum (90%) copper (4%) magnesium (1%) manganese (0.5–1%)	Strong, soft and malleable Excellent corrosion resistance Lightweight	Aircraft structure and fixings, suspension applications, fuel tanks

Table 1.6.7 **Common alloys**

Stainless steel is a ferrous alloy that is regularly used in the manufacture of surgical equipment, catering items and decorative homeware. It has a 18 per cent chromium content, which when combined with the oxygen in the atmosphere produces an oxide that protects it from corrosion. It is extremely hard, corrosion resistant and can be polished to give a shiny finish, which made it the material of choice for this iconic piece of design (Figure 1.6.18).

Figure 1.6.18 **Michael Graves kettle for Alessi (1985)**

ACTIVITY

As we have already mentioned, metals can be found in a huge range of products.
- Using a similar table to the one provided, try to identify the range of metals that can be found in common products. Try to work out if there are hidden metals enclosed within them.
- Why has that material been used? What material requirements did the product have? What desirable property does your chosen material possess?

Product	Low carbon steel	Aluminium	Copper	Stainless steel	What properties does your chosen material have that makes it suitable for the product?
Bicycle					
Cooker					
Kettle					
Pushchair					
Garden gate					

Metal and alloys

Polymers

The popularity of polymers (more commonly referred to as plastics) has continued to grow since their common introduction into consumer products in the 1950s. Their ability to be coloured, shaped and formed, along with their cost and versatile range of working properties, has allowed designers and manufacturers to improve the performance of products and replace the use of more traditional materials.

The majority of polymers are manufactured from the non-renewable resource crude oil. The use of crude oil is not sustainable and chemical engineers are constantly looking for reliable alternatives to petrochemical-based polymers. The increase in popularity means that the environmental impact of polymer use has grown and designers must consider this when selecting materials for use.

Polymers can be categorised into two groups: thermoforming polymers and thermosetting polymers.

Thermoforming polymers

Thermoforming polymers are the most common, used and are found in the manufacture of a huge range of products. They can be moulded into almost any shape have pigment added to them so can be found in a wide range of colours, and most importantly they can be recycled.

Thermoforming polymers can be softened by heating. Once softened, or plasticised, they can be shaped and formed using a wide variety of processes. Once the desired shape has been achieved the polymer cools and maintains its new shape. This process of heating, shaping and cooling can take place over and over again with minimal damage to the properties of the polymer.

ACTIVITY

Gather a selection of polymer products from around your home; they could be children's toys, kitchenware, or packaging. Try to identify what polymers they are made from and why. You may be lucky and find the polymer initials moulded in to the product, otherwise use Tables 1.6.8 and 1.6.9 to help you explore the properties and identify the correct polymer.

Thermoforming polymer	Properties	Common uses
Acrylic (PMMA)	Hard Excellent optical quality Good resistance to weathering Scratches easily	Car light units, bath tubs, shop signage and displays
High-impact polystyrene (HIPS)	Tough, hard and rigid Good impact resistance Lightweight	Children's toys, yoghurt pots, refrigerator liners
High-density polythene (HDPE)	Hard and stiff Excellent chemical resistance	Washing up bowls, buckets, milk crates, bottles and pipes
Polypropylene (PP)	Tough Good heat and chemical resistance Lightweight Fatigue resistant	Toys, DVD & blu-ray cases, food packaging film, bottle caps and medical equipment
Polyvinyl chloride (PVC)	Hard and tough Good chemical and weather resistance Low cost Can be rigid or flexible	Pipes, guttering, window frames
Polyethylene terephthate (PET)	Tough and durable Lightweight Food safe Impermeable to water Low cost	Drinks bottles, food packaging

Table 1.6.8 Common thermoforming polymers

Polypropylene is a polymer that is often chosen for applications where there is a hinge moulded into the product. Polypropylene has excellent resistance to fatigue, so the hinge can be opened and closed without breaking or work-hardening.

Thermosetting polymers

Thermosetting polymers can also be shaped and formed by heat, but in contrast to thermoplastics this process can only occur once. Thermosets cannot be reheated or reformed. This makes them excellent insulators but also means that thermosetting polymers cannot be recycled.

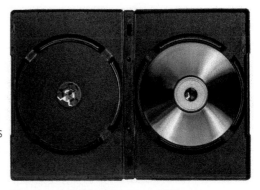

Figure 1.6.19 Polypropylene DVD case

Thermoset	Properties	Common uses
Epoxy resin	Electrical insulator Good chemical and wear resistance	Adhesives such as araldite™, PCB component encapsulation
Melamine formaldehyde (MF)	Stiff, hard and strong Excellent resistance to heat, scratching and staining	Kitchen work-surface laminates, tableware
Phenol formaldehyde (PF)	Hard Heat and chemical resistant Good electrical insulator Limited colours available	Electrical fittings, saucepan handles, bowling bowls
Polyester resin	Brittle but becomes tough when laminated with glass fibre Hard and resistant to UV	GRP boats, car body panels,
Urea formaldehyde (UF)	Stiff and hard Heat resistant Good electrical insulator	White electrical fittings, toilet seats, adhesive used in MDF

Table 1.6.9 Common thermosetting polymers

KEY WORDS

Thermoforming polymers polymers that can be softened by heating, shaped and set over and over again.

Thermosetting polymers polymers that can only be shaped and formed by heat once.

Melamine is found in the manufacture of picnic crockery. It is hard and so it resists scratching, which enables it to maintain its aesthetics while being used and cleaned. It is also more durable than ceramic or glass alternatives.

Polymer additives

The properties of polymers can be further enhanced through the introduction of additives.

- Plasticisers can be added to make the polymer soft and flexible, as often found in PVC.
- Pigments can be added to change the colour of the polymer.
- Stabilisers can be added to help the polymer withstand UV light damage, which is especially useful in products that may be used outdoors.
- Fillers can be used to increase the bulk of the polymer; this can help improve its impact resistance.

Figure 1.6.20 Melamine crockery

Textiles

Fabrics are used to make textile products – the most obvious ones are clothing and home furnishings, but they are also used in many other ways, such as medical applications, car interiors and engines, road and house building, and **safety and security products**. There are many types of fabric available and each will be different depending on the fibre it is made from and the way the fabric is made. When choosing fabric it is important to select one that will perform well for the product being made.

Fibres are very fine hair-like threads and are the basic building blocks of fabrics. Fibres are either **natural** (from plant or animal sources) or **synthetic** (manufactured from oil-based chemicals). A fibre can be short, called a **staple fibre**, or a very long continuous length, called a **filament fibre**. All natural fibres, except silk, are staple fibres. All manufactured fibres are filament fibres. Fibres have different properties according to where they come from.

Natural fibres

Natural fibres can come from plant or animal sources.
- Cotton fibres come from the seed of a bushy plant grown in tropical parts of the world.
- Wool is a hair fibre made from protein and comes mostly from sheep. Some luxury wools come from goats, rabbits and other furry animals.
- Silk is a protein fibre that comes from the cocoon of the silk caterpillar.

Natural fibre	Properties	Common uses
Cotton	Strong, good at absorbing moisture (this means they can take a long time to dry) Can be washed and ironed at high temperatures Creases badly and shrinks unless a special finish is applied Easy to set alight, so can be dangerous	T-shirts, socks and underwear, denim jeans, bed sheets, fishing nets, medical dressings, nappies
Wool	Soft and warm Comfortable to wear; will not crease easily Water-repellent, but also very good at absorbing moisture Takes a long time to dry Most wools will shrink if put in a tumble dryer Does not set alight easily and when it does, it puts itself out Shrinks badly (felting) and therefore difficult to wash unless a special finish is applied to prevent shrinking	Jumpers, coats, socks, blankets, carpet, tennis balls, pool tables, mattresses
Silk	Fibres have a triangular cross section that makes it soft and smooth and gives it a lustre Lightweight, absorbent, warm in cold conditions but cool in hot weather A strong fibre that becomes weak when wet so needs to be washed carefully Has natural elasticity so can crease very badly Expensive and often considered to be a luxury fibre	Evening dresses, ties, lingerie, bedding, wall hangings, parachutes

Table 1.6.10 **Common natural fibres**

1.6 Materials and their working properties

Figure 1.6.21 Cotton is used to produce denim jeans.

Figure 1.6.22 Wool is soft and warm and therefore often used to produce coats, hats and jumpers.

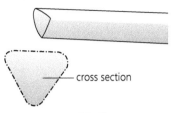

cross section

Figure 1.6.23 Silk fibre

Synthetic fibres

Synthetic fibres are manufactured from oil-based chemicals. Examples of synthetic fibres are polyester, polyamide and elastane.

Synthetic fibre	Properties	Common uses
Polyester	Very strong and resistant to abrasion Does not absorb water so will dry quickly Crease resistant Will soften when heated (it is thermoplastic) and can be heat-set into new shapes that it will maintain when cooled down Finishes can be added easily (for example, permanent pleats and creases) A smooth fibre that does not trap air, so is not very warm to wear	Clothing, bedspreads, sheets, pillows, padding for upholstery, carpets, curtains, ropes, sails for boats
Polyamide (nylon)	Fine and lightweight but extremely strong and abrasion resistant Does not absorb moisture, stays strong when wet Not affected by alkalis but it is weakened by bleach Long exposure to sun will turn white nylon yellow and eventually rot the fabric Thermoplastic so it can be heat-set	Underwear, shoelaces, tights, tents, parachutes, carpets, seatbelts
Elastane (Lycra is the most well-known elastane fibre)	Very stretchy (can stretch by up to six times its length and then return to its original length), allowing it to fit close to the body and give freedom of movement Because it is so stretchy it cannot be used on its own and needs to be blended with other fibres Crease resistant Easily washable Absorbent Resistant to perspiration and quick drying Not very warm to wear	Swimwear, sportswear, leggings, tights

Table 1.6.11 Common synthetic fibres

Figure 1.6.24 **Polyester is very strong and resistant to abrasion, making it ideal for ropes and sails on boats.**

Figure 1.6.25 **Elastane is very stretchy, fits close to the body and allows for movement, and is therefore often used in sportswear.**

Blended and mixed fibres

Most modern fabrics contain more than one fibre. This is because there is no such thing as a perfect fibre so manufacturers include different fibres in a **blend**. Blending is achieved by spinning two or more fibres together to make a yarn. This enables a fabric to be made which will be better suited to the product. Some important reasons for blending fibres are:

→ to help reduce the cost of the fabric
→ to make the fabric stronger
→ to make a fabric easier to care for
→ to enable fabrics to be more crease-resistant
→ to allow fabrics to be heat-set.

Polyester and cotton blends are commonly used to make a wide variety of fabrics. Different percentages of cotton and polyester are included according to what the fabric is to be used for. The polyester helps cancel out the shrinking, creasing and slow-drying of cotton. The cotton makes the fabric better at absorbing moisture and makes the fabric feel nicer next to the skin. But, polyester and cotton blends are very dangerous when they set alight. This is because the cotton burns easily and holds the polyester in place. As the polyester gets hot it starts to melt and drip. The fabric burns very fiercely at high temperatures and gives off a lot of black smoke. Polyester and cotton blends are commonly used to produce shirts, bed sheets, car seat covers and furniture.

Figure 1.6.26 **A polyester and cotton blend is commonly used to make bed sheets**

Elastomers like Lycra are often blended with many other fibres. The Lycra gives the fabric some stretch. Only very small amounts of Lycra are needed to give a lot of stretch. The Lycra also makes the fabric more crease resistant.

Wool is often blended with **nylon** for products such as socks, trousers, jackets and coats. The wool makes the fabric soft and warm, and the nylon gives improved strength and resistance to abrasion, makes the fabric lighter in weight, and helps prevent the wool from shrinking when it is washed.

Woven fabrics

Woven fabrics are produced by interlacing two sets of yarns at right angles to each other on a machine called a loom. The **warp** yarns run the length of the fabric. The **weft** yarns run across the width of the fabric. At the edge where the weft yarns turn round a finished edge called a **selvedge** is formed. The selvedge runs down the length of the fabric.

The warp yarns are usually referred to as the **straight grain** of the fabric and this is the direction that the fabric is strongest (warp yarns need to be able to stand up to the constant stress and strain as they move up and down during the weaving process).

The weft yarns can pull out of the unfinished edge of the fabric – the **raw edge**. This is called **fraying** and will eventually cause the fabric to disintegrate.

The interlacing of the warp and weft yarns makes the fabric strong and stable as the yarns do not stretch much, but the fabric will stretch diagonally – this is called the **bias** of the fabric.

There are three main types of weave: plain, twill and satin. (For more information on these weaves, see Topic 2.4 Sources and origins.)

The **plain weave** is made by passing the weft yarn alternately over and under the warp. On each new row, the weft goes under the warp it went over on the previous row.

Features of plain weave fabrics

- The plain weave is the simplest and therefore the cheapest weave to produce.
- It produces firm, strong, hardwearing fabrics which look the same on both sides, and their smooth plain surface makes a good background for printing.
- Plain weave fabrics include calico, lawn, poplin and chiffon.
- Plain weave fabrics are often used for fashion and furnishing fabrics.

Non-woven fabrics

Some fabrics can be made directly from fibres without being woven or knitted. These fabrics are called **non-woven fabrics** and include felts and bonded fabrics.

Felted fabrics are made from wool fibres and use the natural felting ability of the wool to cause the fibres to matt together using heat, mechanical action and moisture.

Bonded fabrics are made from webs of fibres, which are held together in various ways:
- using a special adhesive
- thermal bonding, which makes use of the thermoplastic properties of some or all of the fibres, to fuse all the fibres together using heat and pressure
- stitching with thread (stitch bonding)
- needle punching, which tangles the fibres together.

Figure 1.6.27 A weaving loom

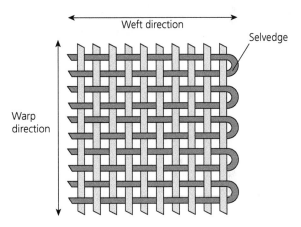

Figure 1.6.28 Warp, weft, bias, selvedge and raw edges of fabric

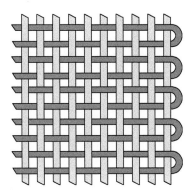

Figure 1.6.29 Plain weave

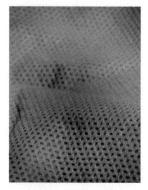

Figure 1.6.30 Bonded fibre fabric

Non-woven fabrics	Properties	Common uses
Bonded	Cheap to manufacture and use Not as strong as woven or knitted fabrics, and do not drape as well Easy to sew Crease-resistant Does not fray	Interfacings and interlinings, disposable items (for example, cleaning cloths and hospital items such as dressings)
Felted	Warm and soft Does not fray Not very strong does not drape well, no elasticity Expensive	Hats, slippers, toys, insulation materials, upholstery

Table 1.6.12 Non-woven fabrics

Knitted textiles

Knitted fabrics are made in a very different way from woven fabrics, as the yarns do not interlace with each other. Instead, the yarns are looped together to make looser, more flexible fabrics. The two main types of looping make **weft knit** fabrics and **warp knit** fabrics. (More information on how these knitted fabrics are produced can be found in Topic 2.4 Sources and origins.)

Knitted fabrics	Examples	Properties	Common uses
Weft knit fabrics	Jersey, rib knits, polyester fleece	Very stretchy but can be pulled out of shape Have a soft drape and do not crease easily Trap air easily and are therefore warm in still air, but cool in windy weather as air can still get through the gaps in the fabric Will ladder easily if snagged	Socks, t-shirts, jumpers, scarves, hats, leggings
Warp knit fabrics	Tricot, knitted lace	Less stretchy than weft knits; firm Do not ladder and cannot be unravelled 'row by row'	Swimwear, underwear, net curtains, industrial textiles and geotextiles

Table 1.6.13 Knitted fabrics

Figure 1.6.31 Net curtains are made from warp knit fabrics.

1.6 Materials and their working properties

Material properties

There are many factors that need to be considered when selecting a material for use. These could include cost and availability, but must include determining the properties that the material must have to successfully perform its function.

Physical properties are those that describe how a material behaves under a specific condition, while mechanical or working properties describe how it behaves when being worked or shaped.

Physical properties

Fusibility can be described as how easily a material's state can be altered to become a liquid. This is an important property in solder, where its low melting point allows it to melt with the heat of a soldering iron.

Electrical conductivity can be described as how easily electrical energy can pass through a material. Gold is an excellent conductor of electricity and, although an expensive material, it is regularly used in small quantities in high-end electrical components.

Thermal conductivity can be described as how easily heat energy can pass through a material. Aluminium is an efficient conductor of heat, which makes it a popular choice when manufacturing pots and pans.

Resistance to moisture can be referred to as a material's ability to prevent liquid and moisture permeating its surface. PVC has excellent moisture-resistant properties and can be added as a coating to fabrics that can then be used to produce waterproof clothing. Other hydrophobic coatings can be applied to fabrics to achieve a water-resistant coating.

Absorbency can be described as a material's ability to soak up and retain liquid, heat or light. Absorbency is an important property when related to fabrics. Natural fibres such as cotton and wool have high absorbency, making them comfortable to wear as they absorb perspiration. Synthetic fibres are virtually non-absorbent so can be very uncomfortable to wear – like wearing a plastic bag.

Mechanical properties

Strength can be described as the ability of a material to withstand a constant force without breaking. Strength is usually associated with one of five forces that can act upon a material. These include: bending, compression, tension, shear and torsion. For more information on these forces see Topic 2.2 Forces and stresses.

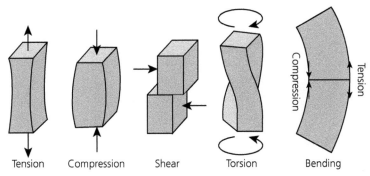

Figure 1.6.32 **Forces**

Hardness can be described as the ability of a material to withstand scratching, cutting and abrasion. Melamine formaldehyde is regularly used for kitchen work surfaces. It maintains its smooth polished finish due to its hardness.

Figure 1.6.33 Child's outdoor toy

Density can be described as a material's mass per unit volume. Expanded polystyrene is used in packaging for protecting expensive or fragile products. It is chosen due to its low density.

Toughness can be described as a material's ability to withstand impact from a dynamic force.

Children's toys are subjected to a huge amount of wear and tear and heavy use. They are often left outside and so need to be durable. HDPE is used in the manufacture of the child's car in Figure 1.6.33 due to its excellent toughness and wide range of available colours.

Malleability can be described as a material's ability to be permanently deformed or shaped by impact, rolling or pressing without breaking. Low carbon steel is used to produce car body panels. It is its malleable property that allows it to be pressed into shape.

Ductility can be described as a material's ability to be drawn or pulled in to a long length or wire without breaking.

When manufacturing electrical cable it is important to select a material that is a good conductor of electricity and which is also a ductile material, so it can be formed into a long wire. Copper possesses both of these properties. PVC is used to coat the wire, as it is a good electrical insulator, is flexible and can be coloured to help with identification.

Figure 1.6.34 Copper wire

Elasticity is a measurement of a material's ability to stretch under force and return to its original shape without deformation when the force is removed.

Figure 1.6.35 Lycra sportswear

Many sportspeople now wear high-performance fitted outfits while training and competing in events. Lycra or elastane can be added to polyester and used where comfort, support and the ability to stretch and move with the body of the athlete is important. In cycling it can also provide an aerodynamic advantage.

KEY POINTS

- Paper is classed as less than 200 gsm.
- Board is classed as over 200 gsm.
- Other materials can be added to paper and card to make it stronger, waterproof or an insulator of heat, etc.
- Hardwoods come from deciduous trees, softwoods come from coniferous trees.
- Hardwoods take approximately ten times longer to mature than softwoods.
- Manufactured boards come in larger sizes than natural timbers.
- Manufactured boards are more stable than natural timber and won't split or twist.
- Veneers and laminates can be added to manufactured boards to improve their appearance.
- Ferrous metals contain iron and are normally magnetic.
- Non-ferrous metals do not contain iron.
- Alloys are combinations of two or more pure metals with other elements.
- Thermoforming polymers can be repeatedly heated, formed and cooled.
- Thermosetting polymers can only be formed with heat once.
- Thermosetting polymers cannot be recycled.
- Natural fibres can come from plant or animal sources.
- Synthetic fibres are manufactured from oil-based chemicals. Examples of synthetic fibres include polyester, polyamide and elastane.
- Different fibres are blended together to make them better suited to different products and to improve a fabric's properties.
- Plain weave fabrics are strong and hardwearing.
- Non-woven fabrics are made directly from fibres without being woven or knitted, and include felts and bonded fabrics.
- Knitted fabrics have yarns that are looped together to make looser, more flexible fabrics.

Check your knowledge and understanding

1 What is a common use for duplex board?
2 Compare the properties of cartridge paper with those of foil-lined board.
3 What are the major differences between a hardwood and softwood? Suggest a product made from one of each.
4 Why do ferrous metals rust?
5 Why are thermosetting polymers less commonly used than thermoplastic polymers?
6 Describe the difference between hardness and toughness.
7 List three properties of cotton and name three products commonly produced from cotton.
8 Describe the differences between woven and non-woven fabrics.

PRACTICE QUESTIONS: core technical principles

Multiple-choice questions

1 A method of funding a project by raising money from large numbers of people is called?
   A Crowd sourcing
   B Enterprise
   C Fair trade
   D Virtual retailing   *[1 mark]*

2 A battery-operated motorised toy will need to work at a minimum of 5 volts. Which **one** of the following are suitable for powering the toy for as long a time as possible, before it needs to be replaced or recharged?
   A 4 × AA Alkaline batteries
   B 4 × AA Zinc oxide cells
   C 2 × Silver oxide coin cells
   D 4 × NiMH rechargeable   *[1 mark]*

3 Which of the following is a technical textile?
   A Teflon
   B Kevlar
   C MDF
   D Shape memory alloy   *[1 mark]*

4 You are designing a t-shirt which should change colour when the wearer becomes hot. Which smart material would you use to make this happen?
   A Photochromic pigments
   B Quantum tunnelling composite
   C Shape memory alloys
   D Thermochromic pigments   *[1 mark]*

5 A designer has created a security system for use in a home. The system is intended to alert the home owner to an intruder. What is the output in this system?
   A Flashing light
   B Light sensor
   C Motion sensor
   D Rocker switch   *[1 mark]*

6 What is the correct order?
   A Fabric → Fibre → Yarn
   B Fibre → Yarn → Fabric
   C Fabric → Yarn → Fibre
   D Yarn → Fibre → Fabric   *[1 mark]*

7 What is the selvedge of a fabric?
   A The bias grain
   B The finished edge
   C The raw edge
   D The straight grain   *[1 mark]*

8 Which of the following is a thermosetting polymer?
   A Acrylic (PMMA)
   B Low-density polythene (LDPE)
   C Melamine formaldehyde (MF)
   D Polypropylene (PP)   *[1 mark]*

9 Which of the following metals will corrode in the form of rust?
   A Aluminium
   B Copper
   C Low carbon steel
   D Stainless steel   *[1 mark]*

10 Hardness is ...
   A the ability to be long lasting
   B the ability to resist a bending force
   C the ability to resist abrasion
   D the ability to resist a dynamic impact.   *[1 mark]*

11 The diagram shows a simple gear train. What is the velocity ratio of
   the gear train?

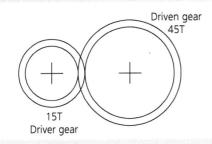

   A 2:1
   B 3:1
   C 1:4
   D 1:3   *[1 mark]*

12 Car windscreen wipers follow which type of movement?
   A Linear
   B Oscillating
   C Reciprocating
   D Rotary   *[1 mark]*

13 If a cam follower rises for 60° of the cam rotation, falls for 60° of the cam
   rotation, and does not move for the remaining rotation of the cam, what will the
   dwell angle be?
   A 60°
   B 120°
   C 180°
   D 240°   *[1 mark]*

14 Which of the following materials is most likely to be used to create a
   working drawing?
   A Cartridge paper
   B Foil-lined board
   C Isometric paper
   D Solid white board   *[1 mark]*

15 Solid white board is often used in the manufacture of architectural models. Which of the following describes solid white board?

   A A heavy type of card that can be printed on

   B A lightweight material that can be cut on the laser cutter

   C A material of 150 gsm that is available in a range of colours

   D A transparent material that can be drawn on with water-based markers *[1 mark]*

Short-answer questions

1 What is the definition of a smart material? *[1 mark]*

2 Name a modern material. *[1 mark]*

3 State two properties of the modern material you have named in the question above. *[2 marks]*

4 Name two sensors that could be used in a system to alert home owners of an intruder. *[2 marks]*

5 Explain the advantages of using a cotton weft knit jersey fabric for a t-shirt. *[6 marks]*

6 What are the benefits of using a manufactured board over a solid timber in the manufacture of an office desk? *[4 marks]*

7 Explain what is meant by 'planned obsolescence'. *[2 marks]*

8 State two ways in which you can prolong the life of a product. *[2 marks]*

9 Which class of lever is a wheelbarrow? *[1 mark]*

10 There are two other classes of levers. Name items that are examples of each type. *[2 marks]*

11 The following product has been made using a type of board. Name the type of board and explain why you think it is suitable for this product. *[4 marks]*

Figure 2 **Box of tea bags**

12 Use notes and sketches to explain how a pumped storage system is able to add electricity to the supply when it is needed quickly. *[3 marks]*

13 Explain the potential advantages tidal power might have over conventional hydroelectric schemes using a blocked valley. *[2 marks]*

2 Specialist technical principles

In addition to the core technical principles covered in Section 1, you will also need to develop an in-depth knowledge and understanding of the specialist technical principles in this section.

You need to know and understand about these principles in relation to at least **one** material category or system. This book focuses on the specialist technical principles in relation to paper and boards; timber, metal-based materials and polymers; textile-based materials; and electrical and mechanical systems and components.

Not all of the properties apply to mechanical or electronic systems, as these are made from the other materials. So, you will need to study at least one of these other areas to make sure you have full understanding and can use this knowledge when completing your assessments.

This section includes the following topics:

2.1 Selection of materials or components

2.2 Forces and stresses

2.3 Ecological and social footprint

2.4 Sources and origins

2.5 Using and working with materials

2.6 Stock forms, types and sizes

2.7 Scales of production

2.8 Specialist techniques and processes

2.9 Surface treatments and finishes

At the end of this section you will find practice questions relating to specialist technical principles.

Selection of materials or components

What will I learn?

In this topic you will learn about:

→ the influences that have an impact on the design of products
→ the factors to consider when selecting materials or components.

There are many different materials and components available and choosing the right ones for a product will be important in ensuring its success with consumers. The influences on designers' choices will vary according to the type of product and the intended target market. The following factors are some of the most important to be considered.

Functionality

The functionality of a product is about it being suited to serve the specific purpose for which it was designed and how easy it is to use for that purpose. For example, a chair that is too small to sit on comfortably or does not stand evenly on the floor is not functional as it cannot easily be used.

Aesthetics

Aesthetics is to do with concepts of beauty and taste. It is concerned with the study of sensory values. Humans have five senses: sight, smell, sound, taste and touch, and all of these are important elements that contribute to making an experience or design *aesthetically pleasing*. For example, when choosing a chair for the home a person will consider its shape, size, proportions and colour as well as the texture and finish of the materials used to make it. Aesthetics is about personal taste, and what one person finds aesthetically pleasing may be unattractive to another.

Figure 2.1.1 **Waste in landfill is a major problem in today's world.**

Environmental factors

All products affect the environment to a lesser or greater degree in terms of pollution, the use of scarce resources, energy use and disposal of the product at the end of its useful life. Choosing and using materials carefully may mean that less waste is sent to landfill, leading to a reduced impact on the environment. Using recycled and biodegradable materials will have less damaging impact on the environment than ones that take many years to break down in landfill sites or that use new materials from finite sources such as coal and oil. Less energy is often needed to recycle materials than to manufacture new ones, and this will also help to protect the environment. Designing products that can be taken apart so that materials can be reused reduces environmental impact (for example, reusing metal from electronic components).

Availability

Some materials are easier to get hold of than others because they are manufactured locally, and some are only available in limited quantities because they are hard to make or take a long time to grow naturally. Certain materials can only be obtained in limited sizes to suit mass production. If materials need to be manufactured, especially for a one-off or bespoke product, then they will take longer to produce than those available for mass manufacture.

Cost

The selling price of a product will include the cost of raw materials, manufacturing costs, packaging and distribution and sellers' costs, with a percentage of the total added for profit. The costs will vary depending on the type and complexity of the product and the target end-users, and manufacturers need to research the price that potential customers are prepared to pay. Some products will use scarce or very expensive materials, some are made for budget or 'value' markets so will use cheaper materials and manufacturing processes. The demand for a product may also influence its selling price.

Manufacturers use spreadsheets to show all the likely costs involved in the design and manufacture of a product when working out the cost of manufacturing each unit. The scale of production will affect the costings, as generally, the more products that are made, the cheaper each unit will cost. This is because there are often discounts for buying larger quantities of materials (known as **bulk buying**) and the costs of setting up the manufacture are the same no matter how many are to be made. Therefore, the more that are made, the less each unit will cost.

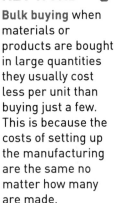

> **KEY WORD**
>
> **Bulk buying** when materials or products are bought in large quantities they usually cost less per unit than buying just a few. This is because the costs of setting up the manufacturing are the same no matter how many are made.

Social factors

Social factors are things that affect our lifestyle and include things such as wealth, gender, family and religion. These can change over time and designers need to be aware of these changes in order to make sure that their products meet the changing needs of consumers. Today, our society is multicultural, with people of many different religions and ethnicities, often with different traditions, which will influence the type of products they will want.

The source of some raw materials can have a big impact on the communities of the people who live nearby. Deforestation to provide land for crops or industrial activity is destroying the Earth's forests on a massive scale. This is thought to contribute to global warming, which can cause flooding of local areas which can have a devastating effect on communities.

Figure 2.1.2 **This envelope carries the FSC® logo and a recommendation for recycling.**

The Forest Stewardship Council® (FSC) is an international, non-governmental organisation dedicated to promoting responsible management of the world's forests, and their logo can be seen on many different forest products, such as those made from wood, paper, cork and natural latex.

Using recycled materials whenever possible can help reduce the need for making new ones and the consequent impact on people's lives.

Cultural factors

Cultural factors are the values of an individual or a specific community that determine the way that a person behaves. Cultural factors have an important effect on the products that people buy, and a product acceptable in one culture may be looked at as offensive or less desirable in another. The use of colour and colour schemes is a good example of cultural influences on design. Red is associated with good luck in China, but is the colour of mourning in South Africa, therefore, red products would be received very differently in those two countries.

Ethical factors

Ethical factors are things that are seen as being the 'right thing to do' and morally right.

The way that a product is designed and made has an impact on the well-being, safety and comfort of all those people who will be affected by the product.

The world's natural resources are dwindling and designers have a responsibility to use them carefully and to use recycled materials where possible. Manufacturers need to be aware of where materials come from and to purchase them from ethical sources, such as those approved by the FSC, whenever possible. Fair trade methods of manufacture ensure that workers are not exploited and have decent working conditions.

Using land to source raw materials means that less may be available for growing food crops in parts of the world where people do not have enough to eat. Manufacturing activities can cause pollution of land and drinking water, which will affect the health of people living nearby. Building factories and access roads may reduce the land available for housing and schools and also contribute to pollution that affects people's health and well-being.

Check your knowledge and understanding

1 Give three different factors that will affect the cost of materials for a product.
2 What are ethical factors?
3 Explain what bulk buying means.

Examples and applications for papers and boards

Functionality

Different types of paper and board are used for different products and it is important to choose the right material that can do the job for which it is intended. A lightweight paper such as layout paper will not be practical for making a model when a board would be stiffer and stronger. Cartridge paper would not be suitable for tracing as a translucent paper is needed to see the image being copied.

Boards used for packaging need to be strong enough to hold the product that is packed yet light enough to be moved; they may also need to be printed on, so the surface should be suitable for this.

Sometimes a waterproof board is needed, for example, when making drinks cartons.

Figure 2.1.3 **Boards are used for model making.**

Figure 2.1.4 **Boxes for moving house**

Figure 2.1.5 **Waterproof board is used to make cartons for drinks.**

Aesthetics

Sometimes the appearance or texture of the paper or board is more important than its functionality, for example, when making greeting cards or decorative wrapping paper.

Environmental factors

Paper and boards are made from wood pulp, and the cutting down of trees leads to environmental problems associated with deforestation. Much of the paper used today is made from recycled paper and board, so has a reduced environmental impact. New developments such as paperfoam and Potatopak are made from starches and designed to replace oil-based materials used for packaging, so are more environmentally friendly as well as being biodegradable.

Figure 2.1.6 **Greeting cards need to be attractive.**

Figure 2.1.7 Posters for outdoor areas are subject to many environmental conditions.

The sort of environment where the product will be used will also affect the choice of material; a poster for an outdoor area will need to be able to stand up to a variety of weather conditions during its life, packaging for frozen food will be subjected to moisture generated as the ice on the packaged food thaws.

Availability

Papers and boards are manufactured in a range of stock sizes and different thicknesses. If a special paper is required for a product (for example, with specific holograms or embossing), it may take longer for the supplier to process the order and will add to the overall costs.

Cost

Paper and boards are relatively inexpensive when compared with many other materials, but cost is still an important consideration and the price of the material must be appropriate for the intended use of the product. Mounting board is a high-quality expensive material, so would not be used for an advertising flyer that will quickly be thrown away: a foil decoration might be used on a book cover intended to last for many years, but its use might not be justified for packaging perfume for the value market.

Social and cultural factors

Culture and religion have a major impact on products made from paper and boards, particularly inappropriate use of colour and pattern or the use of words that have been wrongly translated from one language to another. Images of people are not allowed within most Islamic cultures, but geometric patterns that repeat in a continuous pattern are used as they are thought to represent the never-ending nature of God. Designers need to consider the traditions and beliefs of different cultures if they want to avoid giving offence and to make sure that their products meet the changing needs of consumers.

Figure 2.1.8 Continuous patterns are typical of Islamic designs.

Ethical factors

Paper and boards are easy to recycle and are commonly reused. But the raw material is not the only source of concern when considering the impact of the product on the well-being of people. In today's world there is a focus on disposable products, internet shopping, personal and company image, and convenience. All of these need lots of marketing and packaging materials, most of which are used only for a very short time before being thrown away.

Manufacturing these materials needs land for factories, machinery and energy, additional materials such as inks, plastic films and metallic foils and people to make them.

Figure 2.1.9 Packaging has many different functions.

As consumers, we need to be aware of our demand for these products to make sure that we are not wasting precious resources.

Packaging performs many functions, including protecting the product and providing important information for the user, but perhaps many manufacturers, retailers and consumers are guilty of using too much packaging.

KEY POINTS
- Boards and papers are lightweight yet often strong.
- Boards and papers are easily recycled.
- Boards and papers can be given different finishes to make them suitable for many uses.

ACTIVITY

Make a collection of:
- a take-away pizza box
- a fruit juice carton
- a greetings card
- a promotional leaflet
- a hardback book cover.

Examine the different papers and boards used. Why are they suitable for these products?

Examples and applications for timber, metals and polymers

Functionality

Timber, metals and polymers have very different characteristics, and what is right for one product may not be suitable for another. A child's toy may need to be light in weight and brightly coloured, so acrylic may be more suitable than metal. Cupboards and tables need to be rigid and able to take the knocks of everyday life, so timber is a good material to use. Radiators and pans need to be able to conduct heat, so are made of metal. Some timbers, such as hardwoods, are better at resisting decay than others. Most metals will corrode over time unless given a protective coating, and polymers can crack and scratch easily with use.

Many products are made from more than one material to allow for different functions. A smartphone or tablet needs a clear material for the screen, a body that will contain the electronic parts and resist cracking and scratching with constant use, yet not be too heavy to carry around. Sometimes the method of manufacture will dictate the material to be used; using polymers allows for quicker manufacture of some products when compared with using timber.

Safety is also an important consideration when selecting the material to use. Why, for example, would a toy bucket and spade be made from acrylic rather than metal and timber?

Figure 2.1.10 **Smartphones are made from more than one material so that they can have different functions.**

Aesthetics

Timber, metal and polymers have very different textures and appearances and can easily be manipulated to make them more appealing to the customer. The shape, colour and texture may be as important as how it functions if a product is to be successful, and it is possible to use each type of material creatively to exploit its characteristics. The use of a polymer to make the toy shown in Figure 2.1.11 allows it to be easily coloured and gives it a smooth surface. Garden furniture has an open appearance, yet is strong and long-lasting thanks to the use of metal. Using timber for a chair gives it rigidity and allows curved shapes to be made. A lampshade needs to be attractive with and without the light shining through.

Figure 2.1.11 **Toy bucket and spade for a child**

Figure 2.1.12 **A colourful children's toy made from polymer**

Figure 2.1.13 **Strong garden furniture made from metal**

Figure 2.1.14 **Timber is often used for chairs.**

Environmental factors

The developed world is using or destroying some resources faster than they are able to be replaced, and some, such as oil and coal, cannot be replaced. Timber from managed forests is a renewable material and metals can be recycled and re-used quite easily, as can most plastics. Managing the world's resources has led to a move towards recycling raw materials such as plastic, metal and glass from products that have reached the end of their useful life and using them for new products. The use of energy to make, package and distribute products is an environmental issue that must also be taken into account.

Designing products that are easy to maintain and repair means that fewer need to be made, so less of the world's precious resources are used. Products that use standard components or

Figure 2.1.15 **Aluminium drinks cans can be recycled.**

modules that can easily be replaced can extend a product's life. Products that can easily be broken down and taken apart so that their parts can be used to make new materials are better than further depleting finite resources.

Designers also need to think about where a product is to be used. Garden furniture needs to be able to withstand extremes of temperature and potential damage from UV light and moisture, so a hardwood or protective finish will be needed. Headphones used with mobile devices need to be able to take knocks and function within a range of different environments. Using good-quality materials and finishes can extend a product's life so that fewer replacements need to be made is better for the environment than making many products that need to be thrown away after only a short life.

Availability

Some materials from natural sources are only available in small quantities as there are only limited quantities. Some trees, such as oak, are slow growing, whereas bamboo grows very quickly so more will be available in a given length of time. The availability of naturally sourced materials may vary with unpredictable events, such as volcanoes and hurricanes, or unusual weather patterns that affect the success or failure of harvests.

Availability of materials may also depend on the manufacturing processes used. For example, timbers are often produced in a set number of standardised plank sizes to suit mass manufacture and distribution, so if a bespoke product requires a different size it may have to be produced specially for that product.

Cost

The initial price of the raw materials will of course affect the end price of a product, but these are often only a small percentage of the costs. Some materials require more complex manufacturing processes or additional finishes that will add significantly to the overall costs. Traditional woodworking skills are time consuming and the end-product will also require a finish, but many polymers are self-finishing and use quicker manufacturing processes, such as injection moulding. When choosing a material, the designer and manufacturer need to be aware of the extra costs involved to make sure that the final price matches the expectations of the customer and that it is appropriate for the intended use of the product.

Social factors

Developments in technology and materials have allowed many new products to become part of our daily lives. For example, mobile phones and tablets give us access to a range of social media that allows people to communicate quickly and easily and keep up with news. But these can also have negative effects, such as cyber bullying and using mobile devices when driving, which can cause accidents.

The use of computers and robots to do jobs that used to be done by humans means that many manufacturing processes can be done more quickly, more accurately and more safely, but fewer people are employed in manufacturing jobs. Many people live in poverty and cannot afford to keep up with the latest trends or to buy the best quality. Elderly people and those with physical handicaps may find it difficult to use some products and designers must consider how their designs can be used by a wide group of people.

Figure 2.1.16 **Robots are used in car manufacture.**

Cultural factors

Designers need to investigate the traditions of the cultural group that the product is being made for so that the design is suitable for their needs and fits in with their beliefs of what is good or bad design. Many products are basically the same in all cultures, but the colour or way they are decorated may have specific meanings. Also, some products are very different – for example, Japanese families traditionally sit on the floor to eat their meals, so the design of dining furniture for Japan is unlike that in the rest of the world. The gender of the person buying and using a product may also need to be taken into consideration as the colour, pattern and style of a product may be more popular with men than women.

Figure 2.1.17 **A Japanese dining table and chairs**

Ethical factors

The use of materials and energy can have a big impact on people's lives, and we all have a responsibility to use them carefully and avoid waste. Careful and economical use of materials and the prevention or reduction of pollution during manufacture can help to keep a safe environment which does not impact unfairly on people's lives. When products are made in countries where labour is cheap, people are sometimes exploited and pollution may be less strictly controlled than in the UK.

Deforestation can lead to global warming and to certain species of animals and plants becoming extinct. Choosing ethically sourced timbers, such as those from FSC managed forests, can reduce the damage to the environment and the lives of those living nearby.

Making sure that those operating machinery and working with hazardous materials are adequately protected and kept safe is an ethical must in today's world.

> **ACTIVITY**
>
> Find three chairs made from different materials and used in different areas of your school or college.
>
> Explain why the materials have been used to make each chair. How has the choice of materials been influenced by the method of manufacture?

Examples and applications for timber, metalsand polymers

Making sure that products are fit for purpose and disposed of carefully at the end of their life is also important in protecting people and the world.

Examples and applications for textiles

Functionality

Strength is an important consideration for some products. Garments that will be worn frequently and are washed a lot, and products that will be subjected to friction through movement need to be strong.

Synthetic fibres such as nylon and polyester are very strong and do not break easily, so are ideal for use in products such as bags, sportswear, and outdoor jackets. A closely woven fabric structure is strong, and when used with a strong fibre such as nylon makes fabrics that can stand up to friction and hard wear.

The **absorbency** of a fabric will also affect the comfort of some garments, and some products need to be absorbent if they are to do their job (for example, bathroom towels).

Figure 2.1.18 Polar fleece and knitted fabrics can trap air so are warm to wear.

In very hot conditions, it is important that moisture can be removed from the skin otherwise the wearer becomes hot and sweaty. Fabrics with large air spaces, such as some knitted structures, allow moisture to evaporate from the body. However, in wet conditions, very absorbent fabrics can cause the wearer to be very uncomfortable – think about wearing cotton denim jeans that have become very wet in the rain!

Warmth may be an important consideration when choosing fabrics for garments. In order to keep the body warm, a fabric must be able to insulate. This usually means that it can trap air in the fibres, yarns or the way in which the fabric is constructed. Trapped air is an insulator as it does not conduct heat; the more air that is trapped, the warmer the fabric. Wool is an excellent insulator as its crimp allows air to be held within the fibres.

To keep a person warm, a fabric also needs to be **windproof**. Closely woven fabrics or those with laminated layers, such as Gore-Tex, do not allow air to pass through a garment, and so help to insulate the wearer from cold.

How easily a fabric **catches fire and burns** is an important consideration when choosing fabrics for children's nightwear and soft furnishings. Fibres such as cotton catch fire easily, whereas synthetic fibres are difficult to set alight, but will melt, and the molten fibre can stick to the skin causing serious burns. Wool fabrics do not catch fire easily and tend to put themselves out very quickly.

Figure 2.1.19 Cotton fabrics burn easily so garments carry labels to warn consumers of the dangers.

Trapped air helps fabric to burn quickly, so fabrics such as brushed cotton can be very dangerous.

Fabrics may also need to be able to resist **chemical and biological attack**. Chemicals such as laundry detergents are strong alkalis and can damage fabrics such as those made from wool and silk. Chlorine is found in bleaches and swimming pools and will remove colour and weaken fabrics over time. Perspiration is slightly acidic and can discolour and weaken some fabrics made from fibres such as silk and nylon.

Biological attack from moths and mould (mildew) can also spoil fabrics. Moths lay their eggs in wool fabrics and when they hatch, the larvae eat the fibres, leaving holes in the fabric.

Mildew grows on cellulosic fabrics, such as cotton, when they are kept in humid conditions, and this mould can be very difficult to remove.

Aesthetics

Aesthetics is about the way a fabric looks and feels, and aesthetic choices can be a very personal thing. Some fabrics, such as satin, have a very smooth surface which reflects a lot of light so that they appear shiny. Fabric can be given texture in many different ways, depending on the fibre and yarn used, and the way the fabric is constructed.

Environmental factors

Textiles can impact on the environment in many different ways. The growing of cotton plants uses fertilisers and pesticides which can pollute the atmosphere and waterways. Synthetic fibres are made from chemicals from non-renewable sources. Lots of land is needed for cotton plants, and this causes drastic changes to the landscape because of the intensive farming and deforestation required.

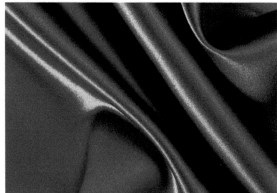

Figure 2.1.20 Smooth satin fabrics made from filament fibres reflect a lot of light so appear shiny.

Cotton can be replaced by newer fibres, such as Tencel and Modal, which come from sustainable sources and use 'clean technology' in their manufacture. Ingeo, a new fibre designed to replace polyester, is made from plant starches and is fully biodegradable. Polyester fibres can be made by recycling plastic bottles – an example of this is Polartec fleece fabric.

The cleaning and dyeing of fabrics and application of additional finishes, such as flame retardancy and stain resistance, during manufacture use many chemicals, and these are often disposed of in rivers in countries where the environmental laws are not as strict as those in the UK. A lot of water and energy are also needed for these processes.

Making components such as zips and buttons also uses energy, as well as plastics and metals, and this can have a toxic effect on the environment.

Textile products are made all over the world before being sent to the country where they will be sold. When choosing materials, designers need to consider whether they can be sourced near to where the products will be made. Using more ecologically friendly forms of transport, (for example, lorries that use biofuels and trains that can transport very large loads) can help reduce carbon emissions.

Figure 2.1.21 Plastic bottles can be recycled to make fleece fabric.

Textile waste is produced when fabrics are made into products and when they are disposed of by consumers, which may end up in landfill sites. Many unwanted textile products can be recycled and reused. Fabrics can be broken down to fibres which can then be spun into yarns to make new fabrics. Some unwanted fabrics can be used for industrial purposes, such as making insulation materials or wiping cloths.

Examples and applications for textiles

Availability

Some fibres are more scarce than others. Cashmere is a luxury hair fibre that comes from the cashmere goat, and is only available in small quantities – one sweater needs the fleece of four to six goats compared to the three to five sweaters from the wool of one sheep.

The availability of many textile materials is governed by trends and fashions, especially colours and patterns on fabrics, so what is freely available one season may be difficult to find in another. Some materials may be produced for specific designers and not available on the mass market, whereas others are mass-produced in large quantities because they are used for many items.

Cost

Cost is an important consideration when designing and making textile products. The cost to the consumer will depend on the fabrics used, the manufacturing processes involved and the quality of the product. Some fibres and fabrics are more scarce or more difficult to make, so will cost more. For example, silk and wool are more expensive than cotton and synthetic fibres, complex weaving and knitting processes will be more costly, and special finishes put on to fabrics will add to the price. Designs that use lots of fabric or have many seams will cost more to make than simpler styles as there are more processes involved. Manufacturers often make a product in many different sizes and will need to calculate the most efficient layout so that fabric is not wasted, as this will add to costs.

The choice of components used on products can affect costs for example, a zip may be cheaper than buttons, which require additional manufacturing processes and materials to make the buttonholes.

Manufacturers are able to save money by buying fabrics and components in bulk as and when they are needed. The style and colour of many textile products are determined by changes in fashion, so buying fabrics as and when they are needed using a JIT system can save money as unused materials are not wasted. Certain components, such as black and white threads, are always needed so it may be possible to make even more savings by buying them in very large quantities.

Social factors

Today's society is multicultural, and designers of textile products need to consider the needs of the specific groups of people who will use the products. Clothing in particular reflects the values of people and makes an important statement about the wearer – what we wear and when we wear it is part of our culture and socially learned way of life, and allows us to communicate who and what we are to others.

Figure 2.1.22 **A patchwork quilt made from reused materials**

In some societies it is traditional to reuse and rework existing products to make new ones. In early American culture, making quilted patchwork quilts was a common social pastime for women, and handmade quilts were often given as a wedding present. Today, it is not uncommon for community quilts to be made to celebrate or commemorate specific events.

How and where a textile product is made will also impact on society in terms of employment opportunities and the impact of manufacturing activity on the local environment.

Cultural factors

The cultural group to which a person belongs will affect their choices, particularly for clothing. For example, many young people are interested in music, and the clothes worn by beatniks, goths and rappers reflect those interests. Peer-group pressure will also impact on clothing choices, along with the influence of what celebrities and fashion idols wear, those who are members of a particular cultural group can recognise each other by the styles they wear. The choice of colours, fabrics and components used for garments will be an important part of that culture. Some people do not want to wear the latest fashions or clothes that bring them to the attention of others, so will want designs that are more conservative.

Figure 2.1.23 **Rappers have a distinctive style worn by members of that cultural group.**

Ethical factors

Ethical factors are concerned with the ways that the design and manufacture of textile products affect the lives and well-being of all the people who might make or use the products.

Many of the issues relating to the environment are also ethical issues. The pesticides and fertilisers used on cotton crops can cause serious health problems for the workers, so choosing fibres like Tencel, Modal and Ingeo are more ethically sound. Certain dyes and fabric-finishing treatments use chemicals that are bad for the workers' health if they are not used carefully, so designers and manufacturers need to know how fabrics have been made and that care has been taken to protect the workers.

Some people are concerned about animal welfare, so will not wear fabrics made from fur or leather, but it is possible to use synthetic alternatives that resemble the real thing.

The manufacture and processing of fabrics and components can have many negative effects on the people and communities where they are made. Designers need to think about the impact of fast and disposable products and whether it is ethically right for some people to be negatively affected in order to replace perfectly useable products to meet these constant changes.

> **KEY POINTS**
> - Many factors affect the functionality of textile products.
> - Choosing fabrics and components carefully can protect the environment and the well-being of people who make them.
> - Styles of clothing are often dictated by social and cultural factors.

> **ACTIVITY**
> Collect a range of different textile products including:
> - a school shirt
> - a party dress
> - a Babygro
> - a decorative cushion cover.
>
> Work in groups to make a table like the one below, listing the fibres and fabrics, and the important functions of each product.
>
> Decide if the aesthetics for each product are very important, important or not important.
>
> List all of the components used in each product.
>
> | Product | Fibres used | Fabric construction | Functions needed | Aesthetics | Components used |
> |---|---|---|---|---|---|
> | School shirt | | | | | |
> | Party dress | | | | | |
> | Babygro | | | | | |
> | Cushion cover | | | | | |

Functionality

Electrical and mechanical systems are built up of interconnecting parts that are required to work together to achieve an output. In order for the product to function as intended, these parts need to be able to work together accurately and efficiently. The different parts of a system need to be held together in some way, so the end-product will need to be an appropriate size and shape with easily accessible controls, and be made from materials suitable for its end-use. The power sources that drive the systems must be appropriate for the product; a large lead-acid battery is fine in a car, but would not be suitable for a portable device such as a tablet or smartphone. The system for holding the batteries within the product so that they are secure yet easily accessible for replacement will affect how efficiently the product functions.

A thermostat needs to be able to control a heating system or an appliance so that it will operate correctly, and sensors must be placed within a product so that they can easily and accurately detect movement, sound, heat or moisture.

Remote controls contain electronic systems that can control televisions and other devices from a distance. They are small enough to be handled comfortably, and the batteries are securely contained and easily accessible for replacement.

Bicycles need an efficient and lightweight system to transfer movement from the pedals to the wheels – the chain and sprocket is the most functional system for this.

Activity tracking wristbands need to be small, lightweight, resistant to moisture and knocks, adjustable in size, not irritate the skin of the wearer, and contain electronic devices that record movement data and transfer it to a computer system.

Figure 2.1.24 **Remote controls**

Figure 2.1.25 **Activity-tracking wristbands**

Aesthetics

The parts that combine to make electrical and mechanical systems are rarely aesthetically pleasing and are often hidden away, but the appearance of products they are used needs to be attractive. The shape, size, styling, colour, decoration and texture are the interface between the system and the user, and will often influence a customer to purchase one brand of product over another. For example, a thermostatic control for a heating system needs to be discrete yet easy to read and set. A smartwatch needs to look and feel good as well as function efficiently. Toys that use movement from cams need to be visually attractive.

Environmental factors

Mechanical and electrical systems require energy in order for them to work, and some of this energy will need to come from finite sources. Many products get their energy from disposable alkaline batteries, which may contain the metals steel, zinc, manganese, potassium, lithium, cadmium and nickel. The use of disposable batteries increases by about five per cent every year.

Figure 2.1.26 **A thermostatic control for a heating system needs to be discrete, yet easy to read and set.**

In Europe, the Waste Electrical and Electronic Equipment (WEEE) directive aims to reduce the amount of waste electrical and electronic equipment that ends up in landfill, and includes the disposal of batteries, which must not be put in with general waste. Instead they must be separated from domestic waste and sent for recycling. Within the EU, most stores that sell batteries are required by law to accept old batteries for recycling.

Some types of batteries contain small amounts of mercury, which can have a negative effect on the environment if they are sent to landfill, as the heavy metal can leech out and poison surrounding land and water courses.

Developing products that encourage the use of rechargeable batteries may have a less detrimental impact on the environment.

Copper is used in electrical systems and it is obtained by mining it from the ground. This process can cause the release of sulphuric acid and heavy metals that contaminate the land, which has detrimental effects on the water, vegetation and biological life in surrounding areas. Designing products that are easily broken down into small parts allows the copper and other metals to be reused for new products.

Availability

Copper, zinc, polymers and timbers used in the manufacture of products containing mechanical and electrical systems are usually freely available. However, as many of these materials come from all around the globe, war and adverse weather patterns may temporarily affect their availability. Many of the components used in electrical systems are made in a range of standard sizes to facilitate their use in a range of products.

Certain products require the use of a printed circuit board, which may need to be made specifically for those products.

The Restriction of the use of certain Hazardous Substances (RoHS) directive aims to eradicate the use of certain materials in new electrical and electronic products, so some materials will no longer be available for use in such products.

Figure 2.1.27 **A collection point for used batteries.**

Cost

The cost of raw materials can vary from one month to another, especially the price of copper, which is essential in the manufacture of so many products; this may affect the price of printed circuit boards and other components used. The cost of products containing mechanical and electrical systems will also include the cost of the additional materials used to make the product itself, and those will vary according to the specific materials used, which may be timbers, polymers, textiles or other metals.

Bulk buying of basic materials, such as copper and standard components, especially when prices are low, may help manufacturers make significant savings. Many of the products that include electric systems reflect new developments in technology, and the price of materials may be dictated by fashion and demand.

The WEEE directive expects producers of electrical and electronic products to meet the cost of the collection and processing of discarded products, and this will need to be factored in to the cost of making new products.

Figure 2.1.28 **Small appliances and electrical equipment are separated at recycling centres, and parts re-used to reduce waste.**

Social factors

Advances in technology, particularly those associated with electrical systems, have led to demands for new products that have become part of everyday life – for example, smartphones, tablets and smartwatches. Many people also want systems that will make their lives easier and give them more leisure time, for example, systems that control and monitor areas of their domestic life, such as switching heating systems and appliances on and off.

During the recycling process, metals from crushed battery cases are separated from the inner mass, which is treated chemically to separate the zinc, manganese and potassium. The resulting liquid is used as a fertiliser to help with production of food crops globally.

Recycling of electrical and mechanical products helps reduce the space needed for landfill. The WEEE directive may require changes in the design of new electrical and electronic products to make them easier to dismantle, recycle and reuse.

Cultural factors

Some groups in society consider it very important to have the newest iPhone, and smartphones and smartwatches and activity tracking wristbands are quickly becoming the latest must-have products.

Most products that use electrical and mechanical systems are the same in all cultures, but the demand for specific products may vary, as will the colour and pattern on the actual products.

Ethical factors

The sourcing of many of the metals used in electrical systems can adversely affect people's health and well-being. The mining of copper starts with extraction of the ore from the ground. The ore is then smelted and converted to copper plates ready for sale on the global market. This process produces chemicals that are harmful to human health (for example sulphuric acid and sulphur dioxide, which can cause breathing and lung diseases).

Alkaline batteries often leak caustic potassium hydroxide, which can cause eye, skin and respiratory irritation.

Although copper is a naturally occurring element and the body requires some for normal and healthy functioning, too much exposure to it can lead to genetic disorders, lung cancer and coronary heart disease. Demand for the latest innovations can have a devastating effect on the health of those who work with some of the raw materials. Some workers may suffer life threatening diseases because of the need to provide more raw materials that satisfy the demand for new products to replace ones that are still perfectly serviceable.

KEY POINTS
- Electrical and mechanical systems are often hidden inside other products
- The disposal of waste electrical and electronic equipment is controlled by law in Europe
- Copper and other metals, the raw materials needed to make many electrical systems, can have an adverse effect on the health of the workers who mine them.

ACTIVITY
Look around your home and make a list of all the products and systems that need batteries to make them work. How many of the batteries can be recharged instead of being thrown away?

2.2 Forces and stresses

What will I learn?

In this topic you will learn about the impact of forces and stresses and the way in which materials can be reinforced and stiffened, including:
→ the five different types of force that can act upon structures
→ how some materials are better at resisting certain forces
→ how materials can be enhanced to resist certain forces.

Have you ever looked at a tower crane or tall building and wondered why they don't fall over? Have you asked why there are so many different shapes of bridges? All structures have to be made to withstand the forces that will act upon them without collapsing and breaking up. Designers therefore need to understand the forces or external loads that can act upon any structure they design and make. There are five different types of force or external load that can act upon a structure: tension, compression, bending, torsion and shear.

Forces acting on materials and objects

Tension forces are pulling forces that cause an object to be stretched or pulled apart. A rope in a tug of war competition is under tension as each side tries to pull the other side.

Compression forces are pushing forces that squeeze an object. An example might be when you stand on a drinks can and squash it. Table and chair legs are under compression when an object is placed on the table or someone sits on a chair.

Shear forces act across a material by acting near to one another but not directly opposite each other. A shearing force cuts the object by pushing it sideways in opposite directions. Scissors and garden shears have a shearing action that causes paper, grass or garden growth to be cut by making one piece slide across the other and create two pieces.

Bending forces act at an angle to an object and make it bend. Placing too many books or very heavy objects on a shelf can apply forces that make the shelf bend.

When an object bends it is under compression and tension at the same time. As seen in Figure 2.2.4, the top of the beam is experiencing compression and the bottom of the beam is under tension.

ACTIVITY

Create a list of objects or products that experience compression under normal use.

Figure 2.2.1 A rope under tension in a tug of war competition

Figure 2.2.2 An object under tension

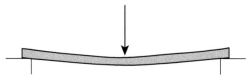

Figure 2.2.3 An object experiencing bending

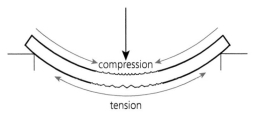

Figure 2.2.4 Compression and tension in a beam that is bending

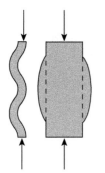

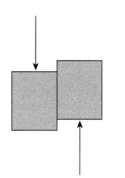

Torsion forces are twisting forces that are applied to an object. When we twist a screw cap off the top of a bottle we are applying a torsion, or twisting, force. Torsion forces act in the drive shafts in cars. These are the shafts that transmit the rotary motion from the engine and gearbox to the wheels.

Figure 2.2.5 **Objects under compression**

Figure 2.2.6 **An object experiencing a shear force**

Figure 2.2.7 **An object experiencing torsion**

Enhancing materials

Materials can be enhanced to resist and work with forces and stresses to improve functionality. A variety of materials are used in many applications to resist tensile forces. Ropes can be made of any long stringy fibrous material, and can be made of natural materials like linen and cotton, or synthetic fibres like nylon or polypropylene. Rope is made by twisting multiple strands or yarns together to form a stronger and more robust material that has good tensile strength.

Concrete is a material that is very good at resisting compressive forces, but not very good when in tension. To overcome this it can be reinforced with steel bars, which are embedded in the concrete before it sets. Reinforced concrete is used extensively in the construction industry in bridges and building construction where it will experience high tensile forces and compressive loads.

Fabrics can be stiffened by having strips of other materials inserted in them, as with shirt collars. Smooth, rigid strips of materials such as horn or nylon, rounded at one end and pointed at the other, are inserted into small pockets underneath the points of a shirt collar to stiffen the collar and prevent bending.

Timber may also be stiffened by the process of laminating. Thin layers, or 'plies', of timber are glued together to shape and stiffen the material in the manufacture of chairs and other furniture. Laminated timber is also used in the construction industry. By laminating a number of small pieces of timber together into a single large beam it is given greater strength to resist the forces that create bending, compressive and tensile loads in structures.

Fabric is given greater tensile strength when it is woven. Car seat belts are made from a flat webbed fabric which is made of synthetic fibres such as nylon and polyester. Fabric interfacing is a technique where another fabric is used to line a garment where additional strength, rigidity or stiffness is required, or to stop a garment stretching.

ACTIVITY
1 Build a bridge using only spaghetti and a hot glue gun to span a distance of 600mm. The bridge should support a road deck made of stiff card measuring 80mm by 600mm. Test your bridge by seeing if a toy car can pass over it.
2 Build a dome using paper straws and a hot glue gun. The centre height of the dome must be at least 300mm high and support a weight of 400 grams.
3 Build a model of a roof for a grandstand of a sports stadium using paper straws, paper and a hot glue gun. The roof must not have any supporting columns that would block the views of spectators.

Figure 2.2.8 **Fibre strands being wound together to make rope**

Figure 2.2.9 **A concrete reinforced structure under construction**

Figure 2.2.10 **A laminated plywood chair**

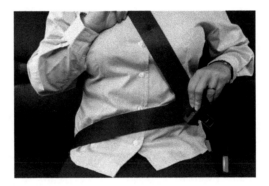

Figure 2.2.11 **Car seat belts are made of webbed fabrics of high tensile strength.**

STRETCH AND CHALLENGE

Describe what can be done to increase the tensile strength of fabrics and how tensile strength might be measured and tested.

ACTIVITY

Carry out experiments to increase the tensile strength of paper using materials commonly available in the technology rooms, without significantly adding to the weight.

KEY POINTS

- There are five main types of force that can act upon any object or structure: tension, compression, shear, bending and torsion.
- Some materials are better at resisting certain forces.
- Materials can be reinforced and stiffened in order to resist certain forces.

Check your knowledge and understanding

1 Give an example of a material that is good in tension.
2 Explain why designers have to understand the forces and external loads that can act upon a structure they design.
3 Give an example of a material that is good in compression
4 Explain how concrete can be reinforced.
5 List materials that rope could be made from.

Figure 2.2.12 A stack of corrugated cardboard boxes awaiting assembly

STRETCH AND CHALLENGE

Describe why corrugated cardboard has greater compressive strength in certain directions compared to others.

Examples and application for papers and boards

Boxes for packaging are often made from corrugated cardboard. By making boxes out of corrugated cardboard the compressive strength of the card is greatly increased because the triangles formed by 'crimping' the inner layer spread any force applied, making the material much stronger. This is extremely beneficial when packaging products for transportation.

ACTIVITY

Create a list of the properties of corrugated cardboard.

Examples and application for timber, metals and polymers

Laminated timber beams are an example of an engineered wood product. They are manufactured from layers of parallel timber laminations, normally of softwoods, such as pine or spruce, although hardwoods can be used. The sawn and planed timbers are selected for strength before being glued together with the grain running in line with the laminates. Knots reduce the strength of timber so they are evenly distributed throughout. This means the beams are able to withstand high stresses.

Figure 2.2.13 Laminated timber beams above a Crossrail station in London

ACTIVITY

Create a list of materials that have high strength when in tension.

Fibre-reinforced polymer is used to increase the strength or elasticity of non-reinforced polymers. These composite materials have a polymer matrix, usually epoxy or polyester, reinforced with fibres, such as glass or carbon. Fibre-reinforced polymer is frequently used in bulletproof armour.

In steel structures, in order to minimise bending caused by loads, the cross sectional shape of a beam is important. The I-shaped beam has been shown to be a very efficient way of supporting bending and shear loads.

STRETCH AND CHALLENGE

Create a table that lists materials that have good resistance to tensile, compressive, torsion, shearing and bending loads, to aid designers in material choice.

Figure 2.2.14 Fibre-reinforced polymer is used in bulletproof armour.

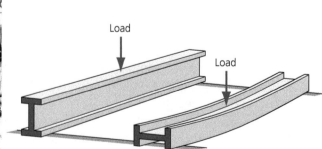

Figure 2.2.15 An I-beam under load

Examples and application for textiles

Figure 2.2.16 **A men's shirt collar will contain a fabric interfacing.**

A fabric interfacing is used to add strength to garments. Products other than garments can also benefit from interfacing, for instance behind buttons or buttonholes on cushions. On garments where certain areas need extra body or strength, such as buttonholes, shirt collars, cuffs, waistbands and pockets, interfacing is used to prevent the fabric stretching. It is also used to stabilise other areas of garments such as shoulder seams and necklines, which may otherwise hang out of shape. Interfacings come in two main types: fusible or sew in, and in different weights and weaves – most interfacings are bonded fabrics, not woven. Fusible interfacing is the easiest to use as it has adhesive on one side and bonds permanently to the fabric when applied with an iron. Where this is not possible the interface has to be sewn in.

Examples and application for electrical and mechanical systems

Figure 2.2.17 **A timing belt being replaced on a car engine**

Many mechanical systems use belts to transfer rotary motion from one shaft to another.

A timing belt, or what is sometimes called a cam belt, is an important part of the internal combustion engine. It synchronises the rotation of the crankshaft and the camshaft to operate the valves that let fuel in and out of the cylinders. Modern timing belts are usually toothed belts (drive belts with teeth on the inside). Timing belts can experience high tensile forces and are typically made of rubber, a material that is very flexible. Rubber however, is not good in tension as it is prone to stretch like a rubber band. To overcome this, timing belts are reinforced with high tensile fibres such as fibreglass, polyester or Kevlar to increase the tensile strength and stop any stretching.

2.3 Ecological and social footprint

What will I learn?

In this topic you will learn about:
→ ecological issues in the design and manufacture of products
→ the six Rs
→ social issues in the design and manufacture of products.

In this section we will discuss the impact that we have on the environment and people due to the choices we make about products. When we discuss 'footprints', we are concerned with the lasting impression we make on the people and places we touch and the effect we have on them.

When designing we need to be aware that every time the materials used in manufacture are either processed or transported energy is consumed. As most energy production also produces pollutants such as smoke or carbon dioxide we need to minimise these effects and improve the way we work.

Reducing distances travelled, in both manufacture and distribution of finished products to be closer to where the final users live would help.

Ensuring the materials used in products are recycled rather than buried or burnt at the end of their useful lives will also help reduce demands on material supply and atmospheric pollution.

Ecological issues in the design and manufacture of products

Mining

Mining is the removal of minerals and metals from the earth. Examples of the type of things that are mined include bauxite (aluminium ore), iron ore, gold, silver and diamonds.

Figure 2.3.1 **Gold mining in the Amazon**

Mining can create a huge amount of employment and income for countries and groups of people, but it is also a concern, as it damages the environment and causes related problems for local people. Mining contaminates water supplies and threatens communities of indigenous populations. Trees and vegetation are cleared and burned in order to make space for mining; this is called **deforestation**. While deforestation is not solely caused by mining (other causes include logging and agricultural expansion), it is important to bear in mind its impact. In order to separate metal from ore, chemicals such as cyanide and mercury are used, and can often end up in rivers and streams contaminating water supplies and poisoning marine life, such as the fish that local communities rely upon for their survival and economic livelihood. These chemicals can cause serious harm to humans, particularly when ingested in large quantities. This affects all of us, irrespective of where we live, due to the fish being collected and sold to other areas of the world – but the people of local communities often use such rivers as their main source of food, so can be exposed to large amounts of these chemicals. People often do not have medical assistance nearby and symptoms can be severe. When the people of the area know that water is contaminated, they cannot use it for the important things that they normally use it for, such as drinking, bathing, washing clothes and so on.

Mining can also be done by land and river dredging. Land dredging involves miners digging a large hole in the ground. They then use a high-pressure hose to expose the layer of sand and clay that holds small pieces of metal, such as gold. The metal is collected from the slurry and the waste is collected in pools, which become stagnant and a breeding ground for mosquitoes. Malaria and other diseases is then an increased concern in areas near to these open pools of water.

River dredging involves moving along a river using a suction hose to collect gravel and mud. This travels down pipes where any metal fragments are collected. The waste gravel and mud is usually deposited further down the river and can disrupt the natural flow of the river, which makes it difficult for fishes to travel down the river, and fish and other marine life often die.

Once the raw material has been mined it has to be melted down. This requires energy, normally in the form of charcoal made from trees that have to be cut down in the local area, which leads to further deforestation.

Mining can also extract coal from under the ground. Coal provides approximately 40 per cent of the world's electricity supply and is an important resource, but as with the extraction of metals such as gold, the effects of mining it are damaging.

Figure 2.3.2 **Offshore oil rig**

Drilling

Drilling is used to extract liquids or gas from underneath the Earth's surface, either on land or in the sea. This is another method which causes a great deal of disruption to the Earth. Exploring an area for oil reserves can cause a great deal of disruption in itself, as roads are created in order to make way for large machinery and camps of people. This newly created access then encourages the arrival of other companies for other purposes, such as logging. Companies see that access has been created, which reduces their initial set-up costs and encourages them to work in the area, thus increasing the environmental impact on the surrounding land and the people who live there. In the example of logging, workers use the roads to get access to areas that might otherwise be expensive to reach and they can then start cutting down trees that might have otherwise been protected.

Once sources of oil have been found, dangerous chemicals are used to extract the material. Unfortunately, as this material is then difficult to dispose of safely, companies have been found to dump these materials in other places. Wells, pipes and other methods of transportation also leak fairly frequently, leading to huge detrimental impact on local areas and those who live there.

Oil can harm wildlife, both when it touches the animal and when it is inhaled or ingested. Damage can occur to the central nervous system, liver, and lungs, and animals often die when they eat contaminated food. Animals can also be affected by breathing in the hydrocarbons left on the water's surface

Figure 2.3.3 **The impact of oil spill on birds**

Ecological issues in the design and manufacture of products

during an oil spill. When physical contact occurs with a bird's feathers, even a small amount of oil can destroy the protective layer on those feathers. This can stop the bird from flying, and can make it sink when in water. Oil also affects an animal's ability to insulate themselves, which can lead to hypothermia and death.

Oil spills can have a huge economic impact on local areas, as fishing has to be stopped when marine life is poisoned. Tourism in the area can be damaged if the oil devastates huge areas and makes water dangerous and unattractive to swimmers.

Drilling for oil and natural gas leads to atmospheric pollution when excess is burnt and carbon dioxide is produced; this may contribute to global warming. This process is called 'flaring'. Methane is also produced as part of the breakdown of biological organisms and is regarded as a greenhouse gas.

Figure 2.3.4 Palm oil plantations leave little space for the orangutan.

Farming

Over 50 per cent of the Earth's useable land has been transformed into **farming** land. This has a huge impact on the Earth's natural resources and habitats. A good example of this is in the current destruction of areas of rainforest for palm oil plantations. This dramatically reduces the natural habitats of the Sumatran rhinoceros and tiger, orang-utan and Asian elephant in places such as Malaysia and Indonesia.

Land is also used in the production of other vegetable or animal products such as leather, wool and dyes. These products all need farm land in order to keep the animals or grow the vegetation.

Farming uses about 70 per cent of the world's useable water supply, which means less is available for other purposes. It also causes large-scale erosion, where large rooted trees that help to hold areas of soil together are replaced by crops. Large areas of nutritious topsoil are often blown or washed away leaving fewer nutrients necessary for successful crop growth. Often soil is eroded on the side of river banks and soil is washed away to silt up rivers and destroy important water sources. This can also lead to flooding.

Pesticides and fertilisers have been used increasingly over the last 50 years. These chemicals are used on crops to make them grow quickly and to kill insects that threaten the produce. They are also used in order to keep animals producing materials effectively, such as sheep dipping in the production of wool. Unfortunately, pesticides can often kill other things and are not restricted to the area in which they are distributed. Chemicals can be sprayed from overhead and are blown into adjoining areas. They can also be transported through water systems after it rains and end up in streams and rivers.

Deforestation

This has a huge effect on animals and people. If there are fewer trees and less vegetation in the forests, wild animals have fewer places to live, and populations decrease. The indigenous people also have fewer places to hunt and less food to eat.

The Amazon alone is losing 200,000 acres of rainforest every year. This is due to a number of reasons, including mining, logging and farming for products such as palm oil or soybean cultivation.

Figure 2.3.5 Deforestation

Illegal logging contributes a large amount to the amount of trees felled each year with an average of between 60 and 80 per cent of all logging activities in the Amazon considered illegal.

The six Rs

Designers need to consider the environment and sustainability when designing products. They must consider how the product affects the environment at every stage of the product from 'cradle to grave'. This means designers must think about the impact of a product from the collection of raw materials to how the product is disposed of at the end of its useful life.

The six Rs of sustainability are often used to remind us of how we can improve the impact our products have on society:

REDUCE – the amount of energy and materials used in the manufacture of a product. This will help to protect valuable resources.

REUSE – the product for something else so you don't need to throw it away.

RECYCLE – take the product apart and separate into the same materials. Convert these materials into another product often by melting the material down. This uses a lot of energy, but considerably less energy than making the material from its raws materials - for example, aluminum uses only 5 per cent of the energy to recycle compared to manufacturing the ore.

RETHINK – products and how we use them. Is there a better way of doing the same job that has less of an effect on the environment?

REFUSE – to buy materials and products that are unsustainable.

REPAIR – products rather than throwing them away. Can you design a product that is easier to repair than throw away?

Figure 2.3.6 **The six Rs**

Social issues in the design and manufacture of products

Earlier in the topic we discussed the negative impact of mining, drilling, farming and deforestation on others. These negative impacts include:

- contamination of water supplies, and therefore loss of livelihood as well as poor-quality or dangerous drinking water
- an increase in the amount of open stagnant water, which can be a breeding ground for malaria carrying insects
- disruption to rivers, which can reduce the ease with which marine life and fishers can travel
- oil leaks and the damage they can do to marine life, people and the economy
- felling of native trees and other vegetation necessary for specific breeds of animals
- soil erosion due to lack of tree roots, which can then lead to the silting up of rivers
- the use of pesticides and their effect on other animals
- the use of fertilisers and the eutrophication of streams and rivers
- deforestation leading to a reduction in hunting areas
- carbon being produced during the manufacturing process.

There are other aspects which we will discuss in more detail below.

Safe working conditions

Many would agree that the designer is partly responsible for ensuring their products are manufactured in a responsible manner. This includes ensuring that manufacture has as little impact on the environment as possible, but also that the people involved in the manufacture are treated fairly and have suitable working conditions. This responsibility demands that designers consider the health and safety aspects of the manufacture of their product and that machinery and environments are maintained effectively. Designers and manufacturers have to consider the legislation put in place to protect workers and also their own ethical guidelines, as some countries differ in terms of their expectations of manufacturers. It is not simply acceptable to stay within legal guidelines, but consumers are increasingly expecting designers, manufacturers and retailers to apply their expected standards to protect those abroad as well as at home.

Designers have a responsibility to ensure the following guidelines are met:
- Employment should be a choice and no one should be forced to work.
- Working conditions should be safe and hygienic.
- Workers should be allowed to join and form trade unions.
- Workers should be paid a 'living wage' which is a wage that is high enough to maintain a normal standard of living.
- Child labour should not be used.
- Working hours should not be excessive.
- Workers do not suffer discrimination.
- Workers do not have to suffer harsh or inhumane treatment..

In the past, companies have been criticised for allowing some of the above to happen. Pressure is increasing to ensure that they take responsibility for the safe working conditions of those involved in the manufacture of their products. More and more consumers are starting to apply this pressure in the hope that companies realise how important these issues are to all involved in the product.

Figure 2.3.7 Tiny microbeads in face wash products can harm marine life and birds.

Figure 2.3.8 Birds have been found with stomachs full of plastic.

Reducing oceanic and atmospheric pollution

Oceanic pollution

Plastic is becoming one of the biggest problems in terms of oceanic pollution. Plastic is a useful and relatively cheap material that can be formed into a range of objects. Unfortunately this material does not degrade quickly, and when disposed of, washes into our oceans where it stays for many years. Plastic products range from large containers to the tiny microbeads found in face wash products. All of these items create huge problems for marine life and birds and can be potentially harmful for us too.

Every year, eight million metric tons of plastic end up in our oceans. That is the same as five supermarket shopping bags filled with plastic for every 300m of coastline in the world. It is estimated that this will double by 2025. It is worrying to think that small particles such as microbeads can get into our waters and can then be consumed by fish and birds. Often this is not only harmful to the animals themselves but also to us, as we then catch and eat them and therefore ingest the plastic ourselves.

In Hawaii, thousands of bird corpses lie dead on the beaches due to plastic poison. In some instances, the body of the birds are entirely disintegrated, yet where their stomachs were piles of plastics remain.

Apart from eating the plastic, marine life can also suffer from the impact of plastic by becoming trapped in it. There are many photographs on the internet of turtles being trapped in plastic rings of different types.

One company has recently developed multipack can rings (known as 'six pack rings') that are 100 per cent biodegradable and can actually be eaten by marine life. They are made of barley and wheat remnants, which are transformed into useful and harmless materials even if they end up in the sea. Rather than being a problem to turtles, these rings can help as they can be a food source, and will disintegrate in time which ensures that they cannot trap animals, as shown in Figure 2.3.9.

Figure 2.3.9 Turtles can become trapped in plastic.

Atmospheric pollution

Atmospheric or air pollution is when a gas is present in the air in a big enough quantity to cause damage to the health of people or animals, kill or damage plants, or effect buildings (for example, make them disintegrate). Sometimes, atmospheric pollution can be caused by natural sources, but most of the time, humans and the decisions they make about their lifestyles and the way they use products is the cause of this problem. Examples of this include:

- Sulphur dioxide is often released when coal is burnt in power stations.
- Carbon monoxide is released when cars are used or when a boiler is not functioning correctly.
- Carbon dioxide is released in dangerous amounts when burning fuels.

The increase of burning fuels for the production of electricity, to power cars and so on has led to a huge increase in the number of dangerous gases that are released in to the atmosphere. This changes the quality of the air we breathe in and can cause problems to our health. It can also have effects on other things, such as farming, as crop yield can be reduced. It also creates problems for our ozone layer, which has an effect on the air temperature and a knock-on effect on the world as we know it.

Reducing the detrimental impact on others

It is a designer's responsibility to consider all of these impacts on society and the environment when considering the production of a product. Designers have to consider all aspects of the life of their product. These include:

- the extraction of raw materials
- the manufacturing processes used
- the treatment of workers
- transportation of materials and products
- the effect the product has on the user and environment
- how easy the product is to repair
- the disposal of the product.

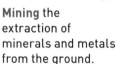

STRETCH AND CHALLENGE

Write an essay about a country of your choice. Explain your thoughts on their policies relating to farming, mining or drilling.

ACTIVITY

Create a poster showing an area of land. Label the possible issues for the environment as a result of modern industrial practices.

KEY WORDS

Mining the extraction of minerals and metals from the ground.

Drilling the process of making a hole in the Earth's surface, usually to extract liquids or gas.

Farming the use of land for growing crops or keeping animals for food.

Eutrophication excessive nutrients in a body of water, often caused by fertilisers.

Deforestation large areas of trees cut down, often due to mining, drilling, farming or logging.

KEY POINTS
- Deforestation is where large areas of trees are cut down, often due to mining, drilling, farming or logging.
- Mining can damage the environment and causes related problems for local people.
- Farming uses about 70 per cent of the world's useable water supply, which means less is available for other purposes.

Check your knowledge and understanding

1 What are the advantages and disadvantages of a mining company arriving in an area of rainforest?
2 Explain the term 'deforestation'.
3 When does deforestation occur?
4 What impact does farming have on the environment?
5 What are the six Rs?
5 How do designers ensure safe working conditions?

Examples and application for papers and boards

The primary source for paper and board products is wood. The management of this resource is important for the continued use of this material, as despite wood being a renewable resource, it is massively depleted. This is due to bad management of forests, a lack of understanding of the effects of deforestation and a lack of responsibility by designers and manufacturers.

Figure 2.3.10 **FSC symbol**

This problem is particularly true of hardwood trees, which take a long time to grow. Many manufacturers now take a greater responsibility for the materials they use and try to replace the trees that are cut down to make their products. They may seek FSC® (Forest Stewardship Council®) certification so that consumers know that they value the responsible management of the forests.

Paper and board products are widely recycled in order to prolong the useful life of the material. In this process, the material is separated into different grades so that they can be recycled together. The paper is pulped and mixed with water and chemicals and then dried out to create new paper. New pulp may be added to the recycled pulp to strengthen it, but the majority of the material is recycled. This reduces the number of trees that need to be cut down for paper and board production.

Examples and application for timber, metals and polymers

Timber

Deforestation is a key concern in the manufacture of wooden products. Hardwoods take many years to grow and have to be managed carefully. Timber is often reused in order to reduce the amount of trees that are cut down. Examples of this are in the building industry, where beams and structural pieces of material may be reused as products such as floorboards. Old railway sleepers are also often used in landscape gardening.

Figure 2.3.11 **Old railway sleepers used in gardens**

It is important in the manufacture of all products for materials to be easily separated. If materials are permanently joined it is difficult to sort them into different groups, and so temporary fixings, such as screws, can be used so products can be dismantled and parts recycled according to their material type. When timber is recycled, it is possible to use this material to make other manmade boards such as MDF. In the production of MDF, recycled wood can be chipped and pressed and combined with glue to make a flat material that is easy to paint on.

Metals

Metal ores that are extracted from the Earth take millions of years to form. These resources will eventually run out, so it is important for us to consider alternative materials or try to recycle and reuse the materials we have already extracted. Mining is the process used to extract metals, and this process is disruptive to the environment surrounding the mine and the atmosphere. Bauxite is the primary source of aluminium, and it is estimated that there is approximately 100 years' worth of Bauxite left in the Earth's crust. This might be surprising, considering it is the Earth's third most abundant element after oxygen and silicon.

Many metal ores are extracted from near the surface of the Earth, and so mining has an impact on vegetation on the surface. Extraction causes erosion and dust and noise in the surrounding areas, and waste has to be managed carefully to avoid pollution.

Bauxite is generally found in tropical forest areas, which are among the most threatened ecosystems. Programmes of reforestation have begun in such areas to try to reduce the impact of mining. It is also important to preserve the reserves we have already extracted. The recycling of aluminium is commonplace due to the ability for aluminium to be recycled time and time again without damaging the material.

Polymers

Polymers are mainly made using oil, which has to be extracted out of the ground using drilling. Polymers are important materials that allow for the mass manufacture of complex shapes that would be expensive or difficult to make in other materials. Unfortunately, not simply in the extraction of oil, polymers have a detrimental impact on the environment.

Polymers take a long time to break down if thrown into landfill rather than recycled. Many plastics are still not recycled, and so much of the plastic used remains in the ground or pollutes the oceans, providing a dangerous environment for birds and marine life. Polymers can be mixed with additives which help them to break down more quickly, or can be made from biodegradable materials such as corn starch. This is an important development in polymer manufacture as the material is not only biodegradable but comes from a renewable source. Unfortunately, this material has also been criticised as less than perfect in terms of environmental impact, as there is not enough land to grow the necessary crops to make the polymer, and it is also argued that these polymers contaminate oil-based plastics in the recycling process and lead to a lower quality recycled plastic.

Figure 2.3.12 **PET symbol. This symbol means that the material is PET (polyethylene terephthalate) and can be recycled.**

Biodegradable plastics are commonly used for disposable items such as plastic cutlery or plates and so on.

Polymers are often burnt to get rid of them at the end of their use, or during the manufacturing process due to flaring during oil production. This leads to a release of harmful gases, including carbon dioxide, which are released into the atmosphere and contribute to global warming.

In order to make recycling more likely, councils have included polymers in roadside collections and manufacturers now have to create a stamp on the plastic to show the type of material. This is so all of the same material is recycled together.

Examples and application for textiles

The most common of the synthetic fibres used in the textiles industry are made from petrochemicals. These materials are difficult to recycle and take many years to decompose. Industrial manufacture in the production of the original material and in the manufacture of the garment have an impact on air quality, and dyeing and printing use huge amounts of water and chemicals which are released into the atmosphere and can be harmful to us and the environment we live in. In Indonesia, the Citarum River is one of the most polluted rivers in the world partly due to the many textiles factories that surround it. The textiles factories were not set up to create the vast amounts of textiles products that they have done and their waste has not been managed effectively, leading to chemicals such as lead, mercury and arsenic being dumped in the river. This has an impact on the human beings that rely on the river for water to drink and the wildlife that uses it as their natural habitat.

Figure 2.3.13 **Oxfam clothing bank**

Even natural fibres have an impact on the environment. Cotton, for example, is seen as a more sustainable material due to it being from the natural resource of the cotton plant, which is renewable. Unfortunately the cotton plant is one of the most chemically dependent crops in the world, meaning that it is often produced using high volumes of chemicals to ensure quick and successful growth. It also takes a large amount of water to grow, which also has an impact on the environment. Companies have started to look at using organic cotton in their garments, but this is still in the early days as there are cost implications of this.

One way that companies are trying to become more sustainable is through the use of waterless or near waterless dyes. These dyes use little water, but instead air to disperse the dye and embed it in the fibres of a garment. This makes the product more resistant to washing, which traditionally removes some dye, which is then fed into the water system.

Textiles are often successfully recycled and reused when products are sent to charity shops or sold to textiles providers for developing countries. Textiles garments can also be reused in the home, where old garments might be used as rags for dusters or cloths and so on. Another way to extend the lifetime of your textiles products might be in repairing them. This method used to be very popular for economic reasons rather than consideration for the environment. However, this is less popular nowadays when the price of clothing and other textiles products is reduced and society is less willing to make alterations, preferring instead to throw something away and buy an alternative.

Examples and application for electrical and mechanical systems

Electrical and mechanical systems are often encased in other materials that have an impact on the environment, as well as the systems themselves. The polymer casing of a remote control car, for example, will be made using oil that is non-renewable and causes air pollution when extracted and made into the stock form. Some mechanical systems may involve the use of metals, such as steel, which have been created using iron extracted from the ground. This iron is smelted in its raw form and then chemicals such as sulphur and phosphorous are added. Both the heating process and the use of such chemicals releases dangerous gases into the atmosphere and uses valuable resources.

Figure 2.3.14 **WEEE symbol**

Batteries are often used in electrical systems as the source of power. These are an important component of the system, but must be disposed of properly to minimise their damage to the environment. Consumers are encouraged to do this at a variety of recycling points, such as supermarkets.

When batteries are thrown away and then sent to landfill they start a long decomposing process in which potentially harmful chemicals are released into the environment. Chemicals such as lithium, lead and mercury may leak out of the batteries and end up in our rivers and streams, and into our water supplies or those of surrounding wildlife.

A photochemical reaction takes place when an object decomposes and this causes gas emissions. Batteries contain harmful chemicals which can leak in the form of gases and contribute to the greenhouse effect.

Batteries and other electrical goods often display the WEEE symbol (Figure 2.3.14). This symbol tells consumers that the product cannot be thrown away at the end of its use and must be taken to a recycling point for safe disposal. This will be because some or all of the product is deemed hazardous, and so must be taken care of in order for the chemicals it contains to be recycled or disposed of without them leaking into our environment.

2.4 | Sources and origins

What will I learn?

In this topic you will learn about:

➡ how cellulose fibres are converted into paper
➡ the source of timber-based materials
➡ how timber is converted into planks of wood
➡ how wood can be seasoned to make it usable and to prevent defects
➡ the source of metal-based materials
➡ the different methods of processing ore into metal
➡ the source of polymers
➡ how crude oil is processed into polymers
➡ the use of additives in polymers
➡ how textile fibres are obtained from animal, chemical and vegetable sources
➡ how fibres are processed and spun into yarns and fabrics.

It is important that as designers and manufacturers we fully understand where the materials we use originate from and that we have a knowledge of the process they have been subjected to before we actually begin to use them. With this information we can make judgements on what materials we should use, based not only on their working properties but also on any financial, moral, ethical and environmental concerns.

Papers and boards

Converting cellulose fibres to paper

Figure 2.4.1 Pulp ready for the paper-making process

Paper is made primarily from wood specifically grown for making paper. Other sources include recycled newspaper and recycled cloth. Softwoods are traditionally used to make paper as their **cellulose fibres** are longer and therefore make stronger paper, but the demand for paper is high, so both hard and softwoods are now used. Recycled fibres cannot be used endlessly, so it is necessary to have some new fibres in the form of wood pulp.

The process of making paper involves 'pulping', which is the process of breaking down wood or recycled materials to extract the fibres.

First, the wood is collected and **debarked** (bark is removed from the wood). It is then chipped and either ground against a stone or cooked with chemicals under high pressure. The pulp is then sent through filters and bleach or other colourings are added.

The pulp is then squeezed and pounded, and materials such as chalks, clays or chemicals are added to alter the **opacity** of the final product. Starch is also added to alter the absorbency of the paper so that it resists the ink and is therefore more effective for writing or printing on.

The pulp is then pumped into automated machines on a moving belt. Normally, the type of machine used is called a Fourdrinier machine. The pulp is pushed through a set of rollers, while excess water is drained off. The belt is shaken, sucked, and blown to remove water from the fibres. A watermark can be added at this stage using a patterned roller called the **dandy roll**. The paper is sent through rollers again and again to become fully dried and then finally pressed to ensure a smooth finish. This process is done with heavy steel rollers called **calenders**. A particular finish, whether soft and dull or hard and shiny, can be imparted by the calenders. The paper is also sometimes given a coating, which can be brushed or rolled on.

Figure 2.4.2 **Paper is often produced in huge rolls.**

The biggest Fourdrinier machines might have between 40 to 70 drying cylinders, and paper is produced at a rate of over 40 miles of paper per hour.

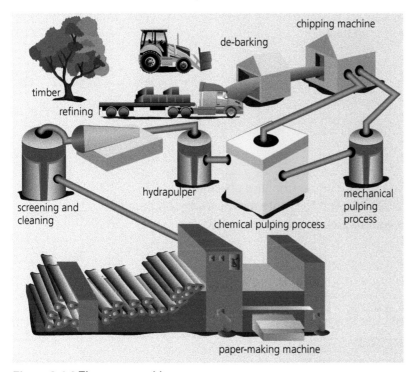

Figure 2.4.3 **The paper-making process**

Life cycle assessment

Life cycle assessment allows us to look at the process of making paper and assess the impact this process has on the environment from 'cradle to grave'. This means we can look at the environmental impact, from the cutting down of trees to the disposal of materials at the end of the product's life.

A great number of trees and other vegetation are cut down in order to make paper. Many paper companies try to replace the trees they cut down, often promising two new trees for every one cut down in order to reassure consumers. Although this is true, it is difficult to replace older trees with younger ones of the same value. New trees take many years to grow and older trees are often established wildlife habitats. They are also more effective

KEY WORDS

Cellulose fibres: fibres found in trees and plants which can be used to make paper and card.

Life cycle assessment a method of assessing the environmental impact of the manufacture and use of materials and products.

Debarked the process of removing bark from timber.

Opacity the transparency or thickness of a material.

STRETCH AND CHALLENGE

Create a life cycle assessment of a specific paper-based product.

at absorbing carbon dioxide. Deforestation also has an impact on other areas of the environment. Trees lost near to rivers can cause banks to fall in, which causes flooding and loss of marine life as streams and rivers become blocked. Paper can be recycled and used in the production of new products, but it can only be used a set number of times. When you recycle paper, the paper is mixed with water and then ground up. This turns the paper back into pulp, but the grinding shortens and weakens the fibres, and so the recycled paper is weaker and easier to tear. The impact of the recycling process can also have an effect particularly, in the de-inking stage where sludge is produced.

When the chemicals used as fillers, colours, etc. are washed into water supplies, they can cause environmental concerns such as the contamination of water supplies for some animals and pollution of the waters where fish live and breed.

See Topics 2.3 and 3.2 for more information.

KEY POINTS
- Paper is made with some new fibres and some recycled materials.
- Trees are cut down and debarked. The wood is chipped and pulped, bleached and dried to produce the final product.
- Life cycle assessment is used to understand the impact of the production of paper on the environment.

Check your knowledge and understanding

1 What are cellulose fibres?
2 What is the pulping process?
3 Name the type of machinery often used for the paper-making process.
4 What are the names of the patterned rollers used to create a watermark on paper?
5 What are the main effects of paper production on the environment?

Timber-based materials

Primary sources

Trees are the primary source of all timber-based materials (including both natural timber and manufactured boards). Trees are grown in forests throughout the world approximately 30% of the surface of the land is covered in forests. Britain was once covered in forests, but now, mainly due to the needs of agriculture, we import most of the timber we use. Different types of timber come from different areas of the world.
- Softwoods mainly come from the cool northern parts of Europe, Canada and Russia.
- Hardwoods are grown in Central Europe, West Africa, Central and South America.

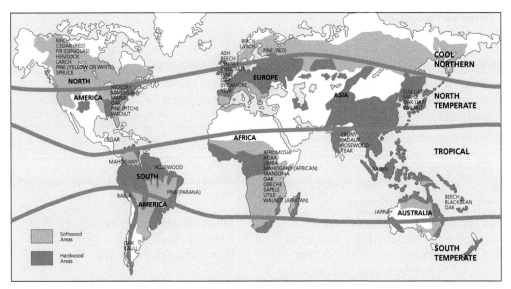

Figure 2.4.4 **The world distribution of forests**

As trees are a renewable resource, it is vital that we only use timber that comes from well-managed forests. The Forest Stewardship Council (FSC) is an international, non-governmental organisation dedicated to promoting responsible management of the world's forests. They ensure that forests are replanted once trees have been cut down, and that the process is done in an ethical and environmentally friendly manner. Always look out for the FSC logo.

Once a tree reaches maturity it can then be **felled** (cut down). The length of time it takes for a tree to reach an age when it can be felled varies depending upon the type of wood. A pine tree grows relatively quickly, only taking around 30 years to be commercially useable, whereas exotic hardwood trees can take considerably longer. Felling is a mechanised process involving logging machinery. A tractor carries a special adapter that can cut the tree, strip off the branches and slice the logs into manageable lengths.

Figure 2.4.5 **Forestry Stewardship Council (FSC) logo**

Figure 2.4.6 **Logging machinery**

Conversion

Once felled, the logs are then transported to the saw mill where they are converted into usable planks. Depending on the type of wood and how it is going to be used, it may go through one of a number of different **conversion** processes, source of which are shown in the table below.

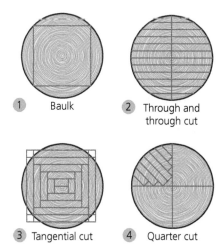

① Baulk

② Through and through cut

③ Tangential cut

④ Quarter cut

Figure 2.4.7 **Conversion**

Timber-based materials

Conversion process	Description	Common uses
Baulk cut	This is the simplest form of conversion. The trunk is simply cut into a square or rectangular section. This removes the bark and evens up the trunk.	Beams in the construction of timber framed buildings.
Through and through cut	This is the most popular form of conversion and involves sawing the trunk into planks. It is a simple and cost-effective method; however, it can lead to a number of problems with warping and twisting. This stock form is most commonly used with softwoods.	Many areas of general joinery
Tangential cut	During this form of conversion the trunk is cut tangentially to the circular trunk. This produces an attractive grain pattern and the wood is less likely to warp and twist. This method is used for both soft and hardwoods.	Where the natural attractive grain is important, such as during the manufacturing of furniture.
Quarter cut	This is a complex method of conversion that produces a lot of waste. It is expensive in terms of its financial cost and its impact on the environment. The trunk is cut radially out from the centre. This stock form of wood is generally used for expensive hardwoods.	To manufacture high quality furniture. Oak furniture that has been quarter cut will display 'figuring'; this shows up as silver markings that catch the light.

KEY WORDS

FSC the Forestry Stewardship Council.

Felling the process of cutting a tree down.

Conversion the process of sawing a tree trunk into planks.

Green timber wood that has not been seasoned.

Air seasoning a natural method of drying out green timber.

Kiln seasoning a relatively quick method of drying out green timber using steam.

Seasoning

Newly converted timber contains a lot of moisture and is known as **green timber**. The high moisture content makes the wood difficult to work, saw or plane. It will twist, warp and split if left in this state. It is also open to rotting and is vulnerable to insect attack.

The high moisture content needs to be reduced by a process known as seasoning. There are two different methods of seasoning: **air seasoning** and **kiln seasoning**.

Air seasoning

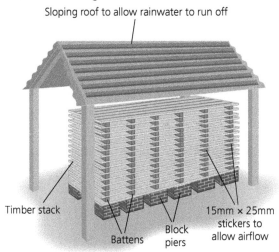

Sloping roof to allow rainwater to run off

Timber stack

Battens

Block piers

15mm × 25mm stickers to allow airflow

Figure 2.4.8 **Air seasoning**

With this method, the aim is to reduce the moisture content of the wood by letting air flow around it in a controlled way. This is a relatively cost effective process but takes a considerably longer time than kiln seasoning – it can take around a year to season a 25-millimetre thick plank of wood.

The planks are carefully stacked inside a building that has a roof, but is open on all sides. The roof will keep off the rain, snow and high sun, and the open sides allow the air to circulate around the boards. The planks are separated by stickers and the stack of boards is kept off the floor by being placed on brick piers. As the air flows around the stack it will slowly, very slowly, dry out the planks.

Kiln seasoning

With this method, the aim is to reduce the moisture content of the wood by gradually reducing the moisture content of steam that is fed into a kiln.

A stack of timber is mounted onto a trolley that is wheeled into a kiln. The kiln is fully enclosed and has steam fed into it. As the moisture content of the steam is reduced it dries out the timber. This is significantly quicker than air seasoning. It is controllable but has a higher financial and environmental cost.

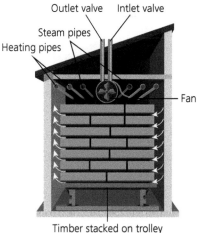

Outlet valve   Intlet valve

Steam pipes

Heating pipes

Fan

Timber stacked on trolley

Figure 2.4.9 **Kiln seasoning**

Manufactured boards

Manufactured board such as MDF, plywood and chipboard are produced by gluing wood layers or wood fibres together. They often we waste wood.
- MDF is made by compressing wood particles and gluing them together.
- Plywood is produced by gluing three or more veneers of wood together at 90 degree angles from the previous layer to increase strength.
- Chipboard is procluced by gluing wood chips together.

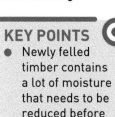

KEY POINTS
- Newly felled timber contains a lot of moisture that needs to be reduced before the timber can be used.
- Trees are a renewable source and it is vital that we only use new timber from managed forests.

ACTIVITY

Ask your groundsman or caretaker for a section from a newly felled tree – a piece around 150 millimetres in diameter or larger and 500 millimetres long will be fine. Get your teacher to convert this into thin planks on the bandsaw using the through and through conversion method. Place it in the school workshop and use stickers to ensure there is an airflow around your stack of short planks. Place a similar size section of baulk cut tree by the side of it and take pictures of what happens over the next few weeks.

Check your knowledge and understanding

1 Describe the work of the FSC.
2 Use notes and sketches to describe the four methods of converting timber.
3 Explain why it is necessary to season wood.
4 Draw a cross-section and fully label a seasoning kiln.
5 Why is air seasoning considered to be environmentally friendly?

Metal-based materials

Primary sources

Most metal is found in rock within the Earth's crust and is known as **ore**. The most popular metals that you will use are steel and aluminium.

- Steel, which we get from the pure metal iron, originally comes from an ore called haematite. Brazil, Australia and Africa have the largest deposits of haematite. Aluminium comes from an ore called bauxite. Australia and Guinea have the largest deposits of bauxite. Metal ores are a finite, non-renewable resource and we are slowly running out of them. It is therefore essential that all metals are recycled and reused.

> **KEY WORDS**
>
> **Ore** a rock which contains metal.
> **Smelting** the process of extracting metal from ore.

Ore is taken out of the earth by a process known as mining. Some ores are quite close to the surface and are relatively easy to mine, but others are buried deep below the surface and require underground mining.

- Surface mining involves stripping away the earth using large excavation equipment to remove earth. The heavy vehicles and equipment used to surface mine use large quantities of fuel and the process leaves large scars on the landscape.
- Underground mining involves digging deep underground tunnels. The ground is often blasted with explosives to release the ore which is then loaded onto conveyor belts to bring it to the surface. Old mines can leave the surface of the ground unstable and lead to subsidence.

Figure 2.4.10 **Mining haematite**

Extraction (smelting)

Once the ore has been extracted from the ground, the metal needs to be extracted from the ore by a process known as **smelting**. There are two main types of smelting: the blast furnace and the reduction cell.

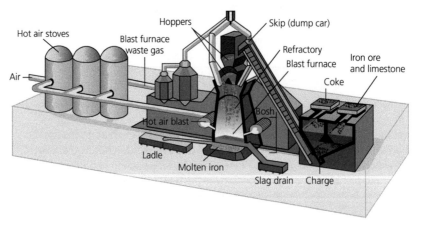

Figure 2.4.11 **The blast furnace**

2.4 Sources and origins

Iron is extracted from haematite in a blast furnace. Crushed haematite, limestone and coke are loaded into the main section of the blast furnace and heated to a very high temperature (1,700 °C). The iron is effectively melted out of the rock and runs out of the furnace into ingots. This is known as 'pig iron'. The manufacture of iron uses large amounts of fossil fuels to heat the blast furnace and creates harmful gases.

Iron smelting is an example of continuous product, as once the furnace reaches the correct temperature it is cost effective to keep it running as long as possible.

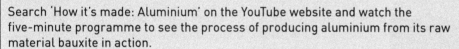

Figure 2.4.12 The reduction cell

Aluminium smelting is done by an electrolytic process in a reduction cell. This involves loading crushed bauxite that has been chemically processed to form alumina, into a chamber filled with molten cryolite. A very high current of electricity is then passed through the mixture via carbon electrodes. This action produces pure aluminium that settles at the bottom of the cell. This process requires large amounts of electricity and is usually undertaken where there is a plentiful resource of inexpensive electricity, such as near a hydroelectric power station.

Pure iron and aluminium are rarely used in that forms. Like many other pure metals they are mixed with other metals to form alloys. The metal we know as aluminium is actually duralumin – an alloy of aluminium, copper, manganese and silicon. Iron is normally alloyed with carbon to form steel.

ACTIVITY

Search 'How it's made: Aluminium' on the YouTube website and watch the five-minute programme to see the process of producing aluminium from its raw material bauxite in action.

KEY POINTS
- Iron ore is known as haematite.
- Aluminium ore is known as bauxite.
- A blast furnace smelts haematite into iron.
- A reduction cell smelts bauxite into aluminium.

Check your knowledge and understanding
1 Explain what ore is.
2 Name the two main processes of mining.
3 What is haematite?
4 What is bauxite?
5 Use notes and sketches to describe what happens in a blast furnace.

Polymers

Primary sources

Figure 2.4.13 **Oil rig**

Figure 2.4.14 **Oil refinery**

Most polymers are obtained from crude oil and are known as synthetic polymers. Deposits of crude oil are distributed throughout the world, but the biggest reserves can be found in the Middle East and in Central and South America. Crude oil is a finite, non-renewable resource and we are slowly running out of it. It is therefore essential that, wherever possible, polymers are recycled and reused.

Crude oil is found deep under the Earth's crust in oil fields. The oil fields are often in very remote locations, such as in the middle of the ocean, in the middle of a desert or in frozen wastelands. Seismic tests (controlled underground explosions) are carried out to find crude oil, and by analysing the results geologists can make a prediction as to whether crude oil is present. Exploratory bore holes are then drilled into the ground to see if the oil is commercially accessible. Crude oil then needs to be pumped to the surface and transported to an oil refinery where it can be made into various oil-based products. Transportation is carried out by pumping the crude oil across vast areas of land in pipelines and by shipping it across oceans in huge oil tankers.

Fractional distillation

Crude oil in its raw state is of very little use. It consists of a mixture of hydrocarbons, of different weights. The lightest of the hydrocarbons are gases, such as liquiefield petroleum gas (LPG) which can be used to fuel cars. The heaviest are thick tar-like substances, such as bitumen which can be used in road surfacing.

The first part of transforming crude oil into these useful petrochemical products is to break it down by a process known as **fractional distillation**. During this process the crude oil is

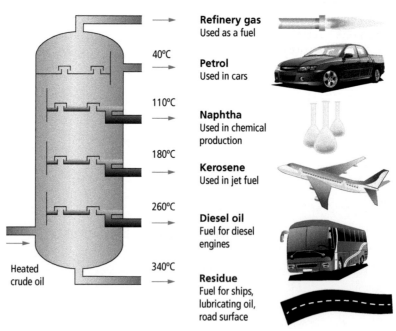

Figure 2.4.15 **Fractional distillation**

heated up until it becomes a gas. It then vents off through a tall column and as it cools it condenses into different petrochemical products, including gas, petrol and oil. One specific petrochemical product that is used to make polymers – naphtha.

A simple monomer

Cracking

Naphtha is still made up of a mixture of hydrocarbons, and needs to undergo another process known as 'cracking' before we can begin to use it to manufacture polymers. The naphtha is heated to break it down further into individual hydrocarbons, such as ethylene, propylene and butylene. These are the building blocks for the manufacture of the polymers that we are familiar with, and we refer to them as 'monomers'.

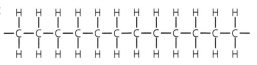

The structure of the polymer polyethylene

Figure 2.4.16 **Monomers and polymers**

Polymerisation

The process of polymerisation takes place in a polymerisation reactor and involves a chemical reaction that links the monomers together into polymer chains. Different monomers linked together in different ways give each polymer its unique properties. The monomer ethylene is polymerised to make the polymer we know as polyethylene (PE) and the monomer propylene is polymerised to make the polymer polypropylene (PP).

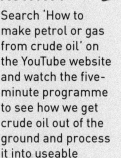

STRETCH AND CHALLENGE

Biopolymers

While many polymers are obtained from crude oil and are synthetic polymers, there are a few polymers that are sourced from plants. These are known as biopolymers. Biopolymers have the advantage of being renewable. However, they are currently expensive and take a long time to manufacture. Polylactic acid (PLA) used in 3D printers is an example biopolymer that is produced from corn starch.

You do not need to know about biopolymers for your GCSE course, but you may find it interesting to do some further research into biopolymers.

ACTIVITY

Search 'How to make petrol or gas from crude oil' on the YouTube website and watch the five-minute programme to see how we get crude oil out of the ground and process it into useable petrochemical products.

KEY POINTS

- Polymers are produced from crude oil.
- Crude oil needs to processed by fractional distillation to produce naphtha.
- Naphtha is the building block of polymers.
- Naphtha must be cracked to produce monomers.
- Polymers are chains of monomers produced by polymerisation.
- Biopolymers are produced from plant material.
- Additives modify the properties of polymers.

Check your knowledge and understanding

1 What is the process of fractional distillation?
2 Name the petrochemical products produced by fractional distillation.
3 What is meant by the term 'cracking'?

Textile-based materials

As you have already learnt in Topic 1.6 Materials and their working properties, fibres are the very fine hair-like threads that are the basic building blocks of fabrics, which are used to make textile products.

Sources of natural fibres

Natural fibres are obtained from plant or animal sources.

Cotton fibres

Cotton is a natural cellulose fibre which comes from a bushy plant grown in the tropical parts of the world. The seedpods of the plant are called **bolls**, and when they ripen the seeds inside the boll become covered with very fine cotton fibres. When the seedpod eventually bursts it becomes a fluffy ball, which looks like cotton wool.

The cotton fibres are like hollow tubes, but as the sun dries the fibres they collapse and become flat like a ribbon and twisted. The fibres in the boll are short, or **staple**, fibres (about two to five centimetres in length). Before they can be used to make a fabric they need to be cleaned and twisted together to make a yarn.

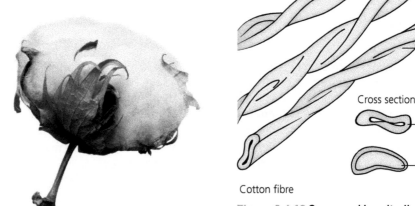

Figure 2.4.17 **Cotton boll**

Figure 2.4.18 **Cross and longitudinal sections of immature and mature cotton fibres**

Wool fibres

Wool is a hair fibre made from protein and comes mostly from sheep (although wool fibres can also come from goats, rabbits, llamas and other furry animals). Wool fibres may be between 4 and 39 centimetres long and are staple fibres.

To obtain wool from sheep, the sheep are sheared. The wool is sent to a mill, where it goes through a process of cleaning and scouring to remove grease and dirt, grading and sorting for quality, carding and combing to straighten the fibres and then spinning to make yarn.

Silk fibres

Silk is a protein fibre that comes from the cocoon of the silk caterpillar. The caterpillar spins the cocoon with two triangular-shaped filaments, one from each side of its mouth. The filaments are held together with a natural gum produced by the caterpillar.

When the caterpillar is ready to turn into a moth, it will break through one end of the cocoon. This spoils the silk thread by breaking it into small lengths which cannot be used as long continuous filament fibres are needed to make the best quality fabrics.

When the cocoon is needed for silk fibres, the caterpillar is killed before it turns into a moth by dropping the cocoons into boiling water. The silk filaments can then be unwound as continuous lengths.

Sources of synthetic fibres

Synthetic fibres are manufactured from oil- or coal-based chemicals.

Polyester fibres

Polyester is a synthetic fibre made from chemicals that come from oil. The chemicals are made into a polymer, which is cut into small pieces. These are melted to make a solution that is spun into continuous lengths of fine fibres (called filament fibres).

Polyester fibres are very smooth and look like glass rods when seen under a microscope.

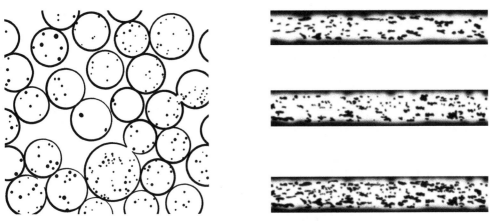

Figure 2.4.19 **Cross- and longitudinal sections of polyester fibres**

Making fibres into yarns (spinning)

Fibres are not used in their raw state and are usually twisted together to make a **yarn** before they can be made into a fabric.

Fibres   →   Yarns   →   Fabrics

Figure 2.4.20 **The process of turning fibres into fabrics**

Staple yarns are made from short staple fibres such as cotton. These have to be **carded**, or combed, so that they all lie in the same direction, before being twisted together to form a yarn. Filament fibres need to be chopped up into short staple fibres if they are to be blended with a staple fibre before being spun into yarn.

Continuous filament yarns are made by lightly twisting together filament fibres (for example, polyester). The twisting holds the fibres together and gives strength to the yarn. If

Textile-based materials

the fibres are tightly twisted together the yarn will be smooth and strong but will not be able to trap air, so it will not be warm.

If the fibres are not twisted so tightly, the yarn will be weaker but also hairier and able to trap air between the fires, so that the yarn will be more insulating. The hairier yarns can also trap moisture between the fibres.

ACTIVITY

Investigating fibres

You will need a five by six centimetre square piece of cotton fabric, such as calico.
1   Pull some of the yarns from the raw edge of the fabric.
2   Untwist the yarns until they become loose.
3   Gently pull the individual fibres from the yarn.
4   Try to count how many individual fibres are in a five centimetre length of the yarn. How long are the individual fibres?
5   What was the purpose of the twisting process in the manufacture of the yarns?

STRETCH AND CHALLENGE

Investigate the different stages of making of cotton yarn.

The YouTube video *How it's made cotton yarn* will help you.

KEY POINTS

- Cotton is a natural cellulose fibre that comes from the ripened seeds found inside bolls, from a bushy plant grown in tropical parts of the world.
- Wool is a hair fibre sheared from a sheep or other furry animal. It is then cleaned at a mill before being spun.
- Silk is a protein fibre that comes from the cocoon of the silk caterpillar. The cocoon is dropped in boiling water and the silk filaments are unwound.
- Synthetic fibres are manufactured from oil- or coal-based chemicals that are made into a polymer and melted to make a solution that is spun into continuous lengths of fibre.
- Fibres are usually twisted together to make a yarn before they are made into a fabric.

Check your knowledge and understanding

1   Describe the origins of cotton fibres.
2   Explain how synthetic fibres are manufactured.
3   Explain how staple fibres are spun into yarns.

2.5 Using and working with materials

What will I learn?
In this topic you will learn about:
→ how to shape and form materials
→ how materials can be modified for a specific purpose
→ how a material's properties influence use
→ how a material's properties affects a product's performance
→ how different properties of materials and components are used in commercial products.
→ how microcontrollers are used in modern systems
→ how the development of modern electronic control systems is changing the way in which we use things
→ how modern control systems are improving performance
→ how to manufacture a printed circuit board (PCB)
→ how materials are modified for specific purposes.

Papers and boards

How different properties are used in commercial products

Materials are chosen based on their properties, and therefore their suitability for a specific job. Many paper- and board-based products are produced in very high numbers. The cost of a material is also a serious consideration, particularly when products need to be mass produced, and what may only be a small difference in cost when you are making one or two products becomes a huge difference when multiplied by thousands.

How properties influence use and affect performance

Properties such as weight, opacity, absorbency, strength and texture are common considerations when deciding which paper or board to use in commercial products. A paper that is resistant to moisture would not be suitable for making kitchen roll as it would not soak up spills effectively. Very thick heavy card is unlikely to be used for bus tickets due to its additional unnecessary cost and weight. Leaflets, brochures and card-based food packaging all require materials that are easy to print on, easy to cut, crease and fold and are affordable. Paper and card products are often disposable so cost is an important factor in the choice of materials.

How to shape and form using cutting and addition
Cutting

When cutting paper or board in a home or school environment, or when working on a one-off project, tools such as scissors and craft knives are used. Craft knives are available in many

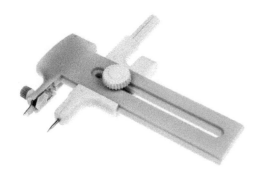

Figure 2.5.1 **A compass cutter**

Figure 2.5.2 **A laser cut paper invitation**

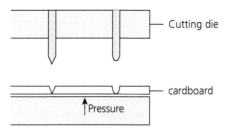

Cutting die

cardboard

↑Pressure

Figure 2.5.3 **Die cutting**

forms, but they often have snap-off blades which can be broken when a new sharp blade is required. They also sometimes have retractable blades making them safer to store. In order to use a craft knife safely, a cutting mat and safety ruler is required.

Scissors often produce poor results when cutting geometric shapes, so crafts knives are best used for straight lines in particular. When cutting circles and arcs, a compass cutter can be used.

Laser cutters are also used in schools to cut paper and card. This can be done effectively if the correct settings are used so that the material does not burn. It is important that both the speed of the laser cutter moving over the material and the power at which the laser is set is accurate, so testing and modifications may be required to get the perfect cut. The laser cutter allows for intricate shapes that would be very difficult and time consuming to cut out by hand.

In industry, it is important for items to be cut very accurately and products are required to be identical every time. Die cutters allow for this and increase the speed at which paper and board can be cut to a high standard. Die cutting allows for cut and fold lines so that products can be assembled quickly. For example, pizza boxes are supplied to fast food restaurants flat and can then be folded along the lines to assemble later. This means that space is saved and therefore more products can be transported at one time. The shapes are 'stamped out' as the paper or card is fed through and products can be cut at high speeds.

Creasing

The die cutter in Figure 2.5.3 comprises of two types of blade. One is used for cutting lines and is very sharp, and another is a more rounded blade which creases the cardboard to create fold lines. These crease the material accurately so that folding is clean and always accurate. This improves the quality of the final product and increases the speed of the process.

Scoring

Scoring is a process often used with thicker boards where the surface of the material needs to be cut in order to fold it effectively. This process is done by hand using craft knives or tools such as bone folders (tools that can be pressed onto the line you wish to hold but will not cut right through the material). Scoring can be done in commercial production or by hand.

Perforating can also be done as part of the die-cutting process. This involves using a serrated edge on the blades so that a dashed or dotted line is created. This is used to produce packaging where part of the product needs to be torn away, such as stamps.

Figure 2.5.4 **Perforated tear strip packaging**

Perforations can be applied to packaging for easy opening. This is an increasingly common method, as more products are bought on the internet and need suitable packaging for delivery. This method is often seen on cardboard envelopes and boxes used for packaging books and DVDs.

Folding

Paper and board can be folded using machinery in order to ensure consistency of the final product. Printed and cut paper and board sheets are fed into a machine where plates fold them at the required places. Commercial paper folding machines often use jets of air to separate each sheet of paper as it enters the machine. This process is quick and produces identical results each time.

Die-cut products can be loaded onto automatic folding and gluing machines to have their sides and bases sealed. This is used when parts of the product need to be glued before assembly. Figure 2.5.5 shows boxes with their sides sealed but their bases left open to allow for folding at their next destination.

Figure 2.5.5 **Flat pack boxes**

Modifying materials for a specific purpose

Materials can be modified to make them more useful for a specific role. Sometimes this is done in the initial manufacturing process.

- Bleach is added to paper in the manufacturing process in order to make the paper white. Different amounts of bleach can be added depending on the required colour of the final outcome.
- Sizing is when a chemical substance is added to paper or board in order to alter its resistance to moisture. This makes the material more suitable for writing and so on.
- Dry strength additives are used to increase the strength of the material under normal conditions. These additives reduce the chance of the paper delaminating and increase the tensile and compressive strength.
- Wet strength additives are used to increase the strength of the material when it is exposed to moisture. This is really important for products such as tissue paper.

Figure 2.5.6 **Foil stamping**

Other modifications take place after the paper or board is formed. These include:

- Clay coating is used for products such as disposable paper plates, where some resistance to moisture and better durability is required.
- Varnishing – this process adds a glossy finish to the material and can be done during printing or afterwards.
- UV coating – this is a clear liquid spread over the material and then cured instantly with ultraviolet light. It creates an extremely smooth surface which looks like a laminate but can be recycled.
- Laminating – this is the process of covering the paper or board in a thin layer of plastic. It provides the material with a protective surface which will prevent damage from water. Unfortunately, once this lamination is applied it is very difficult to remove so recycling is difficult.
- Embossing – the process of stamping paper or board to create a permanent raised surface.
- Foil stamping – this is the process of covering an area of paper or board with a pre-glued metallic coating. A heated die is stamped on to a surface to press the metallic material in place where it permanently adheres to the surface below. This process is usually done in industry with the help of a pneumatic press, although it can be done by hand for smaller print jobs. During the foil stamping process, it is possible to emboss the shape at the same time. In this method, the stamp presses the material into shape while applying the metallic coating.

ACTIVITY

Create an animation showing the stages of cutting card or board with a die cutter. You could do this by acting out the stages, or by using PowerPoint to help you.

KEY POINTS

- Materials are chosen for specific products due to their properties.
- These properties can be modified with additives and finishes.
- Paper and board are cut in industry using die cutting.
- Die cutting allows materials to be cut, folded and perforated.
- Paper and board can be creased using blunt tools on the die cutter.
- Paper and board can be scored using bone folders in small-scale production.
- Paper and board are often folded using machinery, which ensures for a high level of consistency.
- Perforating is the process of creating small dots or dashes in material so part of that material can be removed easily at a later date.
- Varnishing creates a glossy coating on a material, but does not make recycling impossible.
- Laminating creates a layer of plastic on top of the paper or board product. This is difficult to remove.
- Foil stamping is the process of applying a metallic surface to areas of the product.

Check your knowledge and understanding

1 How does the die-cutting process work?
2 Name two ways paper or card can be given a glossy finish.
3 How can paper or card be made more suitable for writing on?
4 Name three ways paper and board can be modified during the paper-making process.
5 Name three ways paper and board can be modified after it is made.

Timber-based materials

How different properties are used in commercial products

There are many factors to consider when selecting a material for use in a commercial product. These can include, cost, availability and the environmental impact associated with sourcing or applying a finish to the material. These are important considerations, but initial selection will relate to identifying the properties the material has that make it suitable for a specific product or application.

How properties influence use and affect performance

Properties such as strength, hardness, toughness and durability are all important things to consider when selecting timber for a commercial product. It is also vital to consider the working environment of the product, as particular woods perform better in certain situations – such as larch being an ideal material for outdoor use, whereas pine would require additional surface treatment to make it suitable for an outdoor application.

Consideration should also be taken about how the product is likely to be manufactured. Some timber is more difficult to machine than others due to the nature of the grain. Equally, certain timbers are more likely to split and warp depending on how they have been machined and prepared.

Brio has been manufacturing children toys since the early 1900s, and although many of their new toys are now making use of polymers, beech remains their most popular manufacturing material. It is an excellent material for the manufacture of traditional Brio trains and has been selected for many reasons, but its physical property of toughness is particularly useful for this application. Children's toys can be subject to a huge amount of use and abuse, meaning that the chosen material must be durable and long lasting. They are dropped and bashed, often stored in large toy boxes where they impact other products and even have to endure being chewed. The toughness of beech and its tight grain mean that it doesn't splinter or break.

Figure 2.5.7 **Child's wooden train**

Chipboard is used in the manufacture of flat-pack furniture. One of the most useful working properties of chipboard that makes it suitable for this application is its density. The manufacturing process compresses the small chipped particles to create a dense and perfectly smooth surface onto which a veneer or decorative laminate can be applied. The consistency of the board and the absence of any grain make it a perfect material to be machined by a CNC router to cut in cavities for hinges and holes for shelf brackets, cross dowels and cam locks. The coarse thread used in the cam dowel knock-down fitting easily cuts a thread into the dense chipboard.

Chipboard has a poor resistance to moisture, which is one of the reasons that flat-pack furniture is usually covered with a veneer or a polymer laminate. This creates an impermeable surface to protect the chipboard from moisture or liquid. If the chipboard does become wet, the particles expand and the board can start to deform and split.

Figure 2.5.8 **Flat-pack furniture**

The modification of properties for specific purposes

Wood can be modified in a number of ways to fit a specific purpose.

When seasoning wood the moisture content can be altered to match the environment to which it is going to be used. More information on seasoning can be found in Topic 2.4

sources and origins. Wood is hygroscopic which means it can absorb moisture, which affects it in various ways. Wood that is drying out can split, cup, warp and twist if the rate that it is drying out is not carefully managed. Wood that is absorbing moisture will swell and be vulnerable to rot.

- Wood that is to be used in a centrally heated home will have a moisture content of around eight to ten per cent.
- Wood that is being used indoors without central heating will have a moisture content of 10 to 15 per cent.
- Wood that is being used outside will have a moisture content of 15 to 20 per cent.

How to shape and form using cutting, abrasion and addition

Marking out on wood

Before you begin to cut or shape your piece of wood it is essential to have some lines to work to. The process of applying the lines to your work is called marking out. Your wood should have a smooth, even surface and have a datum edge. A datum edge is an edge that is flat and smooth and so can be referenced from when measuring.

Marking-out tools for wood

Figure 2.5.9 **A pencil, steel rule and a try square**

Figure 2.5.10 **Marking out on wood**

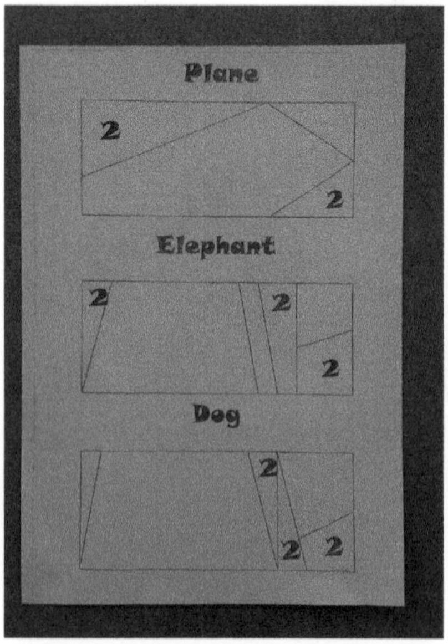

A soft lead pencil is the best tool for marking out on wood as it is easy to see and will not score the surface of the wood. The steel rule is durable within the workshop environment and is less likely to snap or break. A try square will produce an accurate 90 degree line and will improve the accuracy of your work.

Top tips: make sure that your try square is pushed up tight against the side of your wood or you will not get an accurate 90 degree line. Always read a rule by having your eye directly over the measurement or you will get an inaccurate reading.

Templates

A template is often used for irregular shapes where it would be difficult to mark out the profile directly onto the material. Simple templates can be made from paper and could be a print out of a CAD drawing. They are usually stuck down onto the material and then the shape can be cut out following the profile. Paper templates are generally only used as part of a one-off production.

Figure 2.5.11 **A paper template**

Sawing wood

The hand saw is used for cutting thick pieces or large sheets of wood. It cuts relatively quickly, but produces a coarse, rough edge and is generally not as accurate as a 'backed' saw such as the tenon saw.

When sawing with a hand saw it is essential to have your work firmly clamped down. In the picture you can see that the plank of wood has been clamped to the woodwork bench using a G-clamp.

Top tip: place a scrap piece of wood between the clamp and the wood to prevent bruising.

Notice the hand positions: the right hand if you are right-handed (and of course the left if you are left-handed) is positioned through the handle with the index (trigger) finger pointing outwards, this helps to control the saw. The other hand is placed over the top of the handle to provide power for the sawing action. The full length of the saw should be used at a sawing angle of around 45 degrees.

Top tip: support the wood as you are coming to the end of the cut or it will fall off and could leave you with a large splinter.

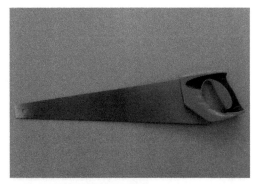

Figure 2.5.12 **A hand saw**

Figure 2.5.13 **Sawing with a hand saw**

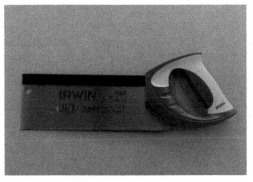

Figure 2.5.14 **A tenon saw**

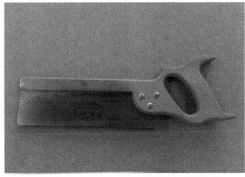

Figure 2.5.15 **A brass-backed tenon saw**

The **tenon saw** is the most commonly used saw in the workshop. It is used for cutting accurate straight lines in wood and produces a relatively smooth cut. The steel or brass back keeps the blade stiff and gives it its accuracy.

Figure 2.5.16 **Sawing with a tenon saw using a G-clamp**

Figure 2.5.17 **Sawing with a tenon saw in a woodworking vice**

Figure 2.5.18 **Sawing with a tenon saw using a bench hook**

Timber-based materials

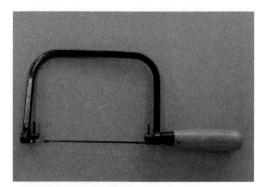

Figure 2.5.19 **A coping saw**

Figure 2.5.20 **A surform**

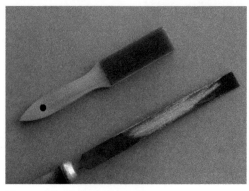

Figure 2.5.21 **Cleaning a file**

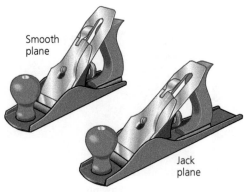

Smooth plane

Jack plane

Figure 2.5.22 **Types of plane**

The work piece can be clamped to the woodworking bench as with the hand saw or it can be held in the woodworking vice, or even held in a bench hook for smaller cuts.

Top tip: you can start a saw cut by dragging the saw backwards three times. This will produce a nick in the wood that will allow you to check that you are about to begin sawing in the correct position. It will also provide a guide for the saw and prevent it from skidding over the surface of your wood.

The **coping saw** is used for cutting curves in wood. It can cut fine, intricate cuts but is not easy to control and it takes practice to be able to use it accurately. The blades are thin and easily broken, but the design of the saw allows the blades to be quickly changed.

Top tip: the blade in the coping saw should face backwards as this keeps it in tension when sawing.

Shaping wood

One of the easiest ways to shape wood is by using a surform or wood rasp. The surform and wood rasp are both similar to a file but have much coarser teeth. Files can be used to shape wood, but will become clogged quite quickly.

Top tip: if you do use a file to shape wood, be sure to keep a file card handy. This will help you keep the file clean.

The **disc sander** is a very effective way of quickly shaping or smoothing wood. It consists of a wheel that is covered with abrasive paper that spins around removing waste wood. You must ensure that you follow all the relevant safety precautions when using this machine. In particular, you should be wearing an apron and safety glasses, and ensure that guards are fitted and that the dust extraction system is switched on.

The **linisher** and **belt sander** work in a similar way to the disc sander but produce a flat surface.

Top tip: make sure that the grain on your wood is facing in the same direction as the direction of the linisher or belt sander.

A **plane** works in a similar way to a chisel in that it slices away thin shavings of wood. All planing should be done in the same direction as the flow of the grain, or the surface will tear. End grain can be planed but must have a waste piece clamped to the end, or it should be planed from the ends to the middle; this will prevent the ends from splitting.

There are two main types of plane as well as a wide variety of special planes.

The **jack plane** is a general-purpose plane used to flatten and smooth the surface and the edges of wood.

The **smoothing plane** is used on small wooden parts or for final cleaning of wooden surfaces.

The image of the **special planes** (Figure 2.5.23) shows four different types of plane and the type of cut that they produce

Drilling will produce a round hole in your wood. Drill bits come in various sizes, and you can get specialist drills that will produce large holes. Whenever you are drilling you must make sure that you follow all the correct safety precautions. Wear the correct PPE (personal protective equipment) and make sure your work is firmly held.

A Forstner bit will produce a large, clean hole with a flat bottom. The drill produces a lot of torque (turning force), and therefore it is essential that the work is firmly clamped down or the work piece will spin out of your hands.

A hole saw will also produce a large hole but the finished hole will not be as clean and it will cut all the way through the material. The drill will, again, produce a lot of torque and therefore it is essential that the work is firmly clamped down.

Top tip: ensure that there is a scrap piece of wood underneath your wood so that the hole saw can drill all the way through, or continue drilling until the pilot drill emerges and then reverse the work.

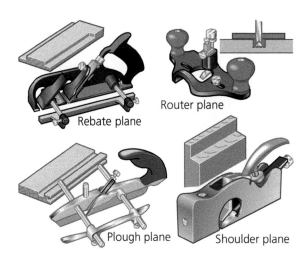

Figure 2.5.23 **Special planes**

Router plane

Rebate plane

Plough plane

Shoulder plane

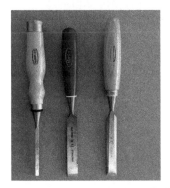

Figure 2.5.25 **Using a hole saw**

Figure 2.5.24 **Using a Forstner bit**

Figure 2.5.26 **A selection of chisels**

Figure 2.5.27 **Using a chisel to cut a housing joint**

Chiselling is used for shaping wood and for producing a variety of wood joints. Your work should be firmly held in a vice or G-clamped to the workbench, with your hands always behind the cutting edge. Horizontal paring will produce a flat horizontal surface while vertical paring will produce a flat vertical surface. Paring involves pushing the chisel through the wood.

Joining

We classify the joining of material into two categories:

- Temporary joints are joints that can be taken apart, such as screws and nuts and bolts.
- Permanent joints are joints that cannot be taken apart, such as glued joints.

Timber-based materials

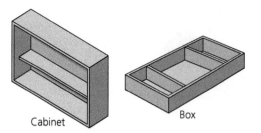

Figure 2.5.28 **Carcase or box construction**

Wood is a very versatile material and can be joined in a variety of ways. There are three main categories of joints:
- carcase or box joints
- stool joints
- frame joints.

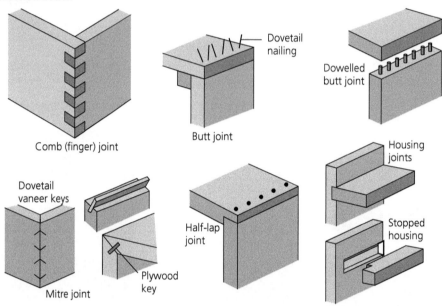

Figure 2.5.29 **Carcase or box joints**

Carcase or box joints all have their own advantages and disadvantages. A butt joint is relatively easy to produce, but is not the strongest joint. A comb or finger joint requires a higher degree of skill to produce, but is far superior in strength. Aesthetically, the mitre joint looks attractive while a nail joint looks inferior.

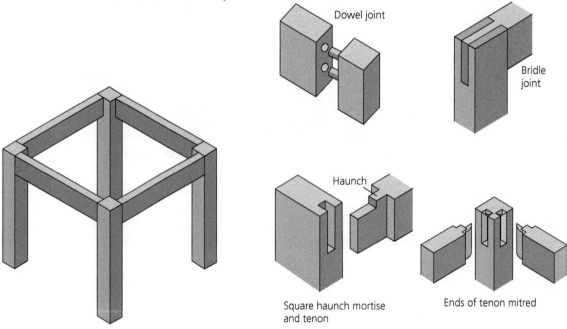

Figure 2.5.30 **Stool construction**

Figure 2.5.31 **Stool joints**

2.5 Using and working with materials

Stool joints should be used when you need to connect a leg to a rail such as on stools, tables and chairs. The dowel joint is relatively simple joint that involves joining wooden pieces together with dowels (a dowel is a cylindrical connecting piece of hardwood that is inserted into a predrilled hole and glued in place). The mortise and tenon joint is a more complex joint that gives greater strength due to the increase in surface area of the joint. This joint consists of a rectangular section of wood that is cut from the rail (the tenon) and a rectangular hole (the mortise) that has been chiselled out of the leg. The tenon is inserted into the mortise hole and glued into place.

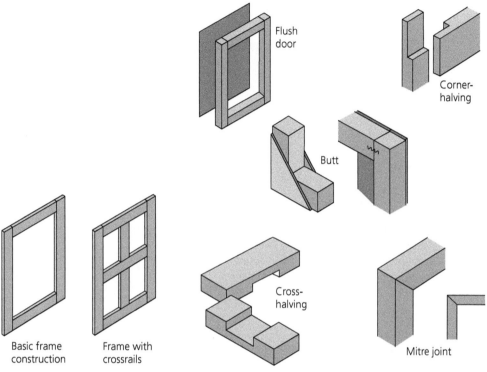

Flush door

Corner-halving

Butt

Basic frame construction

Frame with crossrails

Cross-halving

Mitre joint

Figure 2.5.32 **Frame construction**   Figure 2.5.33 **Frame joints**

Frame construction is used to produce panels, door frames, window frames, mirror frames and picture frames. For greater strength in door and window construction, the mortise and tenon joint would be used.

There are a number of different types of **adhesives** that can be used to glue wood together. Before any gluing takes place all surfaces should be free from dust and dirt.

PVA (polyvinyl acetate) is the most popular woodworking glue. It can be used straight from the bottle and requires no further preparation. It has a good amount of slip time (the amount of time you have for adjusting the parts that are being glued together) and it dries clear, giving a very strong bond. It is available in both exterior (waterproof) and interior versions. Its main disadvantage is that it takes 24 hours before it reaches full strength.

Contact adhesive has the advantage of providing an almost instant joint. It is spread very thinly over both surfaces of the wood then left to dry for several minutes. It is then pushed together providing a quick joint. Its main disadvantages are that it is not as strong as PVA, does not dry clear and has a number of health and safety issues – it is an irritant to the eyes and skin, and is highly flammable.

ACTIVITY

Study the computer games holder shown in Figure 2.5.34. You are to make the games holder from wood using a different wood joint to connect each of the three uprights to the base (e.g. you could use a dowel joint, a glue and nailed joint, and a finger joint). Feel free to reshape the uprights.

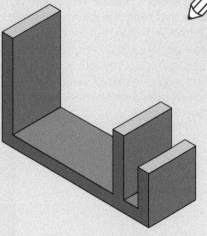

Figure 2.5.34 Computer games rack

KEY POINTS

- Marking out is the process of transferring a design to the material you are working on.
- A template is a profile shape of the part you want to make, like a stencil.
- Tenon saws will cut straight lines in wood.
- Coping saws will cut curved lines in wood.
- Wood joints vary in strength and complexity.

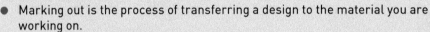

Check your knowledge and understanding

1 Identify and describe how to use marking out tools for wood.
2 Draw and explain how to safely use different types of saws for cutting wood.
3 Show two different ways of jointing a box, a stool and a frame.
4 Explain the advantages and disadvantages of using PVA glue and contact adhesive.
5 Explain the term 'hygroscopic'.

Metal-based materials

How different properties are used in commercial products

There are many factors to consider when selecting a material for use in a commercial product. These can include, cost, availability and the environmental impact associated with sourcing or applying a finish to the material. These are important considerations, but initial selection will relate to identifying the properties the material has that make it suitable for a specific product or application.

How properties influence use and affect performance

Properties such as strength, hardness, ductility and malleability are all important properties to consider when selecting a metal for a commercial product. The range of properties that can be found in different metals is vast, so it is important to have a clear understanding of the intended application.

Stainless steel is often used for cutlery and cooking utensils due to its hardness and its resistance to corrosion, but in applications where heating is involved, aluminium alloys are often used in its place due to them being excellent conductors of heat.

An aluminium alloy has been used in the manufacture of the espresso coffee maker in Figure 2.5.35. It is designed to be used on top of a cooker hob, rather than using electricity, and as such it must effectively conduct heat to boil the enclosed water. The aluminium alloy is lightweight, can be easily formed by die-casting and provides a non-toxic surface for use with food and drink. A thermoset polymer has been used for the handle to insulate the user from the hot aluminium.

The B3 Wassily chair is an iconic piece of design that was designed and manufactured in Germany in 1925 by Marcel Breuer. At that time most pieces of furniture were manufactured from wood, but following the First World War, metal became a more common alternative in furniture. The structure of the chair takes its inspiration from a bicycle frame and is manufactured from tubular low carbon steel. The low carbon steel is malleable and can be easily bent into the curved shape of the chair. It is also an easy material to weld into a strong rigid structure. Low carbon steel is prone to corrosion, but can be easily chrome plated to provide an attractive aesthetic surface finish, which also protects the metal from corrosion.

Figure 2.5.35 **Espresso coffee maker**

The modification of properties for specific purposes

Heat treatment

The properties of many metals can be changed by the influence of heating and cooling. They can be hardened, softened, made brittle or toughened simply by heating them to a certain temperature and then cooling then in a specific way.

Annealing

If a metal is intensely worked on by constant bending or being hammered it can become hard and brittle and is likely to split or fracture. This is known as work hardening. It is possible to remove this work hardening by a process known as annealing. Annealing makes the metal as soft as possible.

The annealing process (steel)

- The steel is placed on the firebricks.
- A brazing torch is used to heat the steel to a temperature above 700 °C (cherry red).
- The steel is then allowed to 'soak' at this temperature for a short while.
- The steel is left to cool as slowly as possible. This is best achieved by burying the work piece in sand.

The annealing process (aluminium)

This process is similar to annealing steel but to control the temperature the surface of the aluminium is marked with soap. When the soap turns black, the aluminium is at the correct temperature for annealing.

ACTIVITY

Heat treat a piece of steel and compare it to one that is not heat treated.

Choose from: soft soldering, hard soldering, brazing, hardening, normalising, annealing or welding.

How to shape and form using cutting, abrasion and addition

Marking out on metal

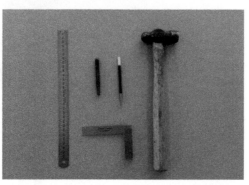

Figure 2.5.36 **Marking-out tools for metal: a scriber, metal rule, ball pein hammer, centre punch and an engineer's square**

Figure 2.5.37 **Marking out on metal**

Figure 2.5.38 **Centre punching metal**

Marking out on metal requires a different set of tools to timber. A pencil would not show up on most metals, therefore a scriber is used instead. The scriber will produce a fine line that scratches the surface of the metal. The metal can be coated with a layer of marking-out fluid which makes the lines easier to see. Marking-out fluid is a thin paint, specially formulated to be quick drying. It can be removed at a later date by rubbing with an abrasive such as emery cloth.

When drilling a hole in metal it is essential to have an indent marked into the metal to prevent the drill from skidding. This is produced by making a dot with the point of the punch by hitting it with a hammer into the metal.

Sawing metal

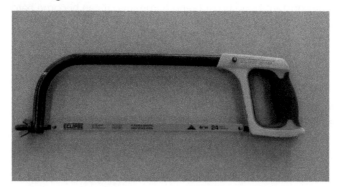

Figure 2.5.39 **Hacksaw**

The **hacksaw** is the most commonly used saw to cut metal. The metal to be sawn is generally held in a metalworking vice. It is important to position your metal carefully in the vice, making sure that the cut is as close to the side of the vice as possible.

Your right hand (if you are right-handed) is positioned through the handle with the index (trigger) finger pointing outwards, this helps to control the saw. The other hand is placed on the curve of the frame and provides power and direction. The full length of the saw should be used.

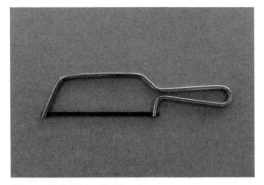

Figure 2.5.40 **Junior hacksaw**

Figure 2.5.41 **Using a junior hacksaw**

The **junior hacksaw**, as the name suggests, is a small version of the hacksaw and is used for sawing smaller metal parts.

Shaping metal

Once the rough shape of the metal has been produced it can then easily be shaped using metalworking files. There are a number files to choose from, but for removing metal a rough-cut file using a cross filing technique is most effective. Hold the handle in your right hand (if you are right-handed) and pinch the end of the file with your other hand. Use the full length of the file and slide along the metal to produce a flat, even surface.

Figure 2.5.42 **Cross filing**

Top tip: the file does not cut on the back stroke so save yourself time and energy, bringing it back through the air and not on the metal.

Files can also be used to smooth and clean the edges and surfaces of metal. A smooth-cut file using a draw filing technique will produce the best results. Hold the blade of the file in both hands and draw the file sideways along the metal to achieve a smooth flat, clean surface.

Drilling metal

Remember, before drilling metal it must be centre punched to prevent the drill from slipping.

A pillar drill can be used to drill holes in metal. The speed of the pillar drill can be altered to match that of the drill and the material it is drilling into. A small drill drilling through a soft metal will required a high speed, while a large drill drilling through a hard piece of metal will require a slow speed. The easiest and safest way to drill metal is in a machine vice. The machine vice will hold the metal securely during the drilling process. Make sure that you are wearing the correct PPE at all times and follow all the relevant safety procedures.

Drilling sheets of metal and irregular pieces of metal can be dangerous. They should be securely held using a clamp with a piece of scrap material underneath to prevent drilling into the machine table.

Figure 2.5.43 **Draw filing**

Figure 2.5.44 **Drilling metal in a machine vice**

Metal-based materials

127

Figure 2.5.45 **Drilling metal using a clamp**

Joining metal

Metal can be joined in a number of permanent and non-permanent ways.

A non-permanent method of joining is a method that can be taken apart. This is particularly useful for maintenance as it means that parts that are broken or worn out can be replaced.

Non-permanent methods of joining metal involve the use of nuts and bolts, and self-tapping screws. More information on the use of non-permanent methods of joining metal can be found in Topic 2.6 Stock forms, types and sizes.

Soft soldering, hard soldering and brazing are all permanent methods of joining metals together that use heat and another metal with a lower melting temperature. As all these processes involve the use of heat, it is essential that you wear the correct PPE and observe all safety procedures.

Soft soldering is commonly used when manufacturing electrical circuits and when plumbing with copper pipework. As with all soldering processes, it is essential that the joint is perfectly clean before the parts are soldered together, as any dirt will prevent the solder from bonding to the metal. When joining electrical components to a circuit board a soldering iron is used to heat the joint, and soft solder is used as the bonding metal. The solder contains a flux that keeps the joint clean and helps the solder to flow into the joint.

When joining copper pipework, a blowtorch is used to heat the metal and a paste flux is applied to help the solder flow.

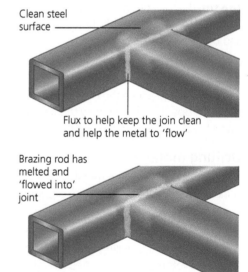

Clean steel surface

Flux to help keep the join clean and help the metal to 'flow'

Brazing rod has melted and 'flowed into' joint

Figure 2.5.47 **Brazing**

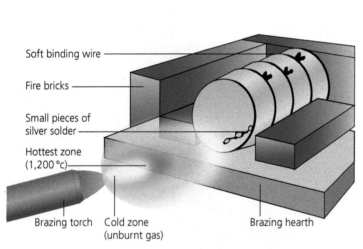

Soft binding wire

Fire bricks

Small pieces of silver solder

Hottest zone (1,200 °c)

Brazing torch   Cold zone (unburnt gas)   Brazing hearth

Figure 2.5.46 **The hard soldering process**

Hard soldering is a similar process to soft soldering, but is used for joining precious metal together. The heat is applied using a blowtorch or brazing torch and the solder is an alloy of silver, copper and zinc.

Brazing is a method of joining steel parts together. This time, heat is applied with a brazing torch in a brazing hearth. The flux (borax) is applied to the cleaned joint before heating – the flux will prevent the steel from oxidising during heating and will help the brass to flow into the joint. Once the steel is red hot, the brazing rod (brass) is introduced. It will then melt and flow into the joint.

Welding is a very strong method of permanently joining metals together. It differs from soldering in that it uses the same metal as that used to produce the joint. There are a number of methods of welding that use either gas, electricity or a mixture of gas and electricity to generate the heat.

Most welded joints require the metal to be prepared by forming a V-gap between the two pieces of metal that are to be joined.

Oxyacetylene welding uses a mixture of oxygen and acetylene gases to produce a very hot flame. (3,500 °C). The flame will melt both sides of the V-cut, and at this time the filler rod (copper-coated steel) is introduced where it will also melt and fill the V-gap.

Electric arc welding uses a low voltage, high current of electricity down a flux-coated filler rod. As the electricity jumps across the gap between the filler rod and the joint it produces a very high temperature arc that instantly melts both sides of the V-gap and the filler rod.

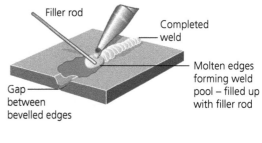

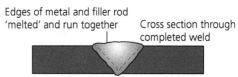

Figure 2.5.48 **Oxyacetylene welding**

Shaping metal by machining

There are a number of other methods of shaping metal that involve the use of machinery.

Turning is carried out using a lathe. The centre lathe is a very useful workshop machine. It is used to make circular components from metal bars. Various stock forms can be held in the chuck while it spins around. A cutting tool can then be used to perform a variety of turning operations, such as facing off (cleaning and squaring the end of a metal bar), parallel turning (reducing the diameter of the bar along its length) and taper turning (producing a taper or cone).

Milling machines are used to carry out a range of machining operations on metal. They can be used to cut slots, grooves, machine edges and to smooth large surface areas. More information on machining processes can be found in Topic 2.8 Specialist techniques.

Casting is a method of producing complex shapes in metal by heating metal to a molten state and then pouring it into a pre-prepared mould. More information on the casting processes can be found in Topic 2.8 Specialist techniques.

KEY POINTS
- A centre punch makes a small indent in metal before drilling takes place.
- A hacksaw cuts straight lines in metal.
- Cross filing is a method of shaping metal with a file.
- Draw filing smooths the edges and surfaces of metal.
- Soldering is a method of permanently joining metal.
- Annealing makes metal as soft as possible.

Check your knowledge and understanding

1 Identify the different types of marking-out equipment for metalwork.
2 Describe the process of brazing, using notes and sketches.
3 Describe how to anneal a piece of steel.
4 Using notes and sketches, show the difference between draw filing and cross filing.
5 Draw three pieces of marking-out equipment used on metal, name them and describe how they are used.

Polymers

How different properties are used in commercial products

There are many factors to consider when selecting a material for use in a commercial product. These can include, cost, availability and the environmental impact associated with sourcing or applying a finish to the material. These are important considerations, but initial selection will relate to identifying the properties the material has that make it suitable for a specific product or application.

How properties influence use and affect performance

The majority of polymers are man-made, and as such, chemical engineers have been able to fine-tune the properties of polymers more than any other material group. This has led to the development of polymers with excellent toughness and durability, chemical resistance and electrical insulation along with the ability to handle extremes of temperatures.

The properties of polymers can also be affected by the addition of additives such as fillers and UV stabilisers. These also need to be considered when selecting a polymer for use.

The office chair in Figure 2.5.49 is a modern replica of a design classic from the 1950s. Once manufactured in fibreglass, this replica has been produced by injection moulding the shell of the chair from high-density polythene (HDPE). HDPE has a number of properties that make it a suitable material for the shell of the chair. It is tough and lightweight, meaning that it can be easily moved around an office and that it will comfortably handle the wear and tear of day-to-day use. It also has the ability to have pigment added to it, meaning that it can be produced in a wide variety of colours. As a thermoplastic polymer it can be easily injection-moulded, which is suitable for large-scale production.

Figure 2.5.49 **Thermoplastic polymer seat**

Thermoset polymers are less common than thermoplastic polymers, but they do possess some properties that make them ideal for applications where resistance to heat and electrical insulation are vital. Urea formaldehyde is a thermoset polymer which is used in the manufacture of electrical plugs and sockets, along with smoke alarm covers. It is a hard material that doesn't soften when exposed to heat, so if there were an electrical error with an appliance, the plug wouldn't melt, protecting the user from electrocution. As a good electrical insulator, it can also be plugged in and removed from the socket without risking any harm to the user.

Figure 2.5.50 **Urea formaldehyde plug**

The modification of properties for specific purposes

There are a number of additives than can be blended with polymers to enhance certain properties.

- Plasticisers are added to polymers to enhance the flexibility of the polymer; phthalate esters are added to polyvinyl chloride (PVC) when used for cables.
- Pigments are added to polymers to change the colour of the polymer.
- Fillers are added to polymers to increase the bulk and reduce the cost of the polymer. Calcium carbonate is a typical filler found many polymers.
- Flame retardants can be added to polymers to prevent or slow down the rate at which they burn.
- Stabilisers can be added that resist UV degradation.

How to shape and form using cutting, abrasion and addition

Marking out on polymers

A spirit-based pen is suitable for polymers as it will mark the surface without scratching it. A plastic ruler is also useful – a metal rule could damage the shiny surface of the polymer.

Top tip: if the polymer comes with a protective coating, keep it on as long as possible to prevent it becoming scratched.

Many of the marking-out tools used on wood and metal can also be used on polymer-based materials. Polymers are ideally suited to the use of paper templates as they also give the polymer an extra protective layer.

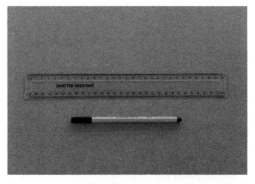

Figure 2.5.51 **Spirit-based pen, plastic ruler**

Sawing polymers

Polymers such as acrylic (PMMA) are ideal for cutting on a laser cutter. However, some cannot be cut by laser and must be cut by a saw or a knife. Most of the metalworking and woodworking saws can be used to cut polymers.

Top tip: when holding your work in a metal vice you can protect the polymer by wrapping it in a paper towel. It is also advisable to keep it as low down in the vice as possible to prevent snapping.

Shaping polymers

Thermoforming polymers can be formed in a number of ways with the use of heat. Line bending is one of the simplest methods of forming a straight line bend in a thermoforming polymer such as acrylic (PMMA).

Figure 2.5.52 **Sawing polymers**

More complex shapes can be achieved using a vacuum forming machine with thermoforming plastics such as high-impact polystyrene (HIPS). More information on shaping polymers by forming can be found in Topic 2.8 Specialist techniques.

Polymers can be shaped using metalworking files. They can also be planed with woodworking planes.

Welding can also be used. Within the school workshop the only convenient method of welding polymers is by solvent welding. This is achieved by using a solvent adhesive such as Tensol cement (dichloromethane). The surfaces of the polymer should be free from dust or dirt and the solvent cement applied with a spatula. The two surfaces are then pressed and held together. The solvent will melt the two surfaces and they will fuse together forming a welded joint.

Commercial methods of welding thermoforming polymers include the use of a hot-air gun to melt the two surfaces or by using hot metal clamps to form a seal on thin polymer sheets.

Casting can be used for certain thermosetting polymers (such as epoxy resin) to form intricate shapes. In a school workshop you could use this process to make pendants, earrings, cufflinks and other types of jewellery.

To produce a casting you must first have a mould. Silicon moulds work well as the epoxy resin will not bond to the surface.

The casting process:
- Pour the required volume of resin into a suitable pot.
- Add a colour pigment if required.
- Add the correct quantity of hardener.
- Mix the ingredients with a spatula.
- Pour the mixture into the mould and allow to set.
- Once set remove the mould. The surfaces can be polished.

(Note: large moulds should be done in layers)

Health and safety: barrier creams and/or latex gloves should be worn together with safety glasses and there should be plenty of natural ventilation in the room.

Drilling polymers

Polymers can be easily drilled with drill bits.

Top tip: it is a good idea to place tape over the area to be drilled to prevent the point of the drill from slipping across the surface.

Joining polymers

Polymers can be joined in a number of ways. They can be treated like metal and bolted or riveted together. They can also be glued and welded together.

Polymers can be glued with a variety of **adhesives**, but there are a number of specific glues that are specially formulated for use with certain polymers.

Tensol cement (dichloromethane and methyl methacrylate) is a clear solvent that gives a very effective bond when used with acrylic (PMMA). It actually melts the surfaces and welds them together. It has a number of health and safety issues, and you must wear the correct PPE before using it and ensure that you follow all the relevant health and safety procedures.

KEY POINTS
- A spirit-based pen is used for marking out on polymers.
- Most marking-out tools can be used on polymers.
- Most saws will cut polymers.
- Tensol cement is used to glue acrylic (PMMA).

(?)

Check your knowledge and understanding

1. Name a saw that can be used to cut acrylic (PMMA).
2. How can you prevent polymers getting scratched when working on them?
3. How do you prevent a drill from slipping when drilling polymers?
4. What additives make PVC flexible?

ACTIVITY

Draw a simple shaped paper template to be used to produce a pendant. Glue it onto a piece of 3 mm acrylic (PMMA) then shape it using saws and files. Clean and polish the edges then drill a 4 mm hole through it and fit a piece of cord.

You could produce smaller shapes to be cut, shaped and polished and glued onto the surface of your first piece.

Textile-based materials

How different properties are used in commercial products

When selecting fabric for a commercially manufactured product the requirements and end-use of the specific product must be considered. Cost is an important consideration when manufacturing in large numbers, and with fashion products colour and pattern will also affect the choice.

The modification of properties for specific purposes

An important **safety finish** is flame retardancy. Some fabrics catch fire and burn easily, especially those made from cotton, viscose, or a blend of polyester and cotton. Some fabrics do not set alight, but in a fire they will melt and may give off toxic smoke (for example, fabrics made from synthetic fibres). Although considered to be inherently flame retardant, fabrics made from synthetic fibres can cause serious skin burns and death by smoke inhalation.

It would not be appropriate to apply a flame retardant finish to all fabrics, but it is extremely important for children's nightwear and furnishing fabrics, especially those used in public buildings. The Nightwear (Safety) Regulations Act of 1985 states that nightwear that does not meet the flammability performance requirements must carry a label with the words 'keep away from fire' in red letters.

Proban and Pyrovatex are chemical flame retardant finishes that can be used on fabrics made from cellulosic fibres, such as cotton and fibre blends. The process is used at the finishing stage of the fabric manufacture and forms a cross-linked polymer inside the fibres. The polymer is insoluble so it will not wash out of the fabric.

When exposed to flame, the finishes decompose to form a local char that stays in the fibre and acts as a protective barrier. These finishes are used in many applications where fire might be a danger, such as safety clothing for workwear.

Many specialised fabrics are used in the manufacture of sportswear. Fabrics with elastane fibres such as Lycra are used for swimwear and other sportswear when great stretch is needed. Polartec fabric is made from polyester fibres, and is used for lightweight, thermally insulating fabrics useful for outdoor sportswear.

Microfibres are extremely fine fibres made from polyester and polyamide and make soft, breathable and lightweight fabrics that are used for many active sportswear garments, such as football shirts.

Moisture management fabrics, such as Coolmax, are developed from synthetic fibres that are engineered so that body moisture can be drawn away from the skin and evaporate on the outer surface of the fabric. These lightweight and easy care fabrics are very popular for active sportswear garments.

Waterproof fabrics that are breathable and windproof are needed for many outdoor sports. Gore-Tex is a fabric that has a waterproof membrane inserted between two layers of fabric, is and used for clothing and footwear.

Kevlar and Nomex are developed from polyamide fibres and used in many products, such as protective clothing worn by motorcycle and racing car drivers, as they are exceptionally strong and can resist very high temperatures.

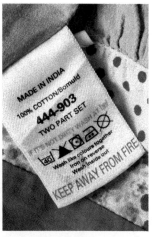

Figure 2.5.53 **Advice to keep away from fire on child's cotton pyjamas**

Figure 2.5.54 **Polyester microfibres are often used to make sports clothing, such as football shirts.**

Figure 2.5.55 **Gore-Tex**

Neoprene is a synthetic rubber fabric used in wetsuits.

Microencapsulated fabrics contain minute capsules containing various substances in their fibres. When subjected to friction the substances are gradually released. Microencapsulated fabrics can contain insect and odour repellents which are useful in sports fabrics.

Newer interactive fabrics incorporate wearable electronics that include fibres made from conductive materials such as carbon, silver and steel. These fabrics allow textile sportswear products to include GPS systems to identify a person's position, solar panels so that mobile devices can be charged, and record physical activity and monitor performance.

How properties influence use and affect performance

The end-use of the textile product will determine the most important properties needed for specific products. Strength, abrasion resistance and easy care are important for products used regularly, such as workwear and household furnishings; the abilities to insulate and repel moisture are needed in wet weather wear; absorbency and smoothness are desirable in garments worn close to the skin.

Figure 2.5.56 **Shirts are commonly made from polyester cotton blends.**

Cotton fabrics are generally cool to wear, absorbent and strong, and so they are used in many different applications, although they can shrink and crease badly. Wool fabrics are insulating and very absorbent, but can be expensive and very difficult to care for. Synthetic fibres are very strong, hardwearing and easy to care for, but are non-absorbent and can pile with regular use. Fabrics made with fibre blends are often used as the good and bad points of fibres cancel each other out. For example, polyester cotton blends are strong, crease- and shrink-resistant and dry quicker than 100 per cent cotton fabrics, while still retaining some absorbency.

Knitted fabrics have more stretch than woven fabrics, and so they are used for garments such as jumpers, t-shirts, socks and other garments where some stretch is needed.

Figure 2.5.57 **Knitted fabrics are used when stretch is important.**

The pattern on the fabric may affect how it is used. Large repeat patterns are fine on large products, but do not work well on small products or those with many small parts and seams. The need to match patterns can add significantly to fabric and manufacturing costs.

How to shape and form using cutting and addition

Cutting fabric

The fabric must be prepared before it can be cut into the shapes needed to make a product. Make sure that creases have been removed and that there are no flaws or dirty marks that could spoil the product. When making garments, the fabric needs to be folded in half lengthways with the two selvedges on top of each other so that both sides are cut at the same time. It is important that this is done carefully or the end-product may be inaccurate.

Before cutting fabric, a template or pattern is needed. When making products at home or school a paper pattern is usually pinned to the fabric.

Figure 2.5.58 **Large patterns work well on large products with little shaping, but are not as appropriate on smaller items.**

Commercial paper patterns can be bought for a wide range of textile products. Many sewing magazines have patterns that you can trace, or you can draft your own pattern template.

Most paper patterns include a seam allowance, which is the amount of fabric added to the pattern for the seam turnings. There are many other markings on patterns to help you make the product; these are explained in the chart shown in Figure 2.5.59.

Pattern Symbol	Meaning	Importance
	Lengthen and shorten lines	These must be adjusted before using the pattern to maintain the proportion of a garment. Fold into a pleat to shorten the length, cut and spread to increase length.
	Straight grain arrow	Must be placed parallel to the selvedge of the fabric so that the grain runs correctly and the product hangs properly or lies flat
	Place on fold arrow	The edge indicated has to be against a fold of the fabric, as the piece is symmetrical and needs to open out.
	Cutting line	This is the line to cut along. Cutting too far in will make the product smaller than intended. Cutting too far outside the line will make the item too big. In either case other pieces may not fit together if not cut accurately.
	Stitching line	This shows where the stitching should be when joining sections of fabric together. Too far in from the edge of the fabric will make the product smaller, too close to the edge will make the product bigger. The usual amount allowed for a seam is 1.5 centimetres.
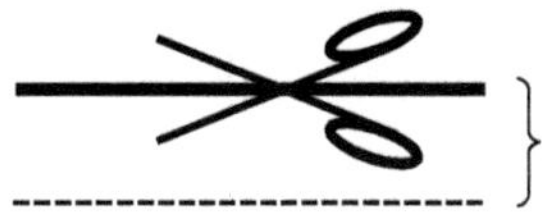	Seam allowance	This is the distance between the cutting line and the stitching line, usually 1.5 centimetres on a commercial pattern.
	Dot	Indicates a position. For example, a dart, gathers, pleats, tucks, pocket, end of a zip.
	Notch	Indicates which pieces fit together and how they need to be aligned. Can also be used to indicate the position of gathers.
	Centre line	Indicates the centre front or centre back of a garment
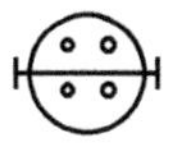	Button and buttonhole position	Shows where to work the buttonhole and stitch the button for correct spacing. These can be adjusted if required.

Figure 2.5.59 **Pattern symbols and their meanings**

Pattern pieces have a straight grain line which must follow the warp grain of the fabric, parallel to the selvedge, to make sure that the fabric hangs correctly in the finished product. Use a tape measure to measure the distance of both ends of the grain line printed on the pattern from the selvedge of the fabric to make sure that they are the same.

It is important that the pattern is pinned to the fabric at regular intervals, with pins about 10 centimetres apart on straight edges and a little closer on curved edges, and pins in all the corners. This will help to stop the pattern from moving as the fabric is cut and help to keep the cutting of the pattern outline accurate. The shape to be cut can also be drawn straight on to the fabric using tailor's chalk.

Figure 2.5.60 A paper pattern is pinned to the fabric.

Figure 2.5.61 The pattern can be drawn on to the fabric using tailor's chalk.

The fabric is cut out following the template using sharp dressmaker's shears or a rotary cutter on a cutting mat. It is important to cut the edges as cleanly as possible as this will help to sew straight lines later.

Figure 2.5.62 Cutting fabric with dressmaker's shears

Figure 2.5.63 Cutting fabric with a rotary cutter

When all the pattern pieces have been cut out, some of the pattern markings need to be transferred to the fabric to help make the product accurately. There are three ways to do this.

Tailor's chalk or a special marking pencil is used to draw on the fabric. This method marks only one side of the fabric and can easily be rubbed off and the markings lost. With a **tracing wheel** and **tracing paper**, the coloured side of the tracing paper is put next to the wrong side of the fabric and the colour is transferred to the fabric using the tracing wheel. This leaves a line of fine dots on the fabric to show the pattern markings. Both sides of the fabric can be marked at once, but the marks can be difficult to remove from the fabric.

Figure 2.5.64 A tracing wheel

Tailor's tacks are made by stitching through the fabric and pattern with a double thread. This method marks both sides of the fabric at the same time and the threads are easy to remove afterwards. This technique can be used when marking the end of a dart. See Figure 2.5.65.

Sewing fabric

A seam is used to permanently join two pieces of fabric together. There are three main types of seam.

A plain, or open seam is the easiest to make. It is not seen on the right side of the product, and gives a flat finish. To make the seam, place the two pieces of fabric together with their right sides facing. Make sure that any pattern notches are matched. Pin the layers of fabric together with the pins at right angles so they can be removed easily as you machine. If necessary, the seam can be held together temporarily using tacking stitches.

Sew the seam on the stitching line – this is usually 15 millimetres from the cut edge – using the guide on the sewing machine to keep the sewing straight and making sure you reverse the stitching at the start and end of the seam to prevent it coming undone.

Remove any tacking and press the seam turnings open.

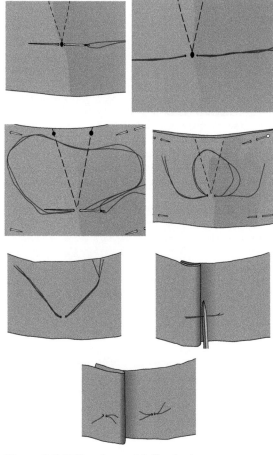

Figure 2.5.65 **How to work tailor tacks**

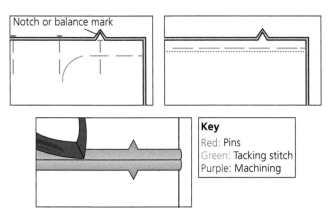

Key
Red: Pins
Green: Tacking stitch
Purple: Machining

Figure 2.5.66 **Making a plain seam**

Neatening the seam edges

The edges of the seam will need to be finished to stop them fraying. The quickest way to do this is by overlocking them. An overlocker has a blade which trims off the surplus seam allowance before the cut edges are covered with looped stitches. This method gives a strong and neat finish and on lightweight and medium weight fabrics both turnings can be neatened together.

The edges of the turnings can be turned under by five millimetres and stitched. This is called edge-stitching and is suitable for medium weight fabrics, but will be bulky on thick fabrics, which can be neatened by using a zigzag stitch on the raw edges. The raw edges can also be neatened using bias binding, but this is a time-consuming method and can make the seam bulky.

A French seam gives a neat and strong finish on fine and sheer fabrics, as all the raw edges are enclosed. The seam is not seen on the outside of the product, but can be bulky if used in thicker fabrics.

Pin the fabrics to be joined with the wrong sides together, matching any notches. Sew the seam 10 millimetres from the cut edge using the guide on the sewing machine to keep the sewing straight, and make sure you reverse the stitching at the start and end of the seam to prevent it coming undone. Remove any tacking and press the seam turnings open.

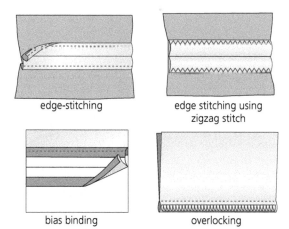

edge-stitching

edge stitching using zigzag stitch

bias binding

overlocking

Figure 2.5.67 **Methods of neatening a plain seam**

Textile-based materials

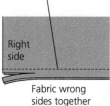

Stitch plain seam 10 mm from edge. Trim seam allowance to 3 mm.

Right side

Fabric wrong sides together

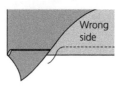

Turn to right side. Press flat Stitch exactly on the seam line 5 mm away.

Wrong side

Figure 2.5.68 Making a French seam

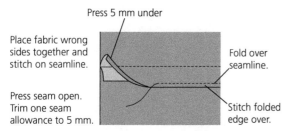

Press 5 mm under

Place fabric wrong sides together and stitch on seamline.

Press seam open. Trim one seam allowance to 5 mm.

Fold over seamline.

Stitch folded edge over.

Figure 2.5.69 Making a double-stitched seam

(a) Knife pleats

(b) Inverted pleat

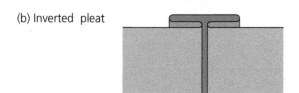

(a) Box pleat

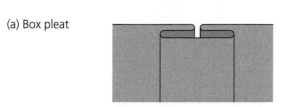

Figure 2.5.70 Knife pleats, an inverted pleat and a box pleat

Figure 2.5.71 Making tucks

Trim the seam allowance to three millimetres and turn the fabric so that the right sides are facing and make sure that the first seam line is exactly on the line of the fold. Pin the seam then sew it again five millimetres from the folded edge opened out. The seam is usually pressed so that it lies towards the back of the product.

A double-stitched seam is a strong and flat seam with all the raw edges enclosed. It is visible on the outside of the product, and is often used as a decorative feature.

Pin the fabrics to be joined with the wrong sides together, matching any notches. Sew the seam 15 millimetres from the cut edge using the guide on the sewing machine to keep the sewing straight and making sure you reverse the stitching at the start and end of the seam. Remove any tacking and press the seam turnings open.

Trim the turning on one side of the seam to five millimetres. Press the second turning over the first, turning the raw edge under by five millimetres. Pin in place then stitch the folded edge to the product, making sure that the stitching is close to the folded edge.

Shaping fabric

Fabric can be manipulated using pleats, tucks and gathering to change it from a flat two-dimensional fabric into a three-dimensional product.

Pleats are folds in the fabric that are stitched or pressed in place. They are used to shape products and add ease (extra room that allow for movement and enhances comfort) as well as texture and decoration. There are three main types of pleat.

Knife pleats are single folds in the fabric and all face the same way.

Inverted pleats are two knife pleats facing towards each other on the right side of the fabric.

Box pleats are two knife pleats facing away from each other on the right side of the fabric.

Pleats can also be heat-set if a fabric containing thermoplastic fibres is used.

Tucks are similar to pleats, but are usually narrower and stitched along their entire length. They are usually used as a decorative feature.

Gathering is made by sewing small stitches in a single layer of fabric then pulling them up to reduce the length of the fabric. They are used to shape products, add ease and give a decorative finish.

A long machine stitch is used and the top tension of the sewing machine should be loosened slightly to help the gathers pull up. Two rows of stitching are made, the first on the stitching line and the second about five millimetres away in the seam allowance. The ends of the stitching must be left loose so that the threads can be pulled up to make the gathers.

The stitching on both rows is then pulled up until the fabric is the right length and the ends of thread wound round a pin put at each end of the stitching. The gathers are then spread out evenly before being joined to another piece of fabric.

Quilting is a functional and decorative technique that gives texture and form to fabric.

The simplest type of quilting involves placing a layer of wadding between two layers of fabric and then stitching through all the layers. The layer of wadding traps air so the fabric acts as an insulator, and the added thickness makes a more protective fabric. This is called English quilting.

The three layers are laid on top of each other, making sure that the grain line runs in the same direction on the top and bottom layers of fabric. The layers are pinned and tacked together, starting in the middle and working outwards to keep the fabric flat.

The layers are stitched together in a pattern, usually based on straight lines, and using a long machine stitch to allow for the thickness of the fabric. The lines can be marked on the fabric using tailor's chalk, or a quilting foot can be used on the machine to guide the stitching.

Quilting can be combined with patchwork to give a more decorative effect.

Piping is used to define a shape, add decoration or add strength to an edge. It is made by covering a cord with fabric, usually cut on the **bias** (against the grain of the fabric), and is sewn between the two sides of a seam. Piping cord comes in a number of thicknesses, and one that is appropriate for the product should be used.

Fold the fabric strip in half lengthways with the wrong sides facing. Place the piping cord inside the fabric strip and pin along the long edges. Stitch the cord in place using a zip foot on the sewing machine.

Pin and tack the piping along the seam on the right side of the product, cover with the second layer of fabric and sewing machine in place close to the cord.

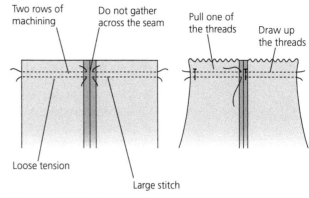

Figure 2.5.72 **Gathering**

Figure 2.5.73 **English quilting**

Figure 2.5.74 **Quilting combined with patchwork**

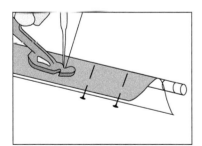

Figure 2.5.75 **Making piping**

ACTIVITY

Make samples of two different seams and three different methods of shaping fabric.

Sketch or collect pictures of some products where each seam and shaping method could be used.

Explain why each is suitable for the products you have chosen.

KEY POINTS

- Fabrics must be matched with their properties and the requirements of the end-product.
- Fabrics must be cut accurately to get a good-quality finish when making a product.
- The type of fabric and appearance must be considered when choosing seams to use on a product.
- Shaping is used to give form and ease to a product.
- Fibre and fabric properties can be modified for specialist applications.

Check your knowledge and understanding

1 Name two methods of transferring pattern markings to fabric.
2 Name two different ways of neatening a plain seam.
3 Name one seam that has its edges completely enclosed.
4 Name two qualities that quilting gives to a fabric.
5 Name three specialist sportswear fabrics and give a specific use for each one.

Electronic and mechanical systems

Properties of materials

Materials are chosen based on their properties, and therefore their suitability for a specific job. When selecting materials and components for commercially manufactured electrical and mechanical products the requirements of the specific product must be considered. Cost is an important consideration along with other factors such as the environmental impact associated with sourcing the material or in manufacturing the components. These are important considerations, but initial selection will relate to identifying the properties that will be required from the material or component to make it suitable for a specific product or application.

Figure 2.5.76 **A modern washing machine containing microcontrollers**

How different properties are used in commercial products

Many electromechanical devices that you come into contact with today are designed to incorporate microprocessor devices that can be programmed to function in different ways. The advantage of using programmable devices is that it takes away the need to produce a range of individual **integrated circuits** for every different product. A common microprocessor is the **PIC** microcontroller. All microcontrollers come with an internal clock, and allow designers to use the same device to control many different situations and processes. A huge number of everyday items such as cars, microwave ovens, washing machines, dishwashers and vending machines contain microcontrollers. One of the properties of a microcontroller is its flexibility in the way it can be programmed or reprogrammed to change its function.

A washing machine using a microprocessor can be programmed to use a range of input components to switch on an assortment of output devices in a variety of sequences. The input components can include different types of switches to input information into the microcontroller. These switches generally include push switches to start different cycles: a switch to sense if the door is closed, a float switch to sense water level and a thermistor to sense water temperature. The microcontroller can then be programmed to turn a variety of output devices on and off in sequence and for different time periods. These output devices include valves to let water into the machine and pumps to remove it, an electric motor to turn the drum, a heater to heat the water as well as a variety of light emitting diodes (LEDs) and usually also a buzzer.

Figure 2.5.77 **Many systems in a car are controlled by microprocessors.**

A modern car can contain as many as 40 microcontrollers controlling the engine management system and fuel consumption, as well as controlling many safety systems.

How properties influence use

How we use machines and equipment is changing rapidly. We can now control many household items in our home from our smartphones when we are not there – we can turn lights on and off, control central heating systems, as well as a continuously increasing number of other devices such as kettles and televisions.

The properties afforded by increasingly complex integrated circuits now allow many changes in the way we use many items of equipment.

Cars now have driver override systems, sensors that not only warn us of danger, actually override the driver's decisions and apply the brakes automatically to slow the vehicle down in dangerous situations. Cars capable of driving themselves are now being trialled. We have seen the switch

Figure 2.5.78 **A self-driving, driverless car being tested in California**

from keys to keyless entry systems in recent years, but in the near future, we may see fingerprint recognition, voice scanning or retina scanners to allow us entry – the same technology as used in many smartphones.

Head-up displays (HUDs) are displays of instrument readings that have been used in aircraft for some time, with the benefit that a pilot can see information from instruments without lowering their eyes – they are generally projected on to the windscreen or visor. Electronic activated glass is now in development that could be capable of displaying instruments on windscreens, bringing the possibility to car drivers of satellite navigation systems projected on the windscreen.

The technology to remotely shut down stolen cars is now available. Stolen cars being tracked and followed by police can be shut down with

Figure 2.5.79 **Head-up display being tested in a car**

ACTIVITY

List household appliances or systems that could be developed to be controlled by smartphone.

their engines slowed and stopped, ending the need for police chases. In-car active health monitoring has also been considered by some motor manufacturers, with sensors in seat belts and steering wheels that monitor the driver's vital statistics; imagine the possibilities created by these technologies!

STRETCH AND CHALLENGE

Analyse developments and benefits that are possible with electronic control systems.

How properties affect performance

The improving properties offered by the electronic control of mechanical systems has enhanced the performance of domestic machinery and cars.

Washing machines have become more efficient with the addition of features that offer wider ranges of programmes, automatic load detection, sensor technologies, and systems that allow the use of less water; these are usually a matter of software, electronics, and sensor systems.

Figure 2.5.80 **The Tesla model X luxury electric vehicle**

The same principles apply to motor vehicles. The increased amount of sensors monitoring vehicle systems can allow software to increase fuel efficiency and performance. We have also seen increasing numbers of electric cars, which has come about through improved battery technology. Developments that allow for longer driving ranges and the introduction of more and more charging points means the number of people using electric cars is likely to continue to grow. Hybrid cars use both technologies: petrol burning combustion engines and electric motors. The goal of hybrid vehicles is to maximise the use of the electric drive motors as they are more efficient and produce no emissions when in use. Electric drive motors use electricity stored in a battery pack. Once depleted, the batteries can then be recharged.

ACTIVITY

List the advantages and disadvantages of electric vehicles.

STRETCH AND CHALLENGE

Write an article that discusses the benefits of controlling modern technology electronically.

The modification of properties for specific purposes

One of the methods of producing printed circuit boards (PCBs) is by using photo-resist board. This is an example of modifying a material to enable the manufacturing method to take place. The copper-coated glass reinforced polymer has another layer of chemical added to it to allow the etching process to be carried out (see below).

Figure 2.5.81 **Anodised carabiners**

Anodising with aluminium is also an example of modification; it is used to increase a material's resistance to corrosion and wear, and to provide better adhesion for paint primers and glues. There are hundreds of components for motor vehicles that are anodised such as trim parts, wheel covers, control panels, and name plates. (See Topic 2.9 for more information on anodising.)

How to shape and form using cutting, abrasion and addition

Electronic circuits in industry and in schools are usually made by using **printed circuit boards (PCBs)**. The non-conductive board is made from glass-reinforced polymer with copper tracks and pads on the surface to conduct the electrical current between the components. Holes are drilled into the board to hold the components in position, where they are soldered to fix them in place and make an electrical connection.

There are two manufacturing methods generally used for making the boards in school: etching using chemicals and milling using a computer numerically controlled (CNC) routing machine. Both methods begin with the circuits being modelled and mapped out using computer software.

KEY WORD

Printed circuit board (PCB) a board that physically supports and electrically connects electronic components using copper tracks and pads on a glass-reinforced polymer board.

Etching

When etching, a piece of photoresist board is used. Photoresist board is a glass-reinforced polymer board with a thin layer of copper laminated onto it, and a layer of photo-sensitive chemical covered by a light-resistant backing on top. The first step when producing a PCB using the photoetch method is to draw and model the circuit using computer software. Some software packages allow you to simulate the circuit at this stage so that you can check if everything functions as it should. Masks can then be printed on to clear acetate sheets; they are drawn from the component side and look down on from above as though the board was transparent.

When you are happy with the PCB design you can then produce the mask.

A piece of photoresist board is then cut to the size of the mask using a guillotine, and the light-resistant backing is removed from the board. The mask is then used to expose the photosensitive copper-clad board to ultraviolet (UV) light in a UV light box. The board is then placed in a developing solution where the protective layer that has been exposed to the UV light is removed. It is then washed in water to stop the developing process and to remove all of the developing chemical. The unwanted copper is then removed from the board by a process called etching in an etching tank using a chemical called ferric chloride. The board is washed and cleaned and is left with only the copper tracks and pads necessary to complete the circuit. The board is now ready for drilling and soldering.

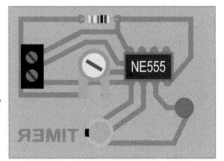

Figure 2.5.82 **A PCB design showing components**

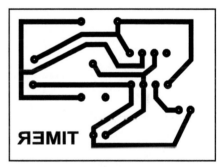

Figure 2.5.83 **A PCB mask**

Drilling

This method of manufacturing PCBs requires the holes to be drilled manually. Drilling PCBs requires a high-speed electric drill that is best used in a drill stand. For most components a one-millimetre drill bit can be used to drill the holes. Care must be taken to make sure that holes are drilled in the centre of pads.

Machining PCBs

A piece of copper-clad board is used as the basis for machining a PCB. The first step when using this method of manufacture is again to draw and model the circuit using computer software. The software determines the areas of copper to be removed from the board so that only the tracks and pads are left on the board where the components can be placed and soldered. Areas of copper are removed as in the etching process, but with this method a CNC milling machine removes the copper mechanically. This method has the advantage of using no chemicals, and it can put the holes into the pads during the milling process.

Figure 2.5.84 **Holes being drilled in a PCB**

Soldering

The method of making sure that there is a good electrical connection between components and copper PCB track and that they stay in place is called soldering. Place components that do not stick out far from the board first, such as resistors and integrated circuit holders. When the components are in place soldering is carried out by:

- placing the soldering iron so that it touches the leg or pin of the component and the copper track – both must be hot
- allowing three seconds for the track and component leg/pin to heat up, and then melting a small amount of solder onto the copper track
- when the solder has run into a shape that resembles a small volcano withdraw the soldering iron.

Figure 2.5.85 A CNC milling machine cutting PCBs

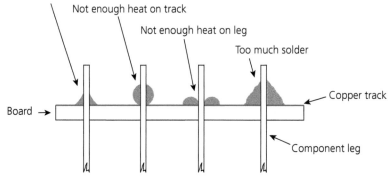

Figure 2.5.86 Good and bad soldered joints

Check your knowledge and understanding

1 Give examples of products that use microcontrollers.
2 Explain the benefits of using microcontrollers in domestic products.
3 Give examples of how modern control systems are changing the ways in which we use machines.
4 Explain what the advantages of making a printed circuit board using the machining method are, as opposed to the etching method.
5 Explain why components are soldered on to printed circuit boards.

What will I learn?

In this topic you will learn about:
- → calculating the amount of material needed when using different stock forms
- → commercially available stock forms of paper and board
- → standard components used with paper and board
- → the stock forms of timber, metal and polymers
- → the commonly available sizes of timber, metal and polymers
- → the standard components that can be used with various material groups
- → the sizes and forms of different textile materials
- → a wide range of standard components
- → buying electrical and mechanical components
- → resistor sizes and tolerances
- → the E12 preferred value series of resistors
- → the different types of capacitors
- → how to recognise integrated circuits and PIC microcontrollers.

Materials are supplied in many common shapes and forms. It is important that you are aware of these **stock forms** when designing and planning your projects and when selecting the most appropriate material for use. The use of an appropriate stock form will keep down material costs and avoid the need for any additional machining or processing before use. Standard forms and sizes are cheaper than special sizes because they are processed in large quantities.

Papers and boards

Often, paper and card is often is sold according to colour, and measured in set sizes such as A3, A4, A5 and so on. Each stock form sizing is half the size of the previous (i.e. A4 is half the size of A3).

Paper and board is also measured in grams per square metre (gsm). This is the density of the material (in other words how heavy it is per square metre). The higher the number of gsm, the heavier the paper or board. Anything more than 200 gsm is considered a board.

Board thickness is also measured. This is different to density and is measured in microns: one micron is one thousandth of a millimetre. Normal paper found in exercise books or used in a photocopier tends to be approximately 80 gsm and 150 microns thick, whereas higher quality paper, such as cartridge paper, is approximately 150 gsm and 600 microns thick.

Paper and card can be sold untreated, but this is of limited use to designers. Untreated paper is used in the production of newspapers to keep costs down, but it is a rough material that absorbs ink and

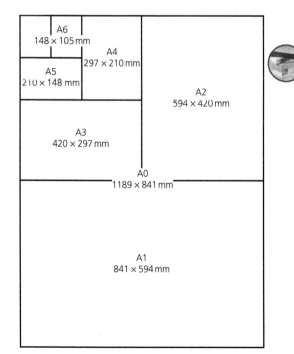

Figure 2.6.1 Typical paper sizes

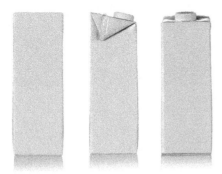

Figure 2.6.2 **Laminated drinks container**

bleeds, which makes it difficult to work with in other situations. For that reason, paper and board is often treated to make it more suited to its use. It is then sold in different forms, such as corrugated card, duplex board, grey board, foil-lined board and so on.

Sometimes card is laminated to give it more useful properties for specific tasks – for example, waterproof layers are added to card to make it more suitable for holding liquids. This card can then be used to package drinks and so on.

In this process, card is covered on one or both sides with a plastic film which creates a waterproof barrier. This can also be done with aluminium foil, which is then used to keep food products warm.

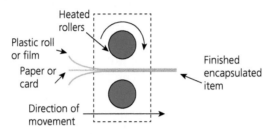

Figure 2.6.3 **The laminating process**

Standard components used in paper and card products

Paper clips, paper fasteners or split pins, binding bars and so on are all **standard components** that are used with paper and board. Standard components are always a set size, which makes it easier for them to be used in a range of different products, and to be replaced in a product if they break or wear out.

- Drawing pins fix paper or board to walls or backing boards.
- Split pins are used in pop-up models and books to allow for movement.
- Rubber bands are used in pop-up cards and models.

Figure 2.6.4 **Split pins**

Split pins or brass fasteners are pushed into holes in paper or board and the legs are separated and bent back behind the material to hold it in place. The legs have a point which makes it easier to punch through the paper if needed. Sometimes, split pins are used instead of staples, but they are most often used because the paper needs to be able to rotate around the joint. This makes tham ideal for use in models such as mock-ups of pop-up cards. Split pins are made of a soft metal, such as brass, and the legs are created in slightly different lengths to make them easy to separate.

Staples are small pieces of folded metal which are used to secure multiple pieces of paper together. They are U-shaped pieces which are placed inside a stapler. When a force is applied the legs of the staple bend inwards securing the material inside. These are a permanent fastening as the paper is damaged during this process and in the removal of the staple.

Figure 2.6.5 **Staples**

Standard components are often manufactured by companies that specialise in that particular component. This means that the product can be produced in high volumes and manufactured at a lower cost. Using standard components brought in from other manufacturers speeds up the process of making the final product.

Choosing the right stock form

When deciding which stock form of material is most suitable it is important to consider the amount of waste that will be created. Shape templates should be positioned on the material as close together as possible to minimise the amount of material that is not used. This is called 'nesting'. Designers will consider the stock forms of materials when they decide the sizes of certain parts of a product. For example, your teacher may ask you to create a net for a gift box. The size of that gift box is likely to be influenced by the fact that the net will need to be printed on A3 paper in order to manufacture it. Rather than creating a product which is marginally too big to fit on the A3 page, it is likely that the size is reduced in order to make the manufacturing simple. This is also true in industry, where the cost of materials is of concern to designers and manufacturers.

Seals and bindings

Seals and bindings are ways of holding multiple pieces of paper together, like a book or a booklet. One of the methods used in schools or colleges is plastic comb binding, using a punch and binding machine. The machine creates holes in the paperwork, the plastic comb is then pushed onto pins on the machine and opened out so they are almost flat. The punched work is then pushed over the combs, which are then closed so that each piece of the comb is inserted into a different hole. The result is a bound document that can be altered if necessary.

Perfect binding is perhaps the most common form of binding in industry. It is the type of binding you often find on paperback books, phone directories and so on. Pages are glued with a flexible adhesive along the spine of the book and wrapped with a heavier printed cover. This is not the strongest form of binding and the book will not open flat.

In section sewn binding, pages are folded together in smaller sections and then sewn down the spine. The spine is then glued together for extra support and the cover then attached. This is the strongest form of binding and allows the book to fold flat.

Screw post binding allows you to add pages at a later date, as the screws can be removed. This type of binding uses screw posts, which are pushed through holes in the material you wish to bind. The screw posts are made of two parts, the barrel post which has an internal thread and the cap screw. The cap screw is twisted into the barrel post over the top of the layers of paper or board and tightened.

Figure 2.6.6 **Plastic comb binding**

Figure 2.6.7 **Perfect binding**

Figure 2.6.8 **Section sewn binding**

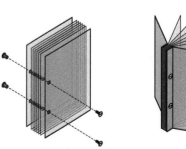

Figure 2.6.9 **Screwpost binding**

KEY POINTS

- Paper and board is usually measured in size, thickness and density.
- Standard components are products that are made to a set size for a variety of different uses.
- It is of benefit to both the manufacturer and the consumer to have materials and products created to a set size.

Check your knowledge and understanding

1 What is the main difference between paper and board?
2 Why are stock forms used by manufacturers?
3 Name three standard components used in paper and card products.
4 Name three different ways of binding paper.
5 Explain the benefits of section sewn binding.

Timber-based materials

Timber

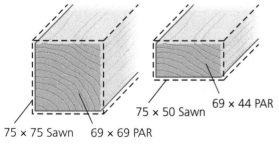

When purchasing timber from a timber merchant, it is important to know the forms that are commonly available. Natural timbers (hardwoods and softwoods) are generally supplied in planks, boards, strips and squares. There is also a wide range of moulded sections available.

Timber generally comes rough-sawn straight from the sawmill, but it's common for the timber merchant to then plane the wood to give it a smooth surface. This can either be planed both sides (PBS) or planed all round (PAR), which is also referred to as planed square edge (PSE). It is worth remembering that the planing process removes around three millimetres off each side of the plank, therefore timber advertised as nominally PAR 100 mm² would actually be 94 mm². Planed timber is more expensive than rough-sawn, but it provides you with a more accurately sized material.

Strip
under 100 mm wide
under 50 mm thick

Square
up to 150 × 150

Board
over 100 mm wide
up to 50 mm thick

Plank
up to 375 mm wide
over 50 mm thick

Figure 2.6.10 **Common timber stock forms**

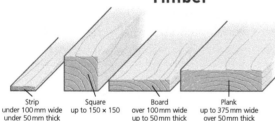

75 × 75 Sawn   69 × 69 PAR

75 × 50 Sawn   69 × 44 PAR

Figure 2.6.11 **Typical planed timber sizes**

In addition to the square-edged forms, timber is available in a variety of shapes and decorative mouldings. These can be used in many areas, but are commonly found in framing applications and architraves. They are manufactured using a spindle moulder and a series of specialised cutters. The waste timber removed can be used in the production of manufactured boards.

The most common timber moulding shape is dowel. It is supplied in a range of sizes from two-millimetre diameter up to 75-millimetres diameter and can come in lengths up to 2,400 millimetres.

All of these timber stock forms are available in both hardwood and softwood, depending on the intended application.

Timber can also be supplied, as large thin sheets known as veneers. They are commonly used to face the external surfaces of manufactured boards, such as plywood and MDF, but they can also be used in laminated products and marquetry, where lots of decorative veneers are combined to make intricate patterns. Veneer is available in a variety of thicknesses and can either be manufactured by rotary peeling, which is then used for plywood, or slicing which is more decorative and used in furniture. Processing timber into veneers is one of the most efficient conversion processes, using as much of the original trunk as possible.

Manufactured boards are readily available in large standard sized sheets of 2,440 × 1,220 millimetres. They are also available in a variety of thicknesses ranging from 3 millimetres up to 25 millimetres. A type of wood called Aeroply is available in even thinner sheets. The thickness of plywood and MDF generally increases in three-millimetre increments (6 mm, 9 mm and so on), whereas particleboard and hardboard board come in a more limited range of thicknesses.

Figure 2.6.12 **Timber mouldings**

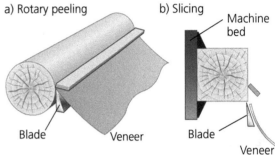
a) Rotary peeling   b) Slicing

Machine bed

Blade   Veneer   Blade   Veneer

Figure 2.6.13 **Producing a timber veneer**

Timber-based components

Components are generally used to assist in the joining of materials. They can be used to increase the strength of a structure or add additional functionality to a product. When used in timber-based products, components usually help create a non-permanent joining method. This means that the product can be taken apart without damaging the individual part. A vast range of common and specialist screws, nails, nuts and bolts, hinges, and knock-down (KD) fittings are available depending on the material and application.

Woodscrews

Woodscrews are used to join two pieces of timber together, or for joining another material to a wooden structure. They can also be removed easily to dismantle the parts. They normally have one or two teeth that spiral around their shank; turning them winds them into the wood. Woodscrews are often made from steel, which is usually zinc coated to prevent corrosion. Where greater resistance to moisture is needed, brass or stainless steel screws are used. There are hundreds of types of screws with special features for particular functions. The modern woodscrew shown in Figure 2.6.14 shows one of the most common types. It has a cross head that allows it to be driven easily with a cordless screwdriver.

Wood screws have different shaped heads depending on the application they are going to be used for. There are specially designed screwdrivers and screwdriver bits that are used with these. The two most common types are Pozidriv and slotted.

Figure 2.6.14 **A modern woodscrew**

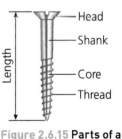
Length — Head
— Shank
— Core
— Thread

Figure 2.6.15 **Parts of a woodscrew**

Straight slot   Phillips   Pozidriv

Figure 2.6.16 **Common screwdriver slots**

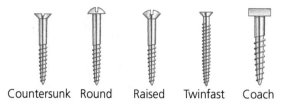

Countersunk　Round　Raised　Twinfast　Coach

Figure 2.6.17 **Common screw heads**

Using screws

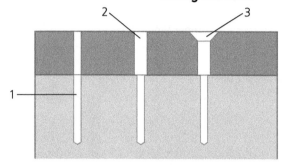

Figure 2.6.18 **Joining two pieces of timber**

When using a wood screw to join two pieces of wood the traditional method is as follows.

1 Drill a pilot hole through the top layer of wood and into the base wood. The diameter should be the size of the core of the screw.
2 Drill a clearance hole slightly larger than the shank of the screw (outside diameter).
3 Countersink the top of the hole to make a recess for the head of the screw to sit in.

Modern screws usually have special features that reduce or avoid the need to go through the stages listed above, saving time (for example the TurboGold screw shown in Figure 2.6.19).

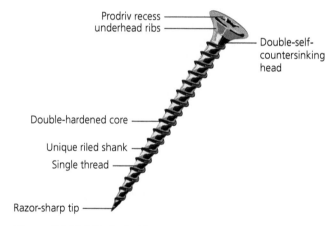

Prodriv recess
underhead ribs

Double-self-countersinking head

Double-hardened core

Unique riled shank

Single thread

Razor-sharp tip

Figure 2.6.19 **A TurboGold screw**

Hinges

Hinges are used where a part needs to move in an arc, for example a door in a house, but they are also used for lids and flaps. A wide variety of hinges are available, some of these are shown in the table below. Different hinges have different uses and are made from materials that are suited to their intended environment. For example stainless steel hinges are more likely to be used on a yacht, where a low-carbon steel hinge would rust quickly when exposed to sea water. Hinges are used in pairs normally, but for larger, heavier parts, three or more can be used together.

Butt hinge	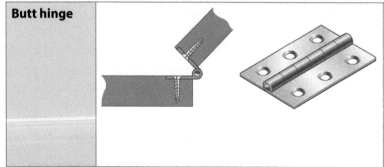	Probably the most widely used hinge. Jewellery boxes (a small pair made from brass) A house door (a large pair made from low carbon steel or stainless steel). Where a door is heavy, e.g. a front door, or fire door, the butt hinges are likely to have ball bearings between the parts to help spread the load of the door. Usually recessed into wooden surfaces. Need to be very precisely aligned.

2.6 Stock forms, types and sizes

Concealed hinge		Kitchen cabinet doors. Usually sprung so that they self close. Often a soft close device is included, so that the doors do not slam shut. Easy to adjust to help with alignment of the doors.
Piano hinge		This is like a butt hinge, but comes in one long length, which you cut to size. Lids that covers the keys on a traditional piano (it gets its name from its common use).
Flush hinge		These are fitted between a door and a frame, like a butt hinge, but are not usually recessed. Because they are not recessed, they are easier to fit, but look less attractive.

Table 2.6.1 **Common hinges**

Knock-down (KD) fittings

Knock down (KD) fittings are those that can be put together easily with a few basic tools (typically a screwdriver and Allen key are all that are needed to join all the parts together). These types of fittings are often supplied with 'flat pack furniture' – furniture sold unassembled in cardboard boxes so that it is more compact and therefore easier and cheaper to transport. The disadvantage is that it needs to be assembled by the customer.

An Allen key is a small hexagonal-shaped tool that can be used with socket heads or hex head fixings. They are available in a wide range of sizes, allow access to hard to reach fixings and are less likely to slip than a screwdriver head.

An Allen key is a small hexagonal shaped tool that can be used with socket heads or hex head fixings. Allen keys are available in a wide range of sizes and are useful as they have a greater contact area with the fixing than a traditional screwdriver head, so are less likely to slip. The shape of the Allen key also allows the user to access hard to reach fixings.

Figure 2.6.20 **Allen key**

Some of the most common knock-down fittings are outlined below.

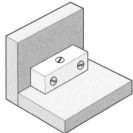

Corner blocks are a simple way to join two boards together at 90 degrees to each other. They are often used to join parts of a cabinet. The block fits in the corner, between the two parts, and three screws are driven in, one into the shelf and two into the side of the cabinet. Although they are easy to use, they are less attractive than other methods, because they are very visible. The limited thickness of the material being joined can also cause difficulties with screw length.

Figure 2.6.21 **Corner block**

The cam lock is a stronger and more attractive way to join parts of a cabinet together. A peg is screwed into a pre-drilled hole on the inside of the cabinet. The other end of the peg is slid into a hole in the

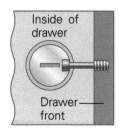

Figure 2.6.22 **Cam lock**

Timber-based materials

end of the shelf. A cam is fitted into a larger hole drilled into the face of the shelf. As the cam is rotated with a screwdriver, it grips the notch in the end of the peg and pulls it and the cabinet side against the end of the shelf.

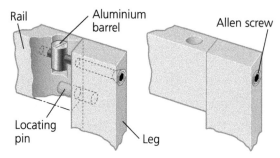

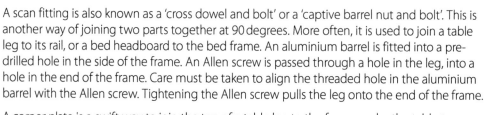

Figure 2.6.23 **Cross dowell**

A scan fitting is also known as a 'cross dowel and bolt' or a 'captive barrel nut and bolt'. This is another way of joining two parts together at 90 degrees. More often, it is used to join a table leg to its rail, or a bed headboard to the bed frame. An aluminium barrel is fitted into a pre-drilled hole in the side of the frame. An Allen screw is passed through a hole in the leg, into a hole in the end of the frame. Care must be taken to align the threaded hole in the aluminium barrel with the Allen screw. Tightening the Allen screw pulls the leg onto the end of the frame.

A corner plate is a swift way to join the top of a table leg to the frame under the table top. The diagram in Figure 2.6.24 is of an upturned table. The plate is screwed onto the two parts of the frame, and the leg is pulled tight against the frame parts by tightening the nut on the studding fitted to the table leg.

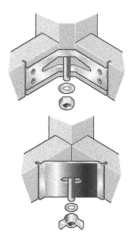

Figure 2.6.24 **Corner plate**

KEY POINTS
- Natural timber is usually supplied with a planed finish – PAR or PBS.
- Manufactured board can be faced with a more expensive veneer.
- Hinges are versatile, but must be fitted accurately in order to properly function.
- Knock-down fittings can be used in the assembly of flat-pack furniture.
- If you are using a knock-down fitting, a jig or a CNC machine should be used to ensure the accuracy of the holes used.

Check your knowledge and understanding

1 What is a dowel?
2 Where would you find a concealed hinge being used?
3 Draw a labelled diagram of the three stages you would undertake when using a countersunk woodscrew.
4 What are the advantages of using knock-down fittings?
5 Where would you find a cross dowel being used?

2.6 Stock forms, types and sizes

Metal-based materials

Metals are also supplied in various stock forms and are available in a variety of different widths and diameters and wall thickness. The range of available forms differs depending on the specific metal used. For example, aluminum and steel can be purchased in a variety of forms such as rods, strips, tubes, bars and angles. Stock forms of metals are generally **extrusions** (a long length of material with a standard cross section), but can also be rolled and shaped out of larger metal sections known as billets. Steel tubes, for example, are rolled and then welded to form a hollow cross section.

Metal suppliers often still list the dimensions of metals in imperial measurements. You will often see metal sheets, or the wall thickness of an extrusion or tube, expressed as 'SWG'. Standard wire gauge (SWG) is a well-established unit of measurement for metals where 12 swg for example is equivalent to 2.642 millimetres and 16 swg equivalent to 1.626 millimetres. The larger the value, the smaller the metric equivalent.

In addition to flat sheets, metal is also available in a wide range of perforated or pressed patterns.

It is important to consider stock sizes of metals when planning what you are designing and making, as metals are harder to machine to a specific shape and size than timber. It is usually more cost effective and less time consuming to adjust your design slightly so that a stock size can be used rather than trying to machine the metal to a bespoke size.

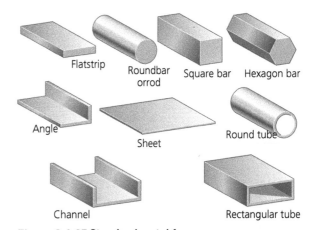

Figure 2.6.25 **Standard metal forms**

Figure 2.6.26 **Perforated metal sheet**

Metal-based components

Components used to join metals include rivets, machine screws, nuts and bolts.

Rivets

Riveting is a widely used technique for joining metals. A hole is drilled through both sheets of metal to be joined and then pass a rivet with the same diameter is passed through the hole.

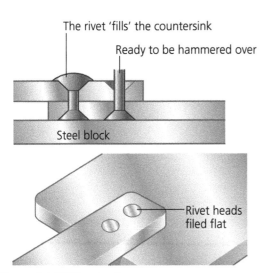

Figure 2.6.27 **The riveting process**

Figure 2.6.28 **Types of rivet**

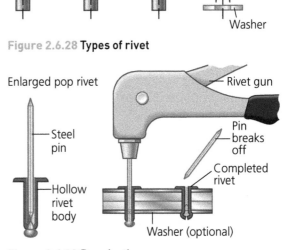

Figure 2.6.29 **Pop riveting**

The head of the rivet butts up against one side of the sheets and the other side is hammered over to hold the second sheet against the first (round or snap head rivets). Riveting does not normally require heat, which can make metals bend and distort (particularly thin sheets). Also, the equipment used for riveting is simpler, cheaper and safer to use.

There are a variety of different types of rivet, which have different uses:
- Countersunk rivets: used for a neater finish. Both metals to be joined need to be countersunk, and then the rivet is filed flat with the surface of the metal.
- Flat head rivets: used for joining materials that are too thin to countersink.
- Bifurcated rivets: used to join soft materials such as fabric and leather.
- Pop rivets: used you cannot access the back of the hole through the materials, for example, when riveting a panel on the outside of a closed box. This rivet is hollow with a pin through it. A pop rivet gun is used to pull the pin into the rivet from the back, squeezing the back of the rivet against the material to be joined. When it is squeezed right against the surface, the pin breaks off. This process is also known as blind riveting.

Machine screws

A machine screw (machine bolt) is designed to fasten an object to an existing tapped hole in a metal surface; there is usually no need for the use of a nut. They are normally referred to by their width and length (e.g. M12 × 50 mm).

Bolts and machine screws are available with a range of heads, designed for different situations, and fitted using different tools. The diagram in Figure 2.6.36 shows common varieties.

Figure 2.6.30 Nut and bolt

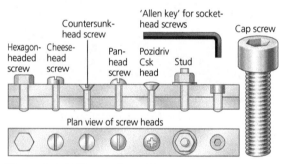

Figure 2.6.31 Types of machine screw

Nuts and bolts

A nut and bolt can be used to join together two materials. A hole is drilled through both materials and then a bolt materials with the corresponding diameter to the hole is passed through them. A nut can then be screwed onto the bolt. A bolt should be chosen that is long enough for the nut to completely screw on the bolt, but short enough so that the bolt does not stick out too far. The bolt and nut are most commonly made with a hexagonal head which can be tightened with a spanner, but they can also have an Allen key or socket head.

There are a range of nuts available.

Most nuts are tightened with a spanner.
- Wing nuts can be tightened by hand using the two metal 'wings' on the side.
- Nylon insert lock nuts have a ring of nylon in one end that is slightly smaller than the screw thread; this grips the thread of the bolt and the friction stops the nut from undoing through vibration. Nylon insert lock nuts are used on folding push chairs. The trade name for them is a 'Nyloc' nut.
- A dome nut is used where a more decorative finish is required, or the exposed end of the bolt needs to be covered for safety reasons.
- When you use a security shear nut, the end snaps off as it tightens, leaving a conical nut that is almost impossible to undo. This type of nut is used where you do not want people to be able to undo the nut.

Figure 2.6.32 Types of nut

Washers

When using a nut, bolt or machine screw, it is common to use a washer to spread the load across a larger area.

- A plain washer is usually slightly bigger in diameter than the head of the bolt.
- If the materials being joined are quite soft, then a larger washer called a penny washer is used.
- If the parts are subject to vibration a lock washer can be used between the face of the nut and a plain washer.

Plain washer Lock washer Grover spring washer Spring washer Tab washer Serrated washer

Figure 2.6.33 **Types of washer**

KEY POINTS
- Metals are sometimes listed using standard wire gauge (SWG) sizes.
- Rivets can be used to join thin sheet material together.
- Washers can spread the load of a joint when using a nut and bolt.

Check your knowledge and understanding

1 What is a billet?
2 Why might you use a cap- or socket-headed machine screw over a hexagonal-headed machine screw?
3 What advantages do wing nuts have over hexagonal nuts?
4 When would you use a 'pop rivet'?
5 What function does a nyloc fulfil?

Polymers

Polymers are available in a more limited range of stock forms than other material groups. Most polymers that you will encounter in school or college will come in sheet form, although extrusions such as tubes and rods are available. All of these plastic stock forms below are available in a wide range of colours, and one of the main advantages of polymers is their ability to be coloured by the addition of a pigment.

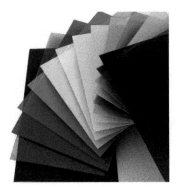

Figure 2.6.34 **A range of acrylic sheet**

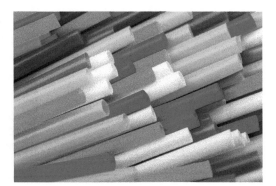

Figure 2.6.35 **Polymer extrusions**

Stock form	
Sheets	Standard sheets of acrylic are usually 1200 × 500 mm or 1200 × 600 mm and are most commonly 3mm in thickness.
	Other thicknesses are available, but the choice of colour becomes more limited in the less common sizes.
	High Impact Polystyrene (HIPS) can also be supplied in large sheets but is usually then processed down into smaller sizes (usually thinner than 3mm) for use in vacuum formers.
Granules	Used in processes such as injection moulding
Powders	Used in fluidising tanks to plastic coat small metal parts.
	Thermoset polymers are also generally found in powder form, but due to the compression moulding manufacturing process, it is unlikely that you will come across them in a school environment.
Foams	These vary in thickness and material depending on the intended application.
	Plastazote foam is a trademark although the material itself is polyethylene foam. It is available in a variety of thicknesses and densities. It is often used in protective packaging, swimming pool floats and safety mattings in gyms and nurseries.
Films	Available in a variety of thicknesses.
	PVC is a popular material that is supplied in film form and can be easily vacuum formed for use in food packaging.
PLA filament	A more recent common stock form of polymer.
	Used in 3D printing.
	Usually supplied on a reel.
	Varies in diameter depending on the brand of 3D printer being used.

Figure 2.6.36 **Polymer granules**

Figure 2.6.37 **Plastazote foam flooring**

Figure 2.6.38 **PVC film**

Figure 2.6.39 **PLA filament**

Polymer-based components

Many of the components used to join metals can also be used successfully with polymers. In addition to these common components, polymer versions of mechanical fixings, including hinges and nuts and bolts, are increasingly becoming available.

Polymer hinges can be used with similar materials, but equally may be suitable for use with metals and woods. The properties of the polymer may be better suited to the application than a metal equivalent (for example, they may be lighter weight and will not corrode).

One advantage of using a polymer hinge with a polymer fabrication is that you can use an adhesive to join the two items together instead of having to add nuts and bolts or similar fixings.

As with metals, the range of nuts, bolts and washers that can be manufactured from a polymer is vast. In applications where weight, conductivity and corrosion resistance may be more important than strength, a polymer fixing may be the most suitable. Most are manufactured from nylon.

Figure 2.6.40 **Polymer hinge**

Figure 2.6.41 **Nylon nuts and bolts**

KEY POINTS
- Most stock forms of polymers are thermoplastic polymers.
- Thermoset polymers are found in powder form.

Check your knowledge and understanding

1 What is an extrusion?
2 Give examples of products that use foam.
3 Why might you not use a polymer nut and bolt?
4 What are the benefits of using a polymer hinge with a polymer fabrication?
5 What process may use a polymer powder?

Textile-based materials

Yarns

Fibres are twisted together to make strands of yarn, but these strands need to be twisted with others to make a strong yarn that can be used to make fabrics or for knitting and craft projects. This twisting is called **plying**.

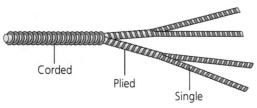

Corded

Plied

Single

Figure 2.6.42 **Multi-ply yarns**

The strands are twisted together in the opposite direction to the one that they were spun – this makes a yarn that has an even thickness with no weak spots. Yarns with more than one ply can also be twisted together to make corded yarns (for example, to make sewing and embroidery threads).

The ply of a yarn refers to the number of strands that make up the yarn, so a 2-ply yarn will have two strands, a 4-ply yarn will have four strands, and so on. Yarns with a higher ply number are more durable than those with a low number as the plies hold the fibres tightly together.

A high number of plies does not always make a yarn thicker or heavier – that will depend on the thickness and weight of the individual strands used to make the yarn.

Fabrics

Fabrics come in a limited number of different widths and are bought by length. The widths of fabrics are approximate, as they will depend on the machinery used to make them. The most common standard widths of woven fabrics are 115 centimetres, 140 centimetres and 150 centimetres, but some fabrics are available in widths of 280 centimetres. The width of a fabric will have an important part to play in determining the length that a manufacturer will need to buy to make a product – narrower fabrics usually need a longer length to be bought.

Components

There are many different types of standard components and trimmings available for use on textile products and they come in many different sizes and formats to allow products to be functional as well as decorative.

Figure 2.6.43 **There is a vast range of embroidery threads.**

Threads are needed to sew products together or to add decoration. Most general-purpose sewing threads used to construct products are made from polyester, but sometimes they are made from cotton or silk. They are available in 100-metre, 250-metre and 500-metre reels, or 1,000-metre and 7,500-metre for commercial use, and are dyed in many different colours so that they will blend in with the colour of the fabric used.

There is a vast range of embroidery threads made from many different fibres (for example, cotton, silk, wool, viscose, lurex and metal). Many are designed to give special effects, including glitter, glow-in-the-dark and lustre. Some are used for specific purposes, such as tacking, overlocking, buttonhole and quilting threads.

Fastenings include buttons, zips, Velcro, press studs and hooks and eyes.

Trimmings include ribbons, bindings, lace edgings, motifs and e-components such as LED lights.

Buttons have two or four holes, or a shank. They come in many different sizes, colours and shapes. Buttons can be used with buttonholes or loops to make functional or decorative fastenings.

Zips have a slider that open and closes the zip teeth. They are available in different types and some of the most commonly used are closed end, open end, invisible and jeans. Most zips have a polyester coil which makes them lightweight. Some zips have metal teeth which are attached to both sides of a zip tape – these zips are most often used on jeans or for decorative purposes. Plastic moulded zips are like metal zips but the teeth are made out of plastic. Open-ended or separating zips are designed to come apart so that a garment can be opened completely (for example, in jackets and coats).

Zips come in a wide range of lengths or can be bought by the metre. They are made in many different colours or can be dyed to match a specific fabric when bought commercially.

Velcro is a touch and close fastener that consists of two fabric tapes with tiny nylon hooks on one, and a soft furry surface on the other. When the tapes are pressed together the two layers stick, making it a quick and easy fastening, often used on children's clothing.

Press studs (also known as snap fasteners) are two circular metal or plastic sections that are stitched on by hand and lock into each other. They are a flat fastening, and useful for openings where there is not much strain as they can pop open easily.

Poppers are similar to press studs, but are applied using a hammer and a special tool. They are a decorative fastening.

Hooks and eyes are a two-part fastening made of metal and applied with hand stitching. The hook fits into the eye, which may be a curved loop or straight bar. These fastenings are meant to be discrete and are often used at the top of a zipped opening or on a waistband.

When choosing components for a product the designer will need to consider the type of product and the fabric it is made from, the colour and weight of the component, safety issues relating to the component, product or user, the cost, how the component is applied to the product, and whether it is to have a decorative purpose or be purely functional.

Figure 2.6.44 Buttons come in many different sizes, colours and shapes

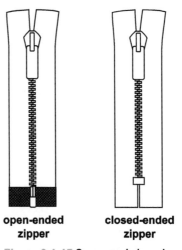

open-ended
zipper

closed-ended
zipper

Figure 2.6.45 Open- and closed-ended zips

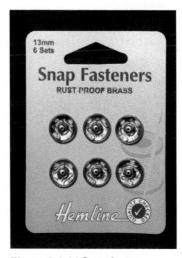

Figure 2.6.46 Snap fasteners

> ## KEY POINTS
> - Yarns are made by plying two or more single yarns together.
> - It is important to know the width of a fabric when calculating quantities.
> - Trimmings and components are used for functional and decorative purposes.

Visit a local shop that sells sewing and knitting resources and research the types of fastenings available.

Make a chart like the one below to show your findings.

Type of fastening	Type	Range of sizes	Where it might be used
Zip 1	Closed end polyester coil		
Zip 2	Open ended		
Zip 3			
Button 1	2-hole round		
Button 2			
Button 3			
Velcro			
Press studs			
Hook and eye			

Check your knowledge and understanding

1   What are the most common standard widths of fabric?
2   Name three different fastenings.
3   What are press studs?
4   What does the 'ply' of a yarn refer to?
5   Choose the best type of zip for a coat and explain your reasoning.

Electrical and mechanical components

There are a range of mechanical components available to buy from suppliers that are useful when prototyping mechanical systems. Cams, gears, pulleys, axles and bearings are all available. Care must be taken when buying these components: it may be easier to manufacture certain parts yourself, such as cams that are straightforward to make or manufacture with the right equipment. If you are buying gears, you will need to know the axle size and the number of teeth required to achieve the necessary speed or torque.

There are many different types of electrical components, including resistors, capacitors, diodes, transistors and integrated circuits. Different components have specific uses within circuits, and they are purchased from electronic component suppliers. It is important when buying components that the correct specification for the component is used. Certain electrical components, such as capacitors, have a maximum working value measured in volts, while other components, like LEDs, are rated for a maximum current. Often, suppliers can offer a better price per component if larger quantities of them are purchased: the greater the number you purchase the cheaper each one will be, so you might see suppliers give the price as:

Sold in multiples of 5:

5+	25+	100+	1,000+
£0.404	£0.33	£0.251	£0.217

This shows how the unit price decreases with the quantity of components bought.

Figure 2.6.47 **Various electronic components**

Resistors

Resistors are very common components in electronic circuits; they limit the amount of current flowing in circuits and set the voltage levels in particular parts of a circuit. When buying resistors you need to know:
- the physical size of the resistor
- the value of the resistor
- the tolerance you require.

Resistance is measured in ohms (Ω) with a single ohm being a very small unit, so you will often see them measured in much larger units:

K (kiloohm) 1 K = 1,000 ohms (10^3)

M (megaohm) 1 M = 1,000,000 ohms (10^6)

Resistor value is indicated by a colour code. To find the value of a resistor it should be placed so that the three bands of colour that are close together are on the left, and the single band of colour is on the right. The first two bands give the first two digits of the value the third band gives the number of zeros to be added. The final band of colour gives the accuracy of the value of the resistor. The stated value cannot be exact so an indication of how close to the stated value is given – its tolerance.

For example a brown, black, red and gold resistor would indicate a 1-K resistor with a tolerance of ±5 per cent. This means its real value could be anywhere between 950 Ω and 1,050 Ω.

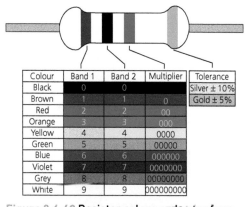

Colour	Band 1	Band 2	Multiplier	Tolerance
Black	0	0		Silver ± 10%
Brown	1	1	0	Gold ± 5%
Red	2	2	00	
Orange	3	3	000	
Yellow	4	4	0000	
Green	5	5	00000	
Blue	6	6	000000	
Violet	7	7	0000000	
Grey	8	8	00000000	
White	9	9	000000000	

Figure 2.6.48 **Resistor colour codes for four band resistors**

Preferred values

It is possible to produce every value of resistor from less than one ohm to millions of ohms, but it would not be practical to make and stock every value. Instead manufacturers produce a range of resistors in a series of preferred values. One such range is the E12 series of preferred values and these values are shown in Table 2.6.2.

It is called the 'E12' series because there are 12 values spaced as evenly as possible across each range in value to the power of ten. If an engineer wanted a resistor for a circuit of a value of 260 Ω they would choose the closest value from the E12 series – a 270 resistor. The nearest value is never more than 10 per cent away from the value required, and this keeps resistor costs as low as possible.

10	100	1K	10K	100	etc.
12	120	1K2	12K	120	
15	150	1K5	15K	150	
18	180	1K8	18K	180	
22	220	2K2	22K	220	
27	270	2K7	27K	270	
33	330	3K3	33K	330	
39	390	3K9	39K	390	
47	470	4K7	47K	470	
56	560	5K6	56K	560	
68	680	6K8	68K	680	
82	820	8K2	82K	820	

Table 2.6.2 **The E12 series of preferred values**

Capacitors

A capacitor is a component that can store electrical charge: the larger its value, the larger the charge it can store. When buying capacitors you need to know:

- which type:
  - electrolytic or polarised capacitor (these capacitors have a positive pole and a negative pole, meaning they must be connected in a circuit the correct way round)
  - non-electrolytic or non-polarised capacitor (does not have positive and negative poles so can be connected to a circuit any way round)
- the physical size of the capacitor
- the tolerance you require
- the value of the capacitor.

Figure 2.6.49 Various electrolytic capacitors

Electrolytic capacitors generally have larger values; the cathode, or negative leg, of an electrolytic capacitor is shorter and the casing is marked with a strip of (–) symbols. Capacitance is measured in farads (F); one farad is a very large unit, so generally you will see them measured in much smaller units:

μF (microfarad), 1μF = 1 millionth of a Farad (10^{-6})

nF (nanofarad), 1nF = 1 thousand millionth of a Farad (10^{-9})

pF (picofarad), 1pF = 1 million millionth of a Farad (10^{-12})

Most capacitors have a maximum working value specified in volts. It is important not to exceed this voltage, but for economy it is best to purchase the lowest voltage that matches the circuit being produced.

Integrated circuits and microcontrollers

Manufacturers stock a range of integrated circuits (ICs). An integrated circuit is a miniature electronic circuit on a semiconductor material. They are recognised and ordered by their size and type and marked with part numbers. One manufacturing format for ICs is dual in-line (DIL). Integrated circuits can be bought with various numbers of legs: 8, 14, 18, 20, 28 and 40 pins, so an 8-pin DIL IC would have two rows of four legs. A common microprocessor IC is the PIC microcontroller. PIC chips can be programmed to respond to input devices and control numerous output devices. They come in a range of DIL packages of varying sizes where different numbers of pins can be allocated to input or output devices. When purchasing PICs users will look for the PIC microcontroller to match the number of input and output components they require.

ACTIVITY

Look online and find a price for buying 200 1K resistors with a tolerance of ±5 per cent.

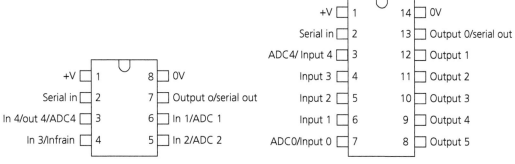

Figure 2.6.50 **Pin layout for an 8-pin DIL PIC**

Figure 2.6.51 **Pin layout for an 14-pin DIL PIC**

STRETCH AND CHALLENGE

Explain how you would recognise a polarised capacitor and how you would know which way round to place it in a circuit.

KEY POINTS

- Electronic components are sold by quantity, volt and current rating.
- Resistors are recognised by colour coding.
- Resistors are not available in every value but are available in a series of values.
- There are polarised and non-polarised capacitors.
- PIC microcontrollers are bought in DIL packages in a range of sizes.

Check your knowledge and understanding

1. Give an example of a component that is rated by a maximum voltage.
2. Explain what is meant by 'tolerance' when referring to a resistor.
3. Give an example of a component that is rated by a maximum current.
4. Discuss the advantages for resistor manufacturers in producing resistors in a series of values like the E12 series.

2.7 Scales of production

What will I learn?

In this topic you will learn about:
- → how products are produced in different volumes
- → why different manufacturing methods are used for different production volumes.

The way products are manufactured depends on the quantity required. For example, if you are producing a prototype for testing, you probably only need one: this is 'one-off' manufacturing. However, if you were making cars for sale across the world then you most likely need to make large batches. There are four main types of production used in manufacturing: one-off (or prototype), batch, mass and continuous. Costs, time, skill needed, efficiency of production and design considerations change as the number of products made increases.

Prototype

A **prototype** is an early sample, model, or release of a product made in order to test specific aspects of a design, such as operation or shape. For example, an electronic control system can be prototyped using computer-controlled circuit boards so that it can be tested to see if it will perform correctly before larger numbers are produced. As the design process continues, increasingly realistic prototypes are often produced to further test the intended elements of the product. The cost of a prototype is high because all the costs of production are absorbed into just one price rather than being spread between lots of the same thing. You can make a prototype in school or college to test aspects of your design, such as how it works or what it looks like. It may not cost a lot of money, but will take a great deal of time – it is labour intensive.

Apart from creating a prototype for testing, one-off products are produced for other reasons. Bespoke production is when a product is made for a specific situation or client, and only one of them is ever made. A bespoke product includes everything that the client wants as well as the specific features that the craftsperson or manufacturer brings to the product. The tools and skills that are used are specific to the end-product and so the eventual cost remains high. Bespoke products include clothing and jewellery – for example, a made-to-order wedding dress, produced to the bride's personal wishes where she chooses the style, fabric and decorations and then a specialised designer or dressmaker completes the garment. Bespoke and one-off production is labour intensive, because every product is different.

Figure 2.7.1 An Arduino prototype board used to test control code

Figure 2.7.2 **Bespoke production of a wedding dress**

Figure 2.7.3 **A one-off motorway bridge under construction in Australia**

One-off production is where just one complete product is produced. It often includes bespoke elements as well as mass-produced parts that are combined to form just one end-product. Typical examples are large objects such as a cruise liner, a power station or a bridge. Where a bridge is being built, even though many parts of it will be similar or identical to other bridges, the whole structure will be unique. Bespoke and one-off production is labour intensive, because every product is different.

The item you produce as part of your non-examined assessment will be a prototype. There will probably only ever be one of the thing you make. You will experience producing something to your own design, appraising it and seeking the advice and opinion of others about how it might be improved. If time allows you may be able to make some improvements.

Batch production

Batch production is where a limited number of the same product is made during a particular period of time. The number of units in a batch is determined by the product that is being made, how much it costs to buy, how often it is sold, the price of raw materials and the costs of running the machines that make it. Jewellery could be made in small batches of as little as two or three, but books like this one are printed in batches of thousands. Other examples of products that are batch produced include clothes, cakes and furniture.

This type of production will make use of moulds, patterns, jigs and formers so that a product can be made over and over again in a series of production runs. Each manufacturing facility has the ability to change aspects of the product very quickly or even change products completely so that the same skilled workers and machinery are making entirely different items.

Figure 2.7.4 **Computer-controlled assembly of car engines**

Computer-controlled machinery is crucial to the success of modern batch production as it allows for a rapid changeover between batches with little **down time** (when the line is not running). In some cases, such as the production of cars, computers can be used to change the operation of a machine each time it is needed. In these situations, parts such as engine blocks can move along a production line and be joined using a sealant applied by a computer-controlled machine.

Figure 2.7.5 **Computer-controlled production of parts for light fittings**

Figure 2.7.6 **Batch production of Windsor chairs**

As the engine part arrives at the machine on a computer-controlled track, information about its shape, position and requirements for the adhesive can be called up and used to control the glue nozzle. This means that a variety of engines can be produced during a single run without stopping the line or changing patterns or jigs. The mould for the product becomes digital and can be instantly changed. Examples of using this digital technology in school or college could include laser cutters and 3D printers, where you can change what is produced just by changing the drawing that controls the machine.

KEY WORD

Mass production manufacturing in large quantities over a long period of time. This typically uses a production line.

Mass production

Mass production is manufacturing in large quantities over a long period of time. This typically uses a production line. Mass production often involves putting together products from sub-assemblies, which are small sections of the whole product, and standardised components, which are simple parts that are produced in large numbers in another factory. Specialist machinery is often used, but the skills required to operate the equipment and assemble the product are low, meaning that the workforce is largely unskilled and can be reorganised quickly when the product changes. Using standardised components and unskilled labour allows the cost of the specialised equipment to be offset, resulting in a large number of cheap products. Examples of mass-produced products are things such as mobile phones, clothes, stationery, printed circuit boards and packaging.

Figure 2.7.7 **Workers assembling smartphones on mass production lines in Guang'an, China**

ACTIVITY

Have a look at products that you have around the house or at school or college. Can you find something that you think is mass produced? What features make it suitable for mass production?

Continuous production

KEY WORD

Continuous production runs constantly and is highly automated.

Continuous production is where an item is made continuously for 24 hours per day, seven days per week with infrequent shutdowns for cleaning and maintenance. Shutdowns happen, for instance, annually or biannually, as stopping the process and restarting it later is more costly than keeping it going. Highly specialised equipment is needed to run the facility, and to offset huge investment costs very high numbers of the same thing are made for long periods of time. The process is hard to change because of the costs of stopping production, and so continuous production is used where the demand for the same product is expected to be the same for a long time. The whole process of continuous production can be fully automated, meaning that no workers are needed. Examples of continuous production include paper, metals such as cars and electronic components.

Figure 2.7.8 **Continuous production of steel at the Llanwern Steelworks in Wales**

ACTIVITY

Take a look at a product such as an aluminium drinks can and identify where different scales of production have been used. Start at the raw material stage and follow the product through to disposal and recycling.

KEY POINTS

- Production can be categorised by the amount that is produced – this is called scales of production.
- One-off production is chosen when a single product is made.
- Batch production is chosen when several of the same product are required before the design of the product needs to change.
- Mass production is used when a large number of the same product are needed. It uses a well-organised production line.
- Continuous production is good when the cost of starting the business and running the necessary equipment is high. The product needs a constant and steady demand.

Check your knowledge and understanding

1 Why is the end-cost of a bespoke product higher than something that has been mass produced?
2 Why do designers make prototypes?
3 Give two reasons why shoes need to be batch produced.
4 Complete the following table.

Labour	Equipment	Unit cost	Scale of production/ manufacturing method used
Unskilled workforce	Specific to the task with some flexibility	Low	
Highly automated with a small workforce	Highly specialised	Very low	
High skills, traditional craftsperson	General purpose with some specialism	Very high	
Skilled and flexible	General purpose with specific adaptations for the type of product being made	High	

5 What type of products benefit from continuous production methods?

Figure 2.7.9 **A wedding invitation that has been produced in a batch**

Examples and application for papers and boards

Programmes for events that are only available for a limited period of time are batch produced. Programmes for football matches, music concerts and wedding invitations are all examples of batch-produced printed products.

STRETCH AND CHALLENGE

Explain why there is a need to batch produce some printed products even though the cost per item is higher than if they were mass produced.

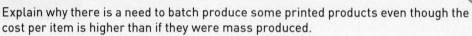

Examples and application for timber, metals and polymers

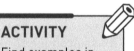

ACTIVITY

Collect two examples of printed products that are normally batch produced.

Timber products such as plywood and MDF are continuously produced. The manufacturers can run their factories continuously because there is a constant demand for these products. The buyers of plywood and MDF want the same product every time they buy it, and so the factory can be kept exactly the same for a long period of time.

The scale of production can affect the cost of a product. Metals and timbers can be used at very low volumes to produce high-quality and expensive products. A professional cabinet maker can use hardwoods to make a bespoke piece of furniture. A blacksmith can use iron and steel to make things such as gates and railings that are specific to a particular building.

ACTIVITY

Find examples in your classroom that have been made at different scales of production. Try to get an example from all four categories. Look for things such as past GCSE projects, shoes, furniture, exercise books, pens and anything that uses continuously produced materials such as MDF.

Figure 2.7.10 **Veneers being layered to produce plywood at a mill**

Figure 2.7.11 **A blacksmith producing a one-off piece of architectural railing**

STRETCH AND CHALLENGE

For each of the examples that you found in your classroom, identify how each part or component has been made. Your chair may have been batch produced, but ask yourself about the screws and bolts. Have they been batch produced too or were they mass produced in a different place?

Examples and application for textiles

Many textile products are a combination of batch and mass production. Screen printed t-shirts, jeans and sweatshirts are mass produced in large factories which are often based in other countries. Each time a new design is made it needs to be produced in different sizes, and so the production is split into batches. High street stores may order large numbers of the same item, but each time the order is delivered it is made up of smaller batches of the same garment in different sizes.

Clothing production is a common form of bespoke or one-off production. Specific items of clothing are made by tailors to the specific needs of individuals. These can be made for particular events, such as weddings, or for particular purposes, such as uniforms. In both cases the garments are made according to the size of the person and their requirements for the finished item of clothing.

Figure 2.7.12 T-shirts of different sizes and in different colours are prepared before being silk-screened with the same design

ACTIVITY

Take a mass-produced item of clothing and analyse it. Find out how many parts and components are used to make the item and list them all next to a photo or drawing of the garment.

Figure 2.7.13 A tailor working on a jacket which will be a precise fit for the client

STRETCH AND CHALLENGE

List the parts of an item of clothing and add the cost of each part to the description. Find out how much it would cost to make just one of whatever you have examined. Compare the price that you would pay in a high street store and the price it would cost to make just one. Explain why they are different and say where the extra costs or the extra savings have come from.

Examples and application for electrical and mechanical systems

Electronic products that are available to consumers, such as phones, computers and tablet devices, are only possible because of mass-production techniques. All of these kinds of product have existed as one-off products in the past, and as new products are designed, new prototypes are made.

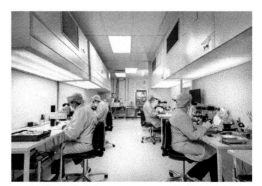

Figure 2.7.14 An electronics factory will have completely clean rooms where the products are assembled.

ACTIVITY

A prototype of a new computer or phone will be very expensive when compared to the price that consumers pay. Explain why there is a difference in price.

STRETCH AND CHALLENGE

A new electronic product will have many different prototypes and early versions which are very expensive, before it is mass produced and available to consumers. Explain why manufacturers make these early versions when they are so costly.

2.8

Specialist techniques and processes

What will I learn?

In this topic you will learn about:

→ production aids used in the manufacture of products
→ the tools and equipment used in the manufacture of products
→ how to cut and shape materials to a tolerance
→ a variety of commercial processes
→ how quality control is used during manufacture.

This chapter covers working with a range of materials and covers commercial practices in addition to school and college workshop procedures. It is essential that you have detailed knowledge and understanding of at least one specific material and that you are aware of the processes and techniques associated with that material. This information will be tested in Section B of the examination paper.

Papers and boards

The use of production aids

Templates and patterns are used in both commercial and one-off manufacture of paper and board-based products. Patterns can be copied both by hand and through computer-aided design so that there is a consistency to designs. This helps to ensure the quality of the final outcome. Templates can be used to show clearly where to put information, again to ensure both the item is accurately made and, in the case of more than one product, so that it can be replicated exactly and the consumer can expect a similar quality throughout.

Templates can be used when designing graphic products, such as leaflets, to help the designer ensure that information is placed in the correct places and that the format looks effective. This type of template is often used in schools to allow students to format their work and easily input information without having to spend too long on layout.

Other templates include those that are used to draw around. This might be done in a school or college environment where multiples of the same item need to be cut out.

Figure 2.8.1 **Using a template to draw around**

Sometimes production aids can actually be part of the cutting process and will ensure that identical products are cut out every time. This is true of die cutting, in both industry and on a smaller scale, which uses a die to shear webs of low-strength materials.

Figure 2.8.2 **Die cutters for use at home**

Measuring

Marking out on paper and board can involve a number of different tools and aids depending on the shapes and materials used. A **ruler** is a basic tool which allows you to measure distances on any flat material. It is also used to help you draw straight lines. **Protractors** are used to measure angles and help you to draw to a specific angle when used with a ruler or a T-square.

Drawing boards are used to give a clean flat surface in order to draw and write. This is a convenient tool in a workshop where surfaces may be dusty or pitted, so this tool is often used in schools and colleges. Some drawing boards have guidelines and paper clips to secure material. They can also include a fitted T-square which can slide up and down the board and allows you to draw vertical and horizontal straight lines quickly and accurately.

Set squares help you to draw lines at specific angles. They are made from clear plastic with 30-degree angled edges for use in isometric drawings or 45-degree edges that are used for oblique or orthographic drawings. Isometric drawing is way of presenting ideas using 30-degree angles to draw the sides. This method produces a three-dimensional representation of the object. Orthographic drawings usually show a front, side and a plan view of the object separately and so are two dimensional. These drawings allow for measurements of all parts to be shown so that there is enough information for third party manufacture.

Compasses are used to create circles or arcs. One side of a compass is the point and should be located at the centre of the circle you wish to draw. The other side usually holds a pencil but attachments can be used to allow other pens to be held in place. Compasses can also be extended with different attachments to enable you to draw larger circles.

French curves are used to draw curves of varying sizes. They are usually plastic templates and feature different curves which can be drawn around to reach the desired effect.

Lightboxes are used for tracing images. The lightbox has a flat surface over the top of a light which shines through the surface of the paper and makes it much easier to see the lines clearly.

Craft knives are used for cutting and scoring paper and board. Craft knives are useful when trying to cut internal shapes as the cut does not need to start from the edge of the paper. They produce an accurate and precise

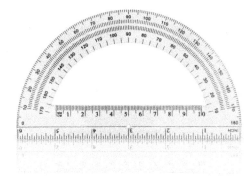

Figure 2.8.3 **Protractor for measuring angles**

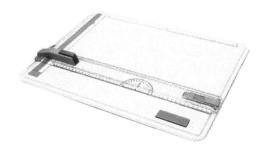

Figure 2.8.4 **Drawing board**

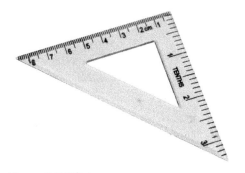

Figure 2.8.5 **Set square**

Figure 2.8.6 **Compasses**

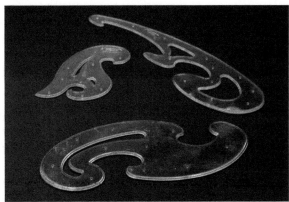

Figure 2.8.7 **French curves**

Papers and boards

171

Figure 2.8.8 Picture mounts are cut on an angle to frame the pictures.

Figure 2.8.9 Compass cutter

Figure 2.8.10 Small die cutters

edge when used well but can be difficult to use, especially when cutting tight curves. It is possible to use jigs to cut angled edges in boards with craft knives. This is often done when creating products such as picture mounts.

When cutting circles and arcs, **compass cutters** and **rotary cutters** are often used to create an accurate and even cut. Compass cutters are used for thinner card and paper, and rotary cutters are often used for thicker board.

Bonding

Paper and board can be bonded using a number of methods which vary according to the strength of bond and aesthetic qualities required.

Adhesives such as **spray mount** are easy to apply and do not soften or damage the paper and materials can peel away from each other which allows the user to reposition. The disadvantage of this method is that the adhesive is expensive and the materials can separate. Other adhesives include **PVA glue** which can create a strong bond between paper and board materials but can be messy and damage thin materials. Wax seals are a traditional method of bonding paper often used to seal envelopes. This provides an attractive bond and makes it clear to the consumer whether the envelope has been opened.

Tools, equipment and processes

In industry, paper and board products are often cut using die cutting. This process stamps out the required shape using sharp blades. It is a quick process that ensures that all of the products are cut to a high level of accuracy, making it cost-effective for high numbers of the same product. Die cutting can be used with a variety of blades, some of which cut and others that are blunt in order to crease. Perforation is done using blades which cut on a dashed line so that the material can be torn. It is possible to use die cutting on a much smaller scale in the production of products such as handmade cards. Small die cutters are used to cut the same shape out of material again and again, so are available in set shapes and sizes.

These processes are discussed in more detail in Topic 2.5.

Lamination is used for all scales of production to give a shiny, water-resistant surface to paper and boards. It is more expensive than varnishing but it creates a thick durable surface to the material. Paper or board is placed inside a plastic-film sleeve. This is then fed into the laminating machine from a tray at one side the machine. The machine pulls the material through its heater using rollers, and the film is warmed up to almost melting point. The rollers press the film together so that the two surfaces bond together and the product is sealed. When the card reappears in the 'out tray', the finished product has a glossy coating which is resistant to water and makes the product last longer.

How materials are cut, shaped and formed to a tolerance

When products are made, it is important for the manufacturer to check the sizes of various parts of the product to ensure consistency and quality. Often there is a measurement **range** which is acceptable rather than a specific measurement. For example, a greetings card may need to be 200 millimetres in length to fit in the envelope, but if it is out by a millimetre less or more than 200 millimetres, this is still acceptable and the card will still fit. This range of measurements is the **tolerance**. In the example of the greetings card this tolerance would be shown as 200 millimetres +/− 1 millimetre.

In the case of paper and card products, the length and width measurements of an item are often the main use of tolerances, but it could also be used for other things such as weight or the distance between one object and another.

Sometimes tolerances are more critical, so they are smaller than others. For example, in the manufacture of an intricate pop-up card, it may be crucial that the parts slide over each other, and so the tolerance may be +/− 0.2 millimetres, whereas a less complex object such as a cereal packet may have a tolerance of +/− 1 millimetre.

Tolerance is important to ensure that the sizes of products look uniform to the consumer and can continue to do their intended job. It is important for the manufacturer to consider how packaging, for example, will be hung in shops and what level of difference between the intended size and the actual size will make this task difficult or impossible for the shopkeeper.

Commercial processes

The main commercial processes for paper and board are die cutting and offset lithography. These processes enable products to be printed and cut in minimal time, and with a much greater consistency compared to machinery and processes that are used for one-off or small-scale production, such as inkjet printers and craft knives.

Offset lithography is a common process used in printing. It prints in four main colours: cyan, magenta, yellow and black. Aluminium plates transfer each of these colours on to the material in tiny dots. These dots are so tiny that they cannot be seen individually. The final outcome is a high-quality print that can be repeated quickly and to a consistent standard.

Die cutting to cut paper and boards is usually done after printing. It is done by stamping blades onto the material to punch out the required shape.

More information can be found in Topics 2.5 and 2.9.

The application and use of quality control

When mass producing products of any type, it is important to ensure that the products are consistent and their quality is always what is expected. It is important than the manufacturer creates products with as little waste as possible in order to keep production costs down. In the production of paper and card products, three main quality control and quality assurance marks are used to ensure that the product is to the correct standard.

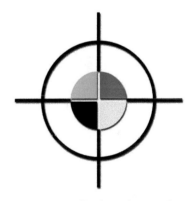

Figure 2.8.11 **Registration mark**

Registration marks

Registration marks are printed using every colour of the four-colour printing process: cyan, magenta, yellow and black. If they are accurate, they should overlap exactly so the mark looks completely black and there is no coloured shadow evident. This mark is used to check alignment, so if there are any other colours it becomes obvious that the product has not printed correctly and therefore amendments need to be made.

Colour bars

Colour bars or densitometer scales are printed outside the trim area and are used to check the quality of the colours. Squares of colour are printed in a strip so that the printer can check colour density and ensure one print is not darker than another further down the print run. These colour bars are usually printed just outside the trim area so that they can be discarded after production and checking has taken place. This checking process is automated by some printers, or handheld devices called densitometers are used. Digital scanners check the colour bars to ensure quality and consistency is maintained.

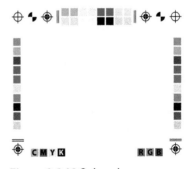

Figure 2.8.12 **Colour bars**

Figure 2.8.13 **Crop marks**

Trim or crop marks

Crop marks are small lines which are positioned in the corners of each page or printed area. They show exactly where the finished page will be cut during the finishing process. Often these crop marks are trimmed off using a guillotine at the end of the production process, although on some products, such as newspapers, the crop marks can still be seen.

ACTIVITY

Collect a breakfast cereal box and a newspaper. Take apart the packaging and look for the printers marks. Make a list of the marks you can see. Use a magnifying glass to study them closely and determine the accuracy of the print.

Check your knowledge and understanding

1 How do you check the quality of a printed product?
2 What would you use to draw a line at 45 degrees to an edge?
3 Name three different production aids when working with paper and board.
4 What are crop marks used for in the production of paper and board products?
5 What is a densitometer?
6 What are tolerances and why are they important?

Timber-based materials

The use of production aids

The first stage in the manufacturing process using timber-based materials is usually marking out. This is when the shape of the part you are making is applied to the prepared material. This is a very important process because any inaccuracies at this stage will result in a part that is the wrong size or shape. This should be avoided, as time and materials will be wasted and there will be a financial and environmental cost.

The timber should have a smooth even surface; this is called a **face side**, and a smooth even edge; this is called a **face edge**. Having these two reference surfaces will allow accurate marking out to take place.

Measuring tools

A 300-millimetre **rule** is the most popular measuring device but they are also available in 150-millimetre (pocket size) and in 600-millimetre and 1,000-millimetre sizes for measuring longer lengths. A steel **tape** measure consists of a coiled steel tape, enclosed in a casing, which is spring loaded for quick retraction. It is used where very long lengths need to be measured and is available in sizes from 5 metres to 20 metres. When using a rule or a tape measure you should always make sure that your eye is directly over the measurement. If your head is slightly to one side then you will get an inaccurate reading. This is called a parallax error. Another frequent error is to forget to allow for the body of the rule when measuring between uprights such as a window.

Figure 2.8.14 **A steel tape measure**

Marking-out tools

A **pencil** is the most obvious and easy to use marking-out tool. A softer pencil lead works best on timber-based materials. A **marking knife** can be used instead of a pencil on wood. It has the advantage of producing a thin cut line that reduces tearing and acts as a guide when sawing or chiselling wood.

A **try square** is used to mark a line 90 degrees to a face edge and can also be used to check an angle is at 90 degrees.

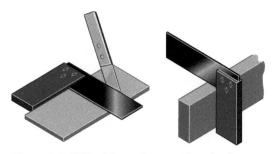

Figure 2.8.15 **Marking out on wood and using a try square to check the face and edge are at 90 degrees**

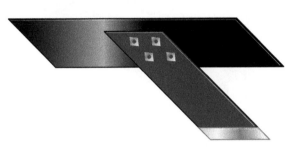

Figure 2.8.16 **A mitre square**

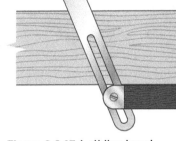

Figure 2.8.17 **A sliding bevel**

A **mitre square** looks similar to a try square but is set to an angle of 45 degrees. It can be used to measure an angle of 45 degrees and can also check an angle of 45 degrees. This angle is particularly important when producing a mitre joint in wood. (Check out the skirting boards around the bottom of walls in your house or the corners of a picture frame and you'll see a mitre joint.) A **sliding bevel** performs the same function as a mitre square but can be set at any angle. This is particularly useful when you have a series of more complex angles to cut, such as cutting the rails to fit into a wooden staircase.

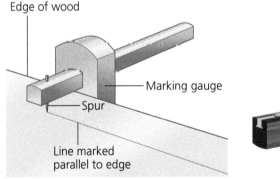

Edge of wood

Marking gauge

Spur

Line marked parallel to edge

Figure 2.8.18 **A marking gauge**

Figure 2.8.19 **A mortise gauge**

A **marking gauge** is used to mark a line parallel to an edge. It is particularly useful to a joiner when fitting a hinge. Initially it can be difficult to use but with practise it can save time when multiple lines need to be marked at exactly the same distance. A **mortise gauge** belongs to the same family of tools as the marking gauge. The mortise gauge has two spurs, and therefore will produce two parallel lines to an edge. It is very useful when marking out a mortise and tenon joint. It will ensure that both the mortise hole and the tenon are marked out to the same size, therefore increasing the accuracy of the joint.

Compasses are used to create a circle or an arc on wood.

A **template** consists of a profile shape of a part that is to be manufactured. They are often used for irregular shapes where it would be difficult to mark out the profile directly onto the material. Simple templates can be made from paper and could be a printed out as a CAD drawing. They are usually stuck down onto the material and then the shape is cut out following the profile. Paper templates are generally only used as part of a one-off production.

When a batch of identical profile shapes are needed, the template may be made from a durable material. Drawing around a cardboard template enables a number of shapes to be produced. If a larger volume of identical shapes are required then the template may be made from MDF, or even aluminium.

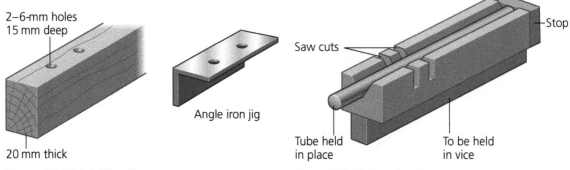

Figure 2.8.20 **A drilling jig**

Figure 2.8.21 **A sawing jig**

A **jig** is a device that is specially made to perform a specific part of the manufacturing process. Jigs are extremely useful when the process has to be carried out multiple times. They can be used when cutting, drilling, sawing, and gluing. They have a number of very important advantages.

- They speed up the manufacturing process.
- They reduce the risk of human error.
- The reduce the unit cost of a part.
- They make the process safer to carry out.
- They increase the accuracy of the process.
- They increase the consistency of the process.
- They reduce wastage.

It should be noted that there are disadvantages of using jigs:

- They are only cost-effective when large numbers of similar parts are required.
- They increase the initial cost of the part.
- They require a high level of skill to produce.

Tools, equipment and processes

Sawing

Once you have marked out your shape or set up your template or jig, you are ready to cut it out. On wood the most popular method of doing this is by sawing, and you will work safely, efficiently and accurately if you choose the correct saw for the process that you wish to carry out.

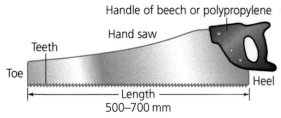

Figure 2.8.22 **A handsaw**

A **hand saw** is used for cutting thick pieces or large sheets of wood. A joiner will use a hand saw for cutting sheets of plywood. The blade is made from hardened and tempered steel that make it strong and durable, so the teeth remain sharp. Modern hand saws frequently have 'hardpoint' teeth, which you can easily identify as they are black. These teeth have undergone a special hardening process. Although hardened teeth stay sharper for longer, when

STRETCH AND CHALLENGE

Investigate a range of timber based products that have an element of curved construction. Try to correctly identify the forming process that has been used and explain why it was most the appropriate method to achieve the desired shape.

they eventually do become blunt, the saw must be thrown away and replaced. Some manufacturers harden the entire blade.

This type of saw is particularly useful for sawing manufactured boards as the extra hardness of the 'hardpoint' teeth prevents them losing their sharpness due to the adhesives found in the manufactured boards. The handle is bolted to the blade and can be made from a close-grained hardwood such as beech or a tough polymer such a polypropylene (PP).

A **tenon saw** is the most frequently used saw in the workshop. It is used for cutting accurate straight lines in wood and polymers. A cabinet maker will use a tenon saw for cutting joints in wood. The blade is made from hardened and tempered steel and it has a stiff piece of metal, called a 'back', running along the length of the blade to improve the accuracy of the saw cut. The 'back' can be made from steel or brass. Modern tenon saws can also be bought with 'hardpoint' teeth. The handle can be made from a close-grained hardwood such as beech, or a tough polymer such a polypropylene (PP), and is bolted to the blade.

A **coping saw** is used for cutting curves in wood or polymers. The hardened and tempered steel blade is held and tensioned in a sprung steel frame. The blade in the coping saw should face backwards as this keeps it in tension when sawing. A cabinetmaker may use this saw when sawing out the waste in a dovetail joint. The blades are thin and easily broken, but the design of the saw allows the blades to be quickly changed. The handle can be made from a close-grained hardwood such as beech or a tough polymer such a polypropylene (PP).

A **scroll saw** is a mechanised version of the coping saw. It allows you to concentrate on following the shape of the part you are cutting while the reciprocating blade does all the hard work. This speeds up the process and usually produces a more accurate cut than using the coping saw.

A **jigsaw** is a very versatile saw as it can be fitted with a wide variety of blades to allow it to saw different types and thicknesses of natural timber and manufactured boards. It can be used 'freehand' to cut around curves and irregular shapes or used with a 'fence' to produce straight cuts.

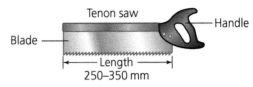

Figure 2.8.23 **A tenon**

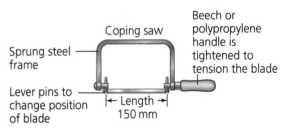

Figure 2.8.24 **A coping saw**

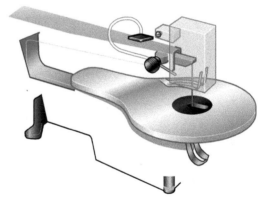

Figure 2.8.25 **A scroll saw**

Figure 2.8.26 **A jigsaw**

Shaping

Once you have sawn your wood then you will want to shape it accurately. Again, knowing which tool to use and how to accurately use it is essential.

There are three main types of **chisel** that are used to shape wood by slicing off shavings.

A **firmer chisel** is a general-purpose chisel that can accept light blows with a mallet.

A **bevel-edged chisel** has edges that are bevelled or angled. This allows the chisel to clean out acute (less than 90 degree) angled corners. This chisel would be used by a cabinet maker when producing a set of hand-cut dovetail joints.

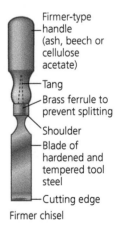

Figure 2.8.27 **A firmer chisel**

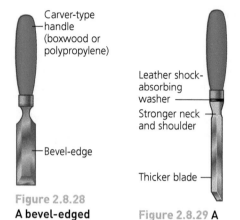

Figure 2.8.28
A bevel-edged chisel

Figure 2.8.29 **A mortise chisel**

A **mortise chisel** has a much thicker blade and a leather washer inserted between the blade and the handle. These two features allow it to be struck quite hard with a mallet. A typical use would be to cut a mortise hole in a handmade table leg.

Horizontal and vertical paring

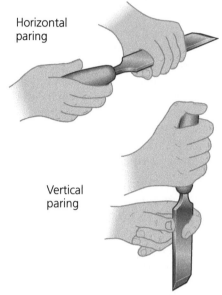

Figure 2.8.30 **Horizontal and vertical paring**

A chisel blade is very sharp and therefore you must take great care when chiselling. Your work should be firmly held in a vice or G-clamped to the workbench, with your hands always behind the cutting edge. Horizontal paring will produce a flat horizontal surface while vertical paring will produce a flat vertical surface.

Planing

A plane is the perfect tool for flattening and smoothing the surface or the edges of a piece of wood. A plane works in a similar way to a chisel in that it slices away thin shavings of wood. All planing should be done in the same direction as the grain or the surface will tear. End-grain can be planed, but must have a waste piece clamped to the end, or it should be planed from the ends to the middle; this will prevent the ends from splitting.

There are two main types of plane, as well as a wide variety of special planes.

A **jack plane** is a general-purpose plane used to flatten and smooth the surface and the edges of wood.

A **smoothing plane** is used on small wooden parts or for final cleaning of wooden surfaces.

Special planes, as the name suggests, are used for special operations.

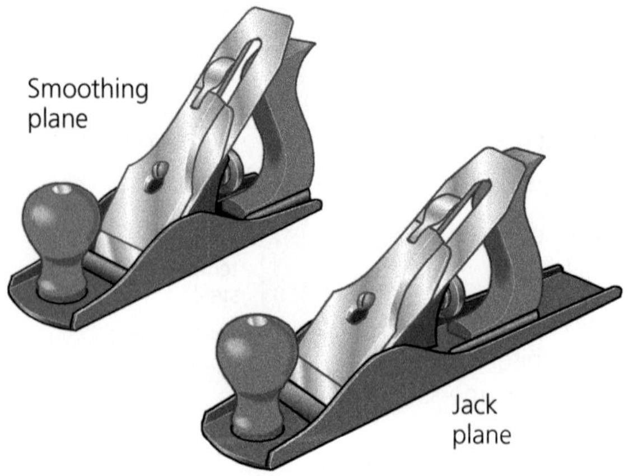

Sanding

A **disc sander** is a very useful, time-saving machine that every woodworker should be familiar with. It will quickly and accurately

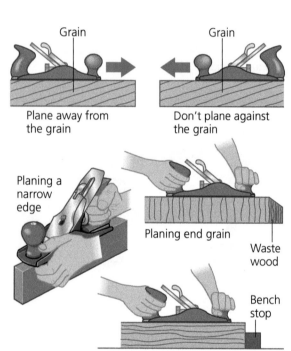

Figure 2.8.31 **Planing**

Smoothing plane

Jack plane

Figure 2.8.32 **A jack plane and a smoothing plane**

clean, shape and true the end-grain of a piece of wood. This is often a difficult, tedious and time-consuming process for any woodworking project. The disc sander does have a number of health and safety issues that must be taken into consideration. The sander will remove skin just as easily as it will remove wood, and therefore you must not get your fingers near the disc. The disc spins at a high speed and so all loose clothing and hair must be tided away. The disc sander produces high volumes of dust, and it is essential that it is fitted with dust extraction equipment and that the user is wearing safety glasses or goggles.

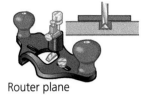

Router plane

Figure 2.8.33 A router plane

A **belt sander and linisher** will clean and smooth the surface of your wood. The linisher is mounted on a bench and can be used on relatively small pieces of work. The belt sander is a portable version that can be used on much larger surfaces such as tabletops.

Drilling

Drill bits will produce a round hole in your wood. It is important to be able to select the most suitable type of drill bit and to be able to use it safely, accurately and efficiently.

Figure 2.8.34 A disc sander

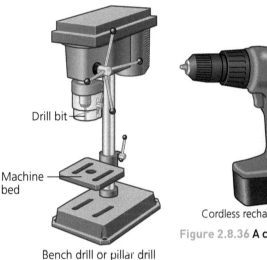

Drill bit

Machine bed

Bench drill or pillar drill

Figure 2.8.35 A bench drill

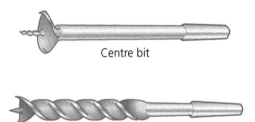

Cordless rechargeable drill

Figure 2.8.36 A cordless drill

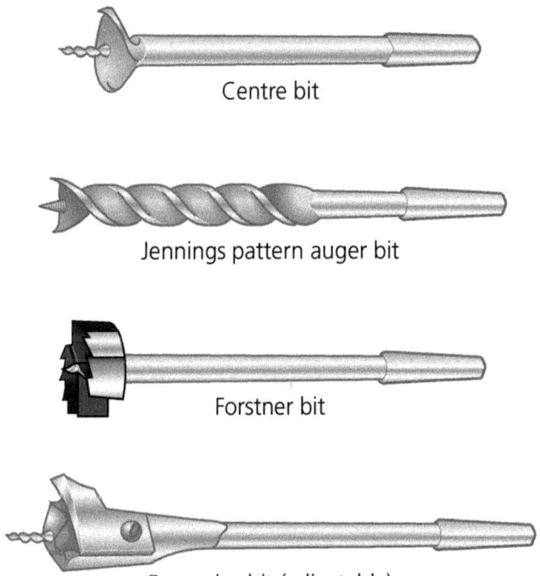

Centre bit

Jennings pattern auger bit

Forstner bit

A **bench or pillar drill** is an essential piece of workshop machinery. Its speed can be changed to suit the size of drill bit and the type of material being drilled. In general, a large drill bit and hard materials require a slow speed while small drill bits and soft materials need a fast drill speed. The bench or pillar drill is normally fitted with a depth-stop so that holes can be drilled to a predetermined depth, and it has numerous safety devices including a chuck guard and an emergency stop button.

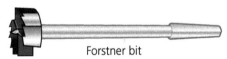

Expansive bit (adjustable)

Figure 2.8.37 A selection of drill bits

Cordless drills are now very popular among most tradespeople as they do not need to be connected to a power source and can therefore be used in remote locations. They cannot only be used to drill holes, but can be also used as a power screwdriver. You can even set the torque (turning force) of the drill so that small screws don't receive too much power.

Twist drills are the most popular type of **drill bit**. They are made from hardened and tempered high-carbon steel.

Countersink bits are used to open the top of a predrilled hole to accept a countersunk screw. This makes the screw level or slightly recessed below the surface of the material.

Flat bits are used for drilling holes in wood. They are available in larger diameters and, as the name suggests, leave a flat bottom.

Forstner bits are similar to a flat bit but have greater accuracy. A kitchen fitter fitting a kitchen cupboard door hinge would use a Forstner bit.

Hole saws are used for drilling larger holes. The cutting edge is actually a saw that is formed into a circle. A plumber wanting to fit a waste pipe through the back of a cupboard would use a hole saw.

ACTIVITY

You are part of a young enterprise team who have decided to make a batch of tealight holders to sell at the school summer fair.

You are to:
- design an interesting shape for your tealight holder
- produce a template
- cut out the shape
- drill the holes
- sand and apply a finish.

How materials are cut, shaped and formed to a tolerance
Dimensional tolerance

It is very difficult, and it would be very expensive and time-consuming, to make something exactly the correct size every time. Most parts that are manufactured do not need to be made with this amount of precision. Therefore, most parts are manufactured to what is known as a tolerance. A tolerance is given as an acceptable difference in size between an upper limit and a lower limit.

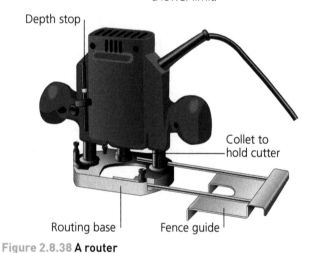

Depth stop

Collet to hold cutter

Routing base   Fence guide

Figure 2.8.38 **A router**

For example, a wooden table top that measures 1,000 millimetres in length may well have a **dimensional tolerance** of plus or minus 2 millimetres. That means to say that the table top could measure between 1,002 millimetres and 998 millimetres in length and still pass quality control.

Moisture content

Another form of tolerance that is applied to timber-based products is its **moisture content**. Wood is hygroscopic; this means that it acts like a sponge and can absorb moisture, making it swell, or dry out, making it shrink, depending on the moisture content of the environment it is placed in.

For example, a kitchen cupboard door is produced with a moisture content of between 10 to 15 per cent. Your kitchen is likely to have a moisture content of less than 10 per cent. This means that the cupboard doors may dry out a little bit and shrink, but they should not absorb any more.

Commercial processes
Routing

A router can be used as a handheld power tool or mounted in a router table. It consists of a powerful motor that can accept a wide variety of router bits. You choose the required profile of router bit to match the desired shape that you wish to achieve. It can be used to place grooves and rebates into wood and can produce a range of decorative edges. A kitchen fitter will use a router fitted with a straight bit to cut out the hole in a kitchen worktop ready to receive the kitchen sink. When used with a specialised jig they can also be used to produce a set of machine-cut dovetail joints or machine-cut mortise and tenon joints. CNC routers are widely used commercially and perform similar tasks to handled routers. The tool paths are controlled by computer numerical control, allowing for consistency, speed and a high quality outcome.

Forming

Forming involves changing a two-dimensional, flat piece of material into a three-dimensional, shape without the need for cutting, sawing or machining. Typically, this will involve some sort of bending or moulding process.

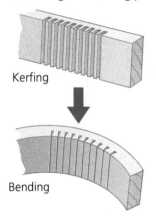

Figure 2.8.40 **An example of kerfing**

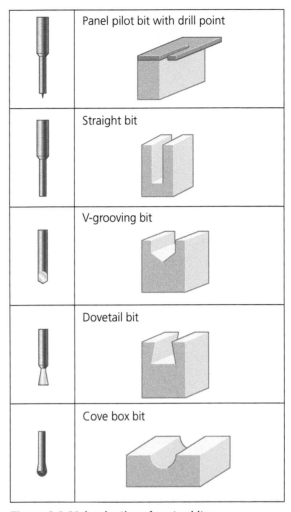

Figure 2.8.39 **A selection of router bits**

Kerfing is a process whereby a simple bend in wood can be achieved by creating a number of even saw cuts (kerfs) on to one side of a piece of wood. This will allow the wood to be bent into a simple curve. If glue is placed into the kerfs before bending, then when it has dried it will hold its new curved shape.

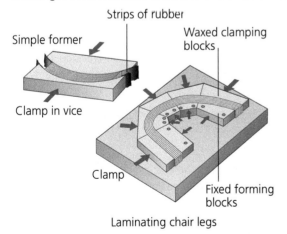

Figure 2.8.41 **The process of laminating**

Figure 2.8.42 **A laminated chair**

Complex bends in wood can be achieved in wood by a process known as **laminating**. It is important to choose a suitable type of wood to laminate; it needs to be a close-grained hardwood such as beech. A furniture maker will use lamination to produce unusual and interesting curved pieces of furniture.

Timber-based materials

First, a former needs to be produced. This is additional work that will take time, money, resources and skill. Because of these factors, lamination is only usually undertaken when batches of products are required. The former will be made from a durable material such as multi-ply and clamped together using G-clamps or wedges.

The wood is firstly sawn into thin strips called veneers. As the veneers are thin they will easily bend round the former. Each veneer should be slightly longer and wider than the final product to allow for finishing. The veneers will be glued and clamped into the former and left until dry. Once dry the lamination is removed from the former and trimmed and finished.

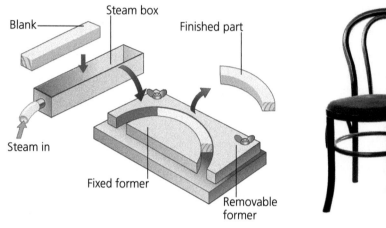

Figure 2.8.43 The process of steam bending

Figure 2.8.44 A steam-bent chair

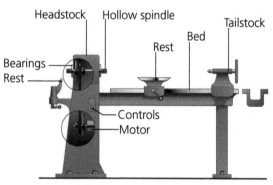

Figure 2.8.45 A woodturning lathe

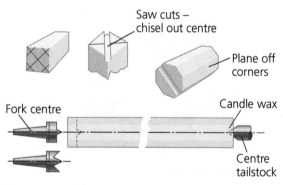

Figure 2.8.46 Turning between centres

Steam bending allows a solid piece of wood to be bent. Again, selection of the type of wood is important; it needs to be a strong wood such as oak, ash or beech.

As with lamination, a former needs to be produced, making this method only really suitable for batch production.

The wood is firstly steamed in a steam box. Typically this will take a few hours to make the wood pliable enough for it to be bent into the former. It is then clamped into position and left to fully dry. Once dry it can be removed from the former and trimmed and finished.

Woodturning is done on a machine called a woodturning lathe. It is used to produce circular forms such as cylinders, cones and spheres in wood. Typical products made by this method include bowls, handles, lamp bases, egg cups, table legs, staircase spindles and salt and pepper mills.

There are two main methods of holding a blank of wood on a woodturning lathe, these are known as 'turning between centres' and 'bowl turning'.

When turning between centres, the wood is first prepared by finding the centre of each end of the section of wood. A light centre punch mark is made in one end and a V-cut is made in the other end. It is useful to plane off the edges of the wood before mounting it onto the woodturning lathe. The tool-rest is then brought up to the wood and the lathe is turned by hand to check that everything is safe before the machine is switched on. A variety of woodturning tools can then be used to produce the desired shape. The finished shape can be sanded and finished on the lathe, which speeds up this often slow process.

When bowl turning, the wood is first prepared by finding the centre and attaching a metal faceplate with a number of screws. Again, it is useful to take off the corners before screwing the faceplate onto the spindle. The tool-rest is then brought up to the wood and the lathe is turned by hand to check that everything is safe before the machine is switched on. When bowl turning, both the outside and the inside of the shape can be formed using a variety of different woodturning tools, such as a gouge, skew chisel or scraper. Again, the finished shape can be sanded and finished on the lathe, speeding up the process.

It must be noted that operating a woodturning lathe is a high-risk activity and all relevant safety precautions should be observed; in particular the need to wear safety goggles or glasses and the need to use dust extraction equipment or a dust mask.

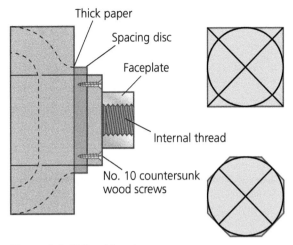

Thick paper
Spacing disc
Faceplate
Internal thread
No. 10 countersunk wood screws

Figure 2.8.47 **Bowl turning**

The application and use of quality control

All manufacturers need to ensure that the products that they produce function as intended, are reliable and are of high quality. **Quality control** is the process of testing and checking a product to ensure that it is fit for purpose. These checks take place on the final product and at various stages throughout the manufacturing process to ensure that a reliable and accurate product is being produced. This is particularly important when manufacturing in volume, where consistency of each product is vital.

Quality control checks should begin when working with timber at the stage when the initial material is being selected. It is important to check that the timber is not bowed, twisted or warped. Starting the manufacture of a product with substandard material will mean that you are likely to have problems when machining or shaping.

When drilling holes in timber you may use many different types of drill bits, including hole saws and Forstner bits. A Forstner bit is used to drill a blind or flat-bottomed hole and can be used when fitting concealed hinges on flat-pack furniture. The depth of the hole here is critical – too shallow and the hinge will not fit properly, too deep and you run the risk of breaking through the front of the panel. A quality control measure in this application could be the use of the depth-stop on the pillar drill. The depth-stop is a mechanical system that can be adjusted and set to the desired depth for the hole. When the depth is achieved the depth-stop prevents the drill bit from going any deeper. The depth-stop is a very useful function, especially if you are drilling a large volume of the same depth hole.

The use of CNC machinery is usually an excellent way to improve and control the quality of a manufacturing process. Quality control measures when using a CNC machine on timber can include ensuring that the tool is correctly calibrated and that the machines datums and reference points are set. This will ensure that when a groove, recess or cut is being made, the depth and width will be accurate and within tolerance.

ACTIVITY

Search for and watch 'How it's made: Laminated wooden beams' on YouTube.

This short programme covers lots of industrial processes.

KEY POINTS
- Accurate marking out is essential if you are to achieve a quality product.
- Jigs and templates will speed up the making process and help achieve consistency.
- There are many saws that will cut wood and it is important to be able to match the saw to the process.
- Chisels and planes help to shape and smooth wood.
- Disc sanders, belt sanders and linishers will mechanically smooth wooden surfaces.
- Wood can be bent by kerfing, steam bending and laminating.
- Wood can by formed into cylinders, spheres and cones by woodturning.

KEY WORDS

Marking out the process of applying a drawing on to a material.

Face side the surface of a piece of wood that is known to be straight and true.

Face edge the surface of a piece of wood that is known to be straight and true.

Template a 2D shape that aids cutting out a shape.

Jig a 3D device that aids a production process.

Dimensional tolerance the difference between the maximum and minimum acceptable size

Moisture content the amount of moisture in a timber.

Laminating a method of bending wood by slicing into thin veneers and gluing back together.

Steam bending a method of bending wood by steaming bending and cooling.

Turning a method of making a wood blank round.

Quality control checks put in place to see if the product meets given standards.

Check your knowledge and understanding

1 Explain the following terms: datum line, face edge, face side
2 Identify and describe the different types of saw used to cut wood.
3 List the advantages and disadvantages of using a cordless drill.
4 List the stock forms and sizes of different types of wood and manufactured boards.
5 Explain how moisture content is calculated.
6 Describe the process of woodturning.
7 Explain how to laminate wood.

Metal-based materials

The use of production aids

Metals are much harder and tougher materials than wood, therefore, it is particularly important to ensure that marking out is carried out with great accuracy, as mistakes are difficult to rectify. The manufacturing of metal-based products is generally slower than manufacturing in wood and so the use of production aids is of greater relevance.

Scriber

A scriber performs the same function as a pencil, but is used on metals. The hard point will score a fine line into metal. It is sometimes useful to apply marking-out fluid to bright shiny metal before using a scriber.

Marking-out fluid

Marking-out fluid is a quick-drying blue liquid that is applied to metals before marking out begins. It produces clean and crisp lines.

Engineers square

The engineers square is used to mark a line 90 degrees to an edge and can also be used to check an angle is at 90 degrees.

Combination square

A combination square can be used for a number of measuring and marking-out processes. It can be used as a rule, it can measure depths and it can be used as a square. It also has the advantage of being able to measure any angle.

Odd-leg calipers

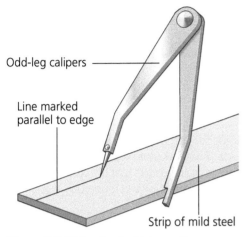

Odd-leg calipers ———

Line marked
parallel to edge

Strip of mild steel

Figure 2.8.49 **A pair of odd-leg calipers**

Odd-leg calipers are used to mark a line parallel to an edge.

Figure 2.8.48 **A combination square**

Outside calipers

Outside calipers are used to measure or check the outside diameter of round bar or rod. They are typically used when turning steel bar on a metal working lathe.

Inside calipers

Inside calipers are used to measure or check the inside diameter of a hole.

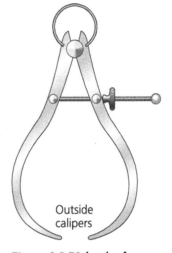

Outside calipers

Figure 2.8.50 **A pair of outside calipers**

Inside calipers

Figure 2.8.51 **A pair of inside calipers**

Dividers

A pair of dividers are used to mark out a circle or an arc on metal. It is useful to first use a dot punch to locate the centre of the circle or arc when working on metal; this will increase the accuracy and will prevent the dividers from skidding and scratching the surface.

Centre punch

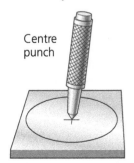

Centre punch

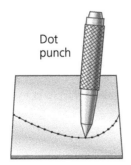

Dot punch

Figure 2.8.53 **A centre punch**

Figure 2.8.52 **A pair of dividers**

The centre punch is a very useful tool in the workshop. When struck with a hammer it will produce a small indentation in metal. It is essential to produce this indentation before attempting to drill a hole in a metal component, as it increases the accuracy of the position of the hole and helps to prevent the drill from skidding off line.

Surface plate, scribing block and vee blocks

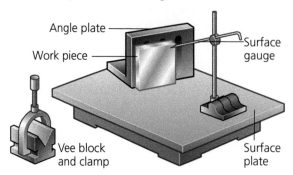

Figure 2.8.54 **A surface plate with scribing block and vee blocks**

This combination of marking-out equipment increases the accuracy of marking out. The surface plate is a very smooth, flat and accurate surface to work from. The scribing block firmly holds the scriber at a pre-determined height, the angle plate ensures that the work piece is held at 90 degrees to the surface plate and vee blocks will steadily hold round bar.

Micrometer and digital vernier caliper

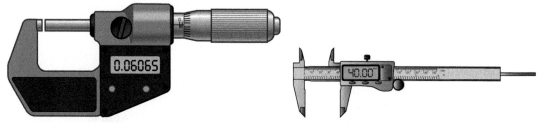

Figure 2.8.55 **A digital micrometer**

Figure 2.8.56 **A digital vernier caliper**

A micrometer is a very accurate tool used for measuring small distances from 0 to 50 millimetres. Larger micrometers are available but the 0 to 50 millimetres is the most common. It utilises a very fine screwthread with a pitch of 0.5 millimetre. Take hold of a 300 millimetre ruler and place your thumbnails either side of a millimetre, now try and imagine dividing that space into one hundred. That is how accurate a digital micrometer can measure!

A digital vernier calliper is another very accurate measuring tool that can measure to an accuracy of one hundredth of a millimetre. It can measure outside dimensions, inside dimensions and depths. Digital vernier callipers normally measure from 0 to 150 millimetres.

Tools, equipment and processes

When you have marked out the metal component that you are wanting to make, the next stage in the manufacturing process is normally cutting out the shape. The type of tool you use will be determined by the material you are using, the profile of the shape and the number of components required. It is important to choose the correct tool and to be able to use it safely, accurately and efficiently.

Sawing

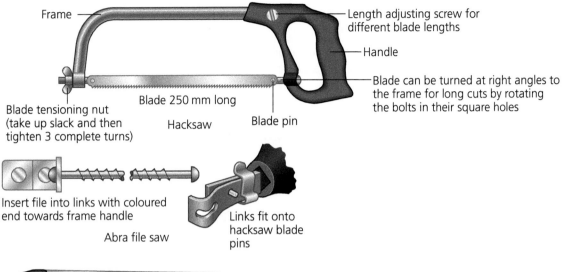

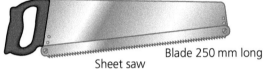

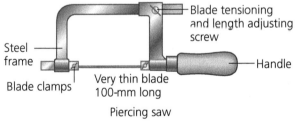

Figure 2.8.57 **A hacksaw**

The **hacksaw** is the most popular saw for sawing metal and polymers. The hardened and tempered high-carbon steel blade is held in a tubular steel frame and tensioned with a wing nut. The handle is usually made from die cast aluminium with a powder-coated finish.

As the name suggests, the **junior hacksaw** is a smaller version of the hacksaw. Again, it is used for cutting metal or polymers. The hardened and tempered high-carbon steel blade is held in a sprung steel frame that can be compressed to allow a worn blade to be quickly changed.

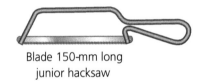

Blade 150-mm long junior hacksaw

Figure 2.8.58 **A junior hacksaw**

Figure 2.8.59 **A Piercing saw**

A **piercing saw** has a very thin, hardened and tempered steel blade. The blade is held in a steel frame and tensioned by moving the handle assembly along the frame. As the blade is very fine and delicate it is easily snapped, however small lengths of blade can be easily accommodated in between the blade clamps.

Filing

Filing is the process of final shaping and smoothing a material once it has already been cut to shape.

There are two basic techniques of filing: **cross filing** and **draw filing**.

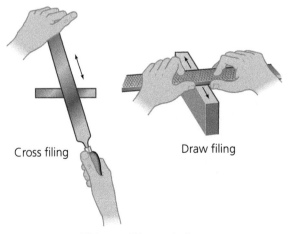

Figure 2.8.60 **Different filing techniques**

Metal-based materials

Cross filing removes the most material and involves pushing the file across the material. You should aim to use the full length of the file and slide along the edge of the material as you push forward. The file does not cut on the backward stroke so save yourself some time and energy and bring it back in thin air.

Draw filing is used to smooth the edge or surface of the material. You hold the blade of a smooth file in both hands and draw it along the edge sideways along the surface. You can also wrap a piece of abrasive paper, such as emery cloth or wet and dry paper, around the file for an even smoother finish.

It is important to keep files in good working order. Check that the handle is secure on the tang and clean it with a file card if it gets impregnated with metal.

Types of file

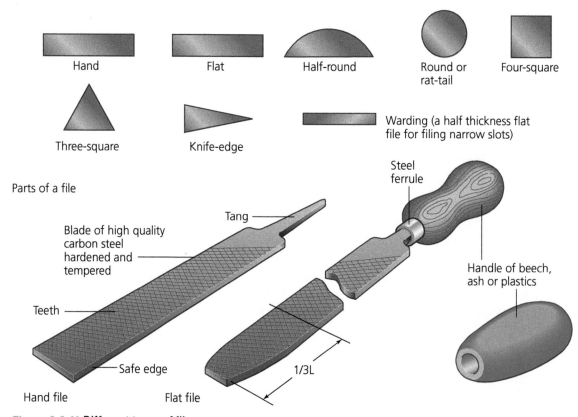

Hand

Flat

Half-round

Round or rat-tail

Four-square

Three-square

Knife-edge

Warding (a half thickness flat file for filing narrow slots)

Parts of a file

Steel ferrule

Tang

Blade of high quality carbon steel hardened and tempered

Handle of beech, ash or plastics

Teeth

Safe edge

1/3L

Hand file

Flat file

Figure 2.8.61 **Different types of file**

There are many different types of file for different purposes. It is important to be able to select the most suitable type of file and to be able to use it safely, accurately and efficiently. Files are classified according to their size, cut and shape.

Forming metal
Bending

The simplest method of forming metal is to produce a 90-degree bend. The metal can be placed into a vice

Bending against the fixed jaw of the vice

Figure 2.8.62 **A simple bend in a metalworking vice**

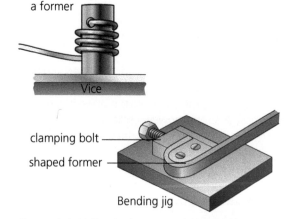

Bending rings round a former

Vice

clamping bolt

shaped former

Bending jig

Figure 2.8.63 **Producing a curve using a former**

and simply hit with a hammer. Where longer bends are required, you can use a set of folding bars, which act as an extension to the vice jaws. For soft metals, such aluminium and copper, a rubber, rawhide or nylon mallet can be used to prevent the surface being marked. Metal can also be bent using commercially produced metal-folding machines.

Joining metals

Brazing

In Topic 2.5 you learned that brazing is a method of joining steel parts together. This time heat is applied with a brazing torch in a brazing hearth. The flux (borax) is applied to the cleaned joint before heating and once the steel is red hot the brazing rod (brass) is introduced. It will then melt and flow into the joint.

Welding

In Topic 2.5 you learned that welding is a very strong method of permanently joining metals together. It differs from soldering in that it use the same metal as to produce the joint. There are a number of methods of welding that either use gas, electricity or a mixture of gas and electricity to generate the heat. Most welded joints require the metal to be prepared by forming a V-gap between the two pieces of metal that are to be joined.

How materials are cut, shaped and formed to a tolerance

As metals are much harder, tougher and more stable than wood-based products, sizes can be defined with a finer tolerance. It is not unusual for a metal component to be manufactured to within 0.01 millimetres. That is to within one hundredth of a millimetre!

Tolerance

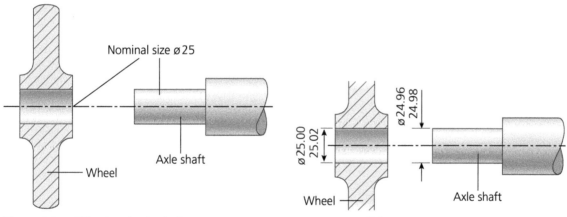

Figure 2.8.64 **Wheel and axle shaft**

Figure 2.8.65 **Tolerance**

The axle shaft shown in Figure 2.8.64 needs to fit into the hole in the wheel. Therefore the hole needs to be at least 25 millimetres but can be no bigger than 25.02 millimetres or it would be too big to function.

The hole must fit around the axle shaft and therefore it must be no bigger than 24.98 millimetres, so it will fit around the largest axle shaft and no smaller than 24.96 millimetres or it would be too small to function.

Measuring tolerance

Measuring to such fine **tolerances** can be quite difficult. Therefore, to help maintain accuracy and consistency, special measurement checking devices are used. For more details, see the section on the application and use of quality control below.

Commercial processes

Machining

The **centre lathe** is a very useful workshop machine. It is used to make circular components from metal or polymer bar. Various stock forms can be held in the chuck while it spins around. A cutting tool can then be used to perform a variety of **turning** operations.

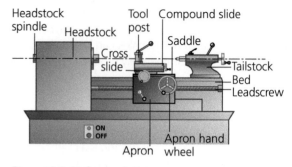

Figure 2.8.66 **Centre lathe**

| Right-hand knife | Left-hand knife | Round nose | Parting |

Figure 2.8.67 **A selection of lathe tools**

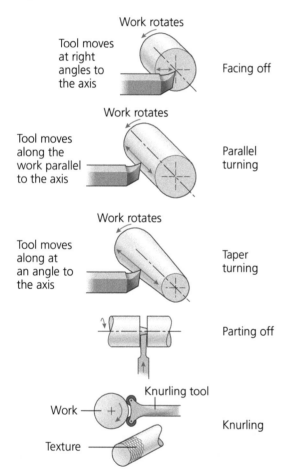

Figure 2.8.68 **Lathe operations**

Knife tools can be used to face off the end of a bar. Facing off a bar involves moving the tool across the end of the bar and produces a flat, smooth 90-degree surface.

Round nose tools can be used to parallel turn along the length of the bar. Parallel turning involves sliding the tool along the length of the bar and produces a smooth, parallel surface. Round nose tools can also be used to taper turn. Taper turning involves setting the compound slide to a predetermined angle and then sliding the tool along an angled path that produces a taper.

Parting tools are used to produce an undercut or to cut off the end of a machined component.

Milling machines are used to carry out a range of machining operations in metal or polymers. They can be used to cut slots and grooves, machine edges and to smooth large surface areas.

Small work pieces can be held in a machine vice that is bolted down onto the milling machine table. Larger pieces of work are bolted directly to the table.

These machines are often used with automatic feeds. This ensures that the correct speed of cut is maintained, giving an accurate surface finish. Due to the large amount of work that the machine can do, the process will generate lots of heat. The heat needs to be controlled by applying coolant or the tool will become damaged and the surface finish will be poor. Many milling machines are now computer numerically controlled; these are usually vertical machines with automatic tools chargers.

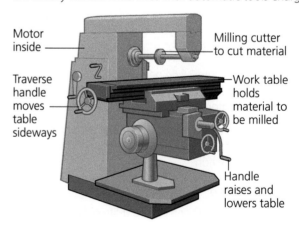

Figure 2.8.69 **Horizontal milling machine**

Horizontal milling

Vertical milling

Horizontal and vertical milling of flat surfaces

Horizontal and vertical milling of vertical surfaces – the horizontal machine is using a side and face cutter which cuts on its side and on its diameter

Horizontal and vertical milling of slots

Figure 2.8.71 **Milling operations**

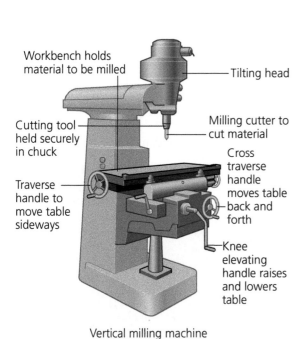

Workbench holds material to be milled

Tilting head

Cutting tool held securely in chuck

Milling cutter to cut material

Cross traverse handle moves table back and forth

Traverse handle to move table sideways

Knee elevating handle raises and lowers table

Vertical milling machine

Figure 2.8.70 **Vertical milling machine**

Casting

When **casting** metal, the metal is first heated until it is in a molten form and then it is poured into a pre-prepared mould. The advantages of casting are that it can be used to produce complex shapes that would be very difficult to make by any other method. Take a look at the picture of 'Eros', a famous cast metal sculpture found in Piccadilly Circus, London. How difficult would it be to cut that out of a solid piece of metal!?

Pewter casting is a relatively quick and simple method of metal casting. Pewter is made mainly from tin (about 90 per cent) and a small amount of copper. The copper increases the hardness of the tin, which is quite soft. Pewter melts at a low temperature of around 200 °C, making it easy to use in the school workshop. It can be heated

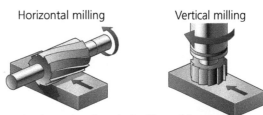

Figure 2.8.72 **'Eros', a famous cast metal sculpture, Piccadilly Circus, London**

in a furnace with a brazing torch, in an oven or kiln, or with a hot air gun. Once molten, the pewter can be poured into a mould where it will cool and set. Although casting in pewter uses a low temperature, it can still cause serious burns, and therefore a full risk assessment must be carried out before casting and the correct PPE (Personal Protective Equipment) must be worn.

Moulds for casting in pewter can be quickly and easily produced in a school workshop. Because of its low melting point, MDF and other types of wood are often used to make the mould. This allows you to hand cut shapes using a coping saw, or use a laser cutter or 3D router to cut out a CAD shape.

Metal-based materials

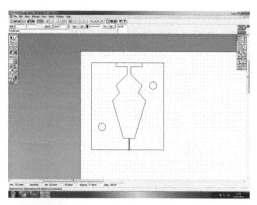

CAD image to be pewter cast

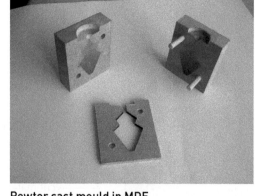

Pewter cast mould in MDF

Heating the pewter

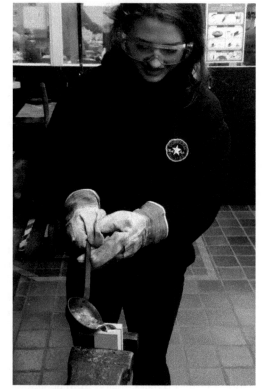

Pouring pewter

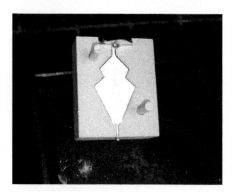

The finished pewter product

Figure 2.8.73 The pewter casting process

Sand casting is a process that can be carried out in a school workshop and is used to produce larger metal casting, usually made from aluminium.

A pattern of the casting is first made from a close-grained hardwood, such as jelutong, or a moulding foam polymer. The pattern must have sloping sides (draft) to allow it to be removed from the mould, rounded corners (fillets) and a smooth surface. A pattern can be in one simple piece for relatively simple castings, it can be split into two for more sophisticated castings, or it can have a core for more complex hollow castings.

Here are the stages of making a mould.

Stage 1

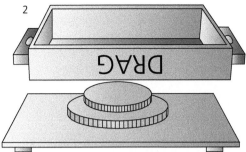

One half of a split pattern is placed onto a moulding board and the bottom half of the moulding box (the drag) is placed upside down over the pattern.

Stage 2

The pattern is cut through the centre and fitted with location dowels.

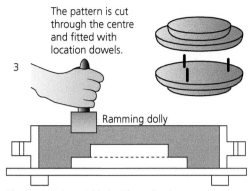

Ramming dolly

The pattern is sprinkled with a releasing agent (parting powder). Sand (petrobond) is then added to the moulding box and rammed around the pattern.

Stage 3

The top is levelled off (strickled) and the whole assembly turned over.

Stage 4

Spure pins

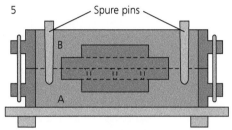

The top half of the moulding box (the cope) is then placed on top of the drag. The top half of the pattern is then fitted to the bottom half. Spure pins are fitted, sprinkled with parting powder, and once again Petrobond sand is rammed around the pattern.

Stage 5

Pour metal via riser

Riser

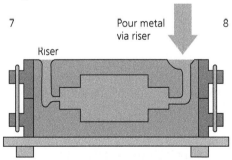

The pins are then removed, the top half of the mould taken off and the patterns are removed. Channels are then cut to connect the runner and riser to the mould cavity, and then the mould is reassembled and is ready for casting.

Figure 2.8.74 **The sand casting process**

For the casting process, the aluminium needs to be heated to a temperature of about 700 °C in a crucible that is housed inside a purpose-built furnace. Once molten, the aluminium can be poured carefully into the mould and left to cool and set. Aluminium casting uses a much higher temperature than pewter casting and there is a real danger of life-changing injuries if the correct health and safety precautions are not taken. Aluminium casting must only be undertaken within a special area of the workshop, under close supervision and using the correct PPE.

The application and use of quality control

The principle of quality control is not material-specific, as it is equally important that metal manufacturing processes and products are tested and inspected in the same way as timbers and polymers.

A quality control process often used in the manufacture of metal products is the use of a **go–no-go gauge**. This is a method of testing and inspecting the accuracy of part of the manufacturing process. A go–no-go gauge is used specifically to test that a part or component is being manufactured within an acceptable tolerance.

Go–no-go gauges

The go–no-go gauge is pre-set to measure a metal component to ensure that it is within a specific tolerance. There are three basic types: a plug gauge used to check holes, a gap gauge used to check external dimensions and an adjustable gap gauge, which is a versatile version of the gap gauge.

Figure 2.8.75 **Go–no-go gauges**

In the case of a plug gauge, the 'go' side of the gauge used to check the axle hole would be set at 25 millimetres, as this is the smallest size that the hole could be. If it was too small then the 'go' side would not fit into the hole and the part would be rejected. The 'no-go' side of the gauge would be set at 25.03 millimetres, as anything at this size or greater would be too large and the component would be rejected.

Many machines that are used to shape and form metals have guides and stops that can be used to increase the accuracy and repeatability of cutting or drilling to a specific dimension.

ACTIVITY

Work out the sizes of a gap gauge used to check the dimension accuracy of the axle shaft.

The depth-stop

The **depth-stop** is a relatively simple method of ensuring holes are drilled to the correct depth. Depth-stops are an integral part of most pillar or bench drills. They can be set to a pre-set limit to ensure that the drill only drills down to a certain distance. You can produce your own drill depth indicator by simply wrapping a piece of tape around the drill bit at the precise distance you want to drill to. Then drill into the material and stop when you reach the tape.

Metal-cutting machines, such as metal-specific bandsaws and chop saws, are fitted with adjustable stops that can be set at a certain distance from the blade. When the piece to be cut is placed up against the stop and clamped, the saw can be used to cut that specific distance. Additional pieces of the same dimension can now be cut without the need for each piece to be marked out, removing the element of human error and providing a reliable quality control measure.

Figure 2.8.76 **Metal-cutting chop saw with adjustable stop**

Check your knowledge and understanding

1. Identify and describe how to use a selection of marking-out equipment.
2. List which saws are used to cut metal.
3. Produce a diagram showing the different filing techniques.
4. Explain what is meant by the word 'tolerance'.
5. List the gauges that are used to check tolerances.
6. Describe an operation carried out on a metal-cutting lathe.
7. Describe an operation carried out on a milling machine.
8. Produce a step-by-step guide to pewter casting.

Polymers

The use of production aids

Polymers tend to share many marking-out tools and equipment with other materials, in particular the ones associated with metal-based manufacture.

Chinagraph pencil and spirit-based pen

Chinagraph pencils and spirit-based pens are used to mark out on polymers. They both have the advantage of being easily removed after the part has been made.

When using compasses, or drilling in to a polymer, it is useful to place tape on the surface to prevent the point from skidding and scratching the surface.

Tools, equipment and processes

Again, polymers tend to share many tools and equipment with those associated with metal-based manufacture.

Forming polymers

Polymers fall into one of two categories: thermoplastics and thermosetting plastics. Thermosetting plastic must be formed into shape from their stock form (powder or granules) and cannot be reformed. Thermoplastics soften when heat is applied and can easily be formed into shape by one of a number of different methods.

Figure 2.8.77 **A strip heater**

Line bending is one of the simplest methods of forming a polymer sheet; acrylic is by far the most common material formed by this method in the school workshop. The **strip heater** consists of a hot wire that is used to heat the acrylic to a temperature of around 160 °C. At this temperature the acrylic becomes pliable and can be bent into shape. The heat can be regulated to accommodate varying thicknesses of acrylic and varying radius of bends. Care should be taken not to overheat the acrylic as it will blister, and you must remember that you are using hot materials and equipment.

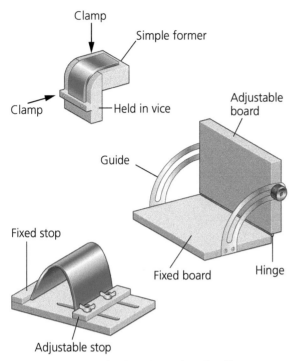

Figure 2.8.78 **A selection of line-bending jigs**

Jigs and formers are often used when bending acrylic to ensure that accurate and consistent shapes are obtained.

Vacuum forming is used to produce three-dimensional complex shapes in a thermoplastic sheet. High-impact polystyrene (HIPS) sheet is the most popular material used for vacuum forming in the school workshop. Before a vacuum-formed product can be made, a mould must be produced.

The accuracy and finish of the mould for vacuum forming is very important, as any imperfections will show up on every product it produces. There are certain key features of a mould.

- It should have sloping sides (usually around 5 degrees) to ensure the HIPS sheet can be removed.
- It should have rounded corners to prevent the HIPS sheet from thinning on the corner and possibly splitting.
- It should have vent holes drilled to allow the air to be removed from inset sections.
- It should have a smooth surface.

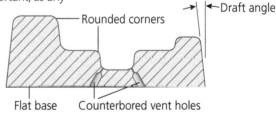

Figure 2.8.79 **A cross-section through a mould used for vacuum forming**

Figure 2.8.80 **A vacuum forming machine**

Figure 2.8.81 **An example of a vacuum-formed product**

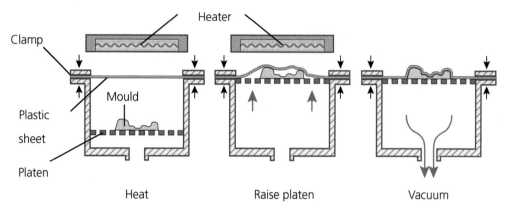

Figure 2.8.82 **The vacuum-forming process**

The vacuum-forming process:
- Once the mould is ready, it is placed on the platen (table) of the vacuum former and lowered into the machine.
- A sheet of HIPS is then clamped over the top of the machine and heat is applied.
- After a short time, the HIPS sheet will become soft. Care should be taken not to overheat the HIPS sheet, as it will not form properly and webbing may occur.
- The sheet is then raised up into the hot HIPS sheet and immediately the air is sucked out to the machine by tuning on the vacuum pump.
- Once formed, the sheet should be allowed to cool then removed from the vacuum former and trimmed.
- Deeper moulds may require the soft HIPS sheet to be blown into a dome before the mould is raised. This gives an even thickness of material around the taller mould.

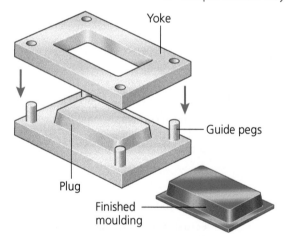

Figure 2.8.83 **A press-forming mould**

Press forming produces similar three-dimensional outcomes to vacuum forming, but is used with simpler shapes in thicker thermoplastic sheets, such as 3-millimetre acrylic.

The press-forming mould comprises of two parts: a **yoke** and a **plug**. The acrylic sheet should be heated up in an oven to make it soft and pliable. It is then placed between the yoke and the plug, and the two parts of the mould are than pressed together. The mould may feature guide pins to ensure that the two parts accurately align with each other. Once cool, the acrylic sheet can then be removed and trimmed.

How materials are cut, shaped and formed to a tolerance

Many of the methods of manufacturing to tolerance that we have already discussed in wood- and metal-based products also apply to polymers. There are some polymer-specific methods that you should know.

The acrylic generally used for laser cutting in the school workshop is normally three millimetres in thickness. This can vary slightly and will have an effect on the performance of the laser cutter. Therefore, it is useful to measure the thickness of the acrylic before it is laser cut and adjust the power or speed settings accordingly. A micrometer or vernier caliper is the best tool to use for accurately measuring the thickness of the acrylic sheet.

Commercial processes

Injection moulding

Injection moulding is used extensively in industry to produce many thousands of objects that we use in our everyday lives. The seat of the chair you are sitting on at this moment, your mobile phone case and the pen in your hand were probably all made by injection moulding.

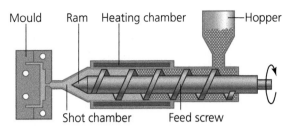

Figure 2.8.84 **An injection-moulding machine**

As with many industrial processes, injection moulding is only a real option when you need to produce products in large batches. The initial set up costs are high, you need to invest in an injection-moulding machine and you need to have sophisticated metal dies produced. However, once set up the injection-moulding machine will produce a very high volume of products, quickly, accurately and consistently. This high-volume production significantly reduces the unit cost of the product.

The injection moulding process:
- Polymer granules are fed into a hopper, which in turn feeds the granules into the heating chamber.
- An Archimedean screw then transports the polymer granules along the heating chamber where they gradually become molten.
- A hydraulic ram then forces the molten polymer through a sprue gate and into the mould.
- The mould is then rapidly cooled with water to set the polymer. The mould is then opened and the injection-moulded product is removed.

Extrusion

Extruding polymers produces a long continuous length of polymer product in a uniform cross section. The guttering around your house, pipework, curtain rails and tubing are all examples of polymer-extruded products.

Figure 2.8.85 **Extruded products**

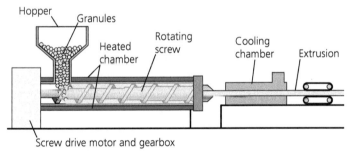

Figure 2.8.86 **The extrusion process**

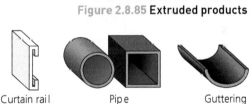

Curtain rail   Pipe   Guttering

The **extrusion process** is similar to the injection-moulding process.
- Polymer granules are fed into a hopper, which in turn feeds the granules into the heating chamber.
- An Archimedean screw then transports the polymer granules along the heating chamber where they gradually become molten.
- The molten polymer is then forced through a die. The die is a mould made from metal that forms the shape of the extruded polymer product.
- As the polymer emerges from the die it is also pulled to keep it in tension to prevent distortion.
- The extrusion is then rapidly cooled with water to set the polymer.
- The extrusion can then be cut to length or coiled.

Rapid prototyping

Rapid prototyping involves the manufacture of a three-dimensional image to produce a model of a product or component. There are three main methods of rapid prototyping: 3D printing, stereo lithography and laser sintering. In each case the process starts with a three-dimensional CAD drawing of the model. This is uploaded onto a computer and is processed into machine code, taking into account the material being used, the density of the mould and introducing any support material that may be required. If you have ever used a rapid prototyping system you may wonder how the term 'rapid' could be applied to this system as it often takes hours to produce a relatively simple model.

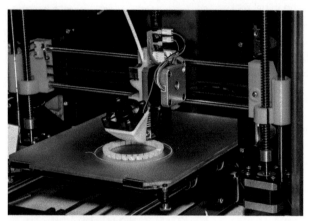

Figure 2.8.87 **3D Printer**

The **3D printer** uses a filament polymer that is fed through a heating unit that converts it into a molten state. The most common polymers used with a 3D printer are ABS and the biopolymer PLA. Once heated, the polymer is extruded and follows a path to produce a three-dimensional object by building up the shape layer by layer, each successive layer fusing to the previous one. The path is controlled by the use of three stepper motors, allowing any three-dimensional shape to be produced. The model can be altered to be solid, hollow or structured.

The **stereolithography printer** works in a similar way to the 3D printer, but here the tool is a laser that follows a similar path that works in a bath of resin. The laser cures the resin to form a solid shape as it moves along, again building up a model layer by layer, which can be solid, hollow or contain an internal structure.

The **laser sintering** method of rapid prototyping is very similar to the stereolithography method, but involves the use of fine heat fusible powder. This powder can be made from a polymer, a ceramic or a metal. The laser, once again, traces the path of the model, but here the laser turns the material into a molten state. Each successive layer fuses to the previous layer, gradually building up the three-dimensional model.

The application and use of quality control

Very few polymer-manufacturing processes are done by hand, instead making use of highly automated machines. Therefore, most of the quality control measures are applied during the manufacture of the moulds and in the setup of the moulding machines.

Three-dimensional printing is now a popular method of manufacturing prototypes or small component parts. A 3D printer simply follows a series of commands that are generated from a three-dimensional CAD drawing, and as such there is little opportunity for quality control during the manufacturing cycle. Measures to ensure that an accurate product is produced include running a computer simulation of the print to see if rafts and supports are correctly placed, which also ensures that there is sufficient filament installed to complete the print. Most simulation programs will provide you with an estimate of the volume and weight of filament needed for a particular job.

Commercial products that are made from polymers tend to make use of manufacturing processes, such as injection moulding. Once an injection-moulding system is set up, there will be very few errors, and quality control measures such as visual inspections will take place at the end of the manufacturing process. Over time, moulds can wear and the tolerance of the moulded product may fall outside the acceptable range.

ACTIVITY

The Year 5 students from your local primary school are coming on a visit to the Technology Department. They have one hour and you need an activity that will get them actively involved in the workshop.

On a CAD software program, produce a template of a key fob that they can stick onto a piece of acrylic. They can then file around the template, clean and polish it and take it home as a souvenir of their visit.

KEY POINTS

- Thermoplastics can be formed by using heat.
- Acrylic (PMMA) can be bent using a strip heater.
- Accurate bends can be produced using a bending jig.
- A sheet of high-impact polystyrene (HIPS) can be vacuum formed using a mould.
- Injection moulding is used to produce high volumes of polymer-based products.
- Extrusion produces long lengths of uniform section polymer.
- Rapid prototyping machines will take a three-dimensional image and turn it into a solid three-dimensional model.

KEY WORDS

Strip heater a machine used to produce bends in thermoplastic sheet.

Vacuum forming a process that uses heat to soften a thermoplastic sheet, then sucks it around a mould.

Press forming a process that forms pre-heated thicker thermoplastic sheet between two moulds.

Yoke the top part of a press-forming mould.

Plug the bottom part of a press-forming mould.

Injection moulding a reforming method of production that involves forcing a molten polymer through a die and into a mould.

Extrusion a reforming method of production that involves forcing a molten polymer through a die to produce a regular cross section polymer product.

Check your knowledge and understanding

1 Describe the process of bending a 3-millimetre acrylic sheet.
2 How do bending jigs improve the manufacturing performance?
3 Draw and label all the features of a mould used for vacuum forming.
4 Produce a flow chart to accurately describe the process of vacuum forming.
5 What is the difference between injection moulding and extrusion?
6 Use notes and sketches to describe the process of injection moulding.

Textile-based materials

The use of production aids

Making textile products uses number of different tools and equipment that are also used with other materials, especially those used for measuring and cutting.

A template is generally used when cutting fabric so that the right sizes and shapes of the product parts are cut correctly. Commercially-printed patterns for a wide range of products are made from tissue paper and often have several sizes printed on the same sheet of paper. These patterns come with instruction sheets that tell you how to make the product.

Figure 2.8.88 Commercial patterns come in a range of sizes.

Figure 2.8.89 Basic blocks for jeans

Figure 2.8.90 Cutting many layers of fabric using a special cutting knife

You can also find patterns in sewing magazines and websites. You could also make your own using CAD software such as Fittingly Sew, by disassembling an existing product and drawing round the pieces, or start from scratch using basic blocks. A basic block is a template for the different parts of a simple garment, including the front and back bodice, sleeve, shirt, front and back trouser and skirt shapes. Other styles of garment are developed from these basic templates.

In commercial clothing manufacture the pattern templates are developed from basic blocks and produced in standardised sizes based on British Standards Institute (BSI) standard sizing. The block patterns are made from a heavy card or plastic so that they can be used many times. The pattern maker uses the basic blocks to develop patterns for a wide range of garments by adding the style details and decorative features of the design.

Tools, equipment and processes

In commercial manufacture the fabric is not folded, but laid out with many layers on top of each other, and a computerised program places the pattern pieces on the fabric as economically as possible. The pattern templates are held in place using suction and the pieces cut out using knives. Some computerised systems send the cutting instructions directly to the cutting machine and use lasers or water jets to cut the fabric.

Pattern markings need to be transferred to many layers of fabric at the same time. Different methods of doing this are used in commercial manufacture. The most commonly used techniques are:
* drill markers, which make small holes in the fabric, such as at the end of darts; these small holes will not be visible after the product is sewn together
* hot notchers, which mark the edge of fabrics with small cuts; these show where different sections of a product need to match up
* fluorescent dye markers, which are used on the surface of the fabric (for example, to show pocket positions); the dye is only visible under an ultraviolet lamp so the marks will not be visible on the finished product.

The sewing is done by a team of machinists who each sew part of the product before it is moved on to another person who will sew a different part of the product. In most factories, an automatic conveyor system delivers work to the machinists as it is required. A small terminal at each work station monitors the work in progress and records any problems.

Specialised computer-controlled machines are used in some factories to carry out identical operations that need to be repeated many times. These include machines that make buttonholes, hems, seams, pockets and embroidered patterns.

In commercial manufacture, a number of special pressing tools are used. A flat press uses steam, heat and pressure to press the product and can be set to suit different types of fabric and products.

A steam dolly is shaped like a body, and the finished garment is placed over it. The dolly is inflated inside the clothes using steam and air, which makes the creases fall out as well as giving it its correct shape.

A tunnel finisher is used to press garments such as shirts. The garment is placed on a hanger and steamed with some parts, such as the collar, pressed manually with a steam iron.

Pressing of completed products can also be controlled by computers that can store different pressing programs in their memory.

How materials are cut, shaped and formed to a tolerance

Different tolerances will apply to different types of textile products and these will be written into the manufacturing specification. The tolerance is the acceptable range within which the size of the product or part can vary. It is almost impossible to make products exactly the same every time, and it is not necessary for textile products. But individual parts of the product will need to be made so that they fit together when the product is constructed. For example, a collar to fit on to the body of a shirt must measure the same as the shirt neckline and be within a few millimetres, otherwise it will not fit well.

Commercial processes

Different types of production systems are used in commercial manufacture as products need to be made quickly and easily with costs kept to a minimum. One-off or bespoke production is used for special orders when only one or a very small number of products is made (for example, a dress for a celebrity). These products are usually unique and very expensive to make.

Batch production is used to make a specific number of identical products and is used in the manufacture of most textile products, as fashions change quickly so it would not be wise to make too many of one type of product. For more information on batch production see Topic 2.7.

When pattern pieces have been pinned to a fabric a range of tools, equipment and processes can be used to shape, fabricate, construct and assemble a textile prototype. The fabric is cut out using dressmaker's scissors or a rotary cutter and pattern markings are transferred to the fabric before it is sewn. (More information on the tools and processes associated with cutting, pattern marking and sewing can be found in Topic 2.4).

Pressing of the product as it is made and when it is complete is important to get a good quality finish. A steam iron is a useful piece of equipment for this.

In modern manufacturing systems, garments are stored and transported on hangers and moveable rails so that they arrive at the shop ready to go on display. The ticketing and tagging of products is now largely done by the manufacturers and not the retailers. The barcode system used in large stores records which items are selling and can help decide when and how many to re-order from the manufacturer. Re-ordering is often done automatically as stocks of a particular product become low.

Commercial processes for dying and printing are explored in Topic 2.9.

Weaving

Fabrics can be woven on small or large industrial scales using a weaving loom, which weaves two yarns at 90 degrees to each other (weft yarns go horizontally across the loom; and warp yarns run vertically). The main actions involved in the weaving process include:
1   Shedding: heddles, through which the warp threads are passed, separate the warp yarns into two layers by raising and lowering to create a space for the weft threads to pass through (called a shed)
2   Picking: the weft yarn (or pick) are driven across the warp threads and through the shed
3   Beating: a comb-like component called a reed pushes the weft yarn into place as it is woven.
4   Let off: the warp yarns are unwound from the warp beam
5   Take off: the fabric is wound onto the cloth beam

Different mechanisms are used to control the shedding process, including a crank, cam or dobby.

The weft can be picked using a rapier, air jet, water jet or shuttle.

Different types of loom are used to produce different types of fabric. For example:
- a shuttle loom is used to create a plain weave
- a jacquard loom can be used to create fabrics with a more complicated pattern

Colour and pattern are then put on to fabrics using dyeing and printing processes in order to meet the client's requirements and to keep up with the demands of fashion. More information about these processes can be found in Topic 2.9.

The application and use of quality control

Quality control is an important part of manufacture and manufacturers need to make sure that the goods they produce are of an acceptable standard.

Different sections of a product need to be cut to the correct dimensions or they will not fit together as intended. For example, patchwork squares need to be the same size or they will not fit together accurately and the finished product will not be aesthetically pleasing; the front and back sections of trousers must be cut to the right size or they will not match up correctly when sewn together.

A repeat pattern printed on to fabric must be checked against the original sample to make sure that the size is correct and that the pattern repeats accurately along both the length and width of the fabric. Small mistakes in the placing of the repeat pattern can cause larger errors along the whole length or width of fabric. The colour must be consistent on each repeat, and care must be taken to place the screens for different parts of the pattern accurately on the fabric so that it is the right size.

One way to check the quality of products is to inspect every one. A quicker method is to inspect a sample of the products as they are made, and collect and analyse information about the samples. This can help the manufacturer to notice that a particular machine is faulty, so it can be fixed before too many defective products are made. If a machine operative is not producing the correct quality of work, that person can be identified and given training to improve the quality of their work.

Manufacturers need to make quality control checks at certain stages during manufacture to make sure the products are of the quality required. These stages are built in to the production specification. Checking at an appropriate stage is important – for example, it may be too late to do anything about a pocket that is not made to the right size once it is sewn on the product.

KEY POINTS
- Pattern templates are developed from basic blocks in standardised sizes.
- Many layers of fabric are cut out at once in commercial manufacture.
- Commercial machinists usually work on specific sections of a product before passing it on to someone else who makes a different part of the product.
- Computer control is used in industry to carry out identical and repetitive processes.
- Quality control is an important part of manufacture to make sure that products are made to the quality specified in the manufacturing and product specifications.

Check your knowledge and understanding
1 What is a basic block pattern?
2 Give three uses for automated sewing machines.
3 Explain why is it important to make quality control checks regularly when manufacturing textile products.

Electrical and mechanical systems

The use of production aids

When making any product or prototype there are many things that can be done to make sure that it is made accurately. One method of ensuring accuracy is to use a jig. A jig is a simple mechanical device that is used during the manufacturing process to make sure that every piece is made to the correct specification. Another production aid that can be used if you are making multiple identical pieces is to make an accurate template.

Jigs

When a number of identical pieces are required for a product or system a jig may be required. A jig allows identical pieces to be made in quantity by making sure that a machine tool moves in the same way to the same place. Some jigs in industry are designed to increase productivity, some are designed to do activities that have to be repeated many times, others are designed solely to make an outcome more accurate.

A jig might be used when 50 PCBs are manufactured using the etching method and the holes then need to be drilled accurately and consistently. It would be designed to hold the PCB in the same position every time on the bed of the PCB drill stand so that every hole was accurately drilled in the correct place.

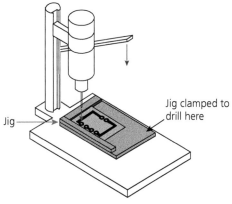

Figure 2.8.91 **A jig being used to drill a hole in a PCB**

Templates

A template can be cut out of a variety of materials, but it is important that the material used is reasonably hard to prevent wear. A template can ensure that the pieces are the same shape and size and can also show where any holes are to be drilled. Unlike a jig, however, a template will simply mark out the position of the holes, instead of drilling through them.

Tools, equipment and processes

Making PCBs

When making PCBs (see Topic 2.5) there are two ways of going about this: etching or machining.

Figure 2.8.92 **A small CNC milling machine**

When machining, a CNC milling machine like the one shown in Figure 2.8.92 is used. These machines remove the copper on the surface of the board to leave just the tracks and pads necessary for the circuit and cut the shape of the board.

> **ACTIVITY**
> Make a list of the benefits of using jigs and templates.

> **STRETCH AND CHALLENGE**
> Consider a product or prototype you have made in school recently. Design a jig that could be used for one of the manufacturing processes if you were going to make a batch of 500.

When etching a PCB, either a bubble etch tank or a spray etching machine can be used. Both methods use a heated ferric chloride solution to remove the unwanted copper, and protective goggles and gloves are required by the operator. Spray etching is quicker than using a bubble etch tank, but in both methods the time taken is governed by the quality of the solution, the more the solution is used the longer the process takes.

Cleaning

When removed from the etching solution, the PCB is washed in water and an abrasive block is used to remove any film left on the pads and tracks.

Drilling

Holes for the components need to be drilled with a suitable drill. One millimetre carbide drill bits are generally used for most components. These drill bits are very brittle and therefore snap easily, so a drill stand should be used for drilling holes.

Populating the PCB

Components are placed on the opposite side of the board to the copper tracks and pushed as close as possible to the board. Leads for components such as resistors and diodes are bent carefully using snipe-nosed pliers and placed in the correct position on the board. Low-profile components (those that do not stick out very far) are placed first.

Soldering

When soldering, make sure the area is well ventilated. Hold the tip of the soldering iron against the leg of the component and the copper pad on the board for about three seconds. Touch the solder against the pad and the leg, letting some encircle both the leg and pad and withdraw the soldering iron.

Figure 2.8.93 **A PCB drill and stand**

Figure 2.8.94 **A populated PCB**

Figure 2.8.95 **Soldering a PCB**

Cutting legs

Finally, to finish off, cut off the remainder of any protruding legs of the component. Wear goggles to protect your eyes when cutting the leg with sidecutters.

Figure 2.8.96 **Cutting component legs with sidecutters**

ACTIVITY

Create a list of all the tools and equipment necessary to etch and populate a printed circuit board.

STRETCH AND CHALLENGE

Write an article that explains which of the two methods of printed circuit board production you consider to be the most appropriate one to use in schools and colleges.

How materials are cut, shaped and formed to a tolerance

If you have ever tried making something to an exact size you might have noticed that you were out by a fraction of a millimetre. The accuracy that components must be made to so that they fit together is known as the tolerance limit. This is the acceptable difference from its perfect size. It is written as two numbers that give the upper limit and the lower limit.

If it were found that components intended to be 100 millimetres in length actually measured a variety of lengths between 99 millimetres and 101 millimetres, and this was acceptable as they still fit together with other components, we could work out the acceptable tolerance. The tolerance is the difference between the upper and lower limit, so in this case we could state that the tolerance is ±1 millimetres. The specification for a product should include criteria regarding tolerance.

Tolerance is important to ensure quality and reliability. Generally, the smaller the tolerance the better, but improving accuracy means that manufacturing costs are higher. As mentioned in Topic 2.6, resistors are bought according to their tolerance – you will pay more for a resistor with a tolerance of ±5 per cent than a one with a tolerance of ±10 per cent.

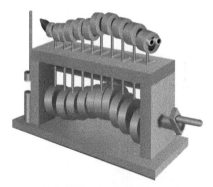

Figure 2.8.97 **A caterpillar toy using cams needs to be made to tolerance so that the moving parts operate correctly.**

When manufacturing a mechanism with moving parts, all the moving parts must have a clearance gap between them so they can move with minimum interference with each other. If making a shaft for an axle, it must be slightly smaller than the hole it goes through so it can turn. Tolerances are used in industry to make sure any shaft of a given size will fit correctly with other parts of the final product and function. The tolerance that is set for a part will depend upon its application.

Commercial processes

Most commercially manufactured PCBs now use surface-mount technology. This is a method of manufacturing electronic circuits in which the components are mounted or placed directly onto the surface of PCBs on the copper or track side of the board. They use component placement systems, commonly called **pick and place machines**, in which robotic machines are used to place surface-mount components onto PCBs. Electronic components, such as resistors, capacitors, and integrated circuits, are placed with a high degree of precision at high speed onto the surface of the PCB. The pick and place machines are part of larger overall systems that work together to supply and pick up and place the components onto precisely the correct place on the PCB.

KEY WORD

Pick and place machine a machine used to place surface-mounted components onto printed circuit boards.

KEY WORD

Flow soldering a method of soldering where solder paste is melted by heating the printed circuit board.

ACTIVITY

List the advantages of surface-mounting electronic components.

STRETCH AND CHALLENGE

Produce a one-sided A2 flyer for a company that manufactures pick and place, and flow soldering machinery, outlining the benefits to industry of pick and place technology and flow soldering.

KEY WORD

Quality control: a process that ensures every product is manufactured to the same standard.

A sticky paste of solder and flux is used as a temporary method of keeping the components in place and attaching them to their contact pads. This process, known as **flow soldering**, then heats the assembled PCB board which melts the solder, making a permanent connection between the components and the board. Heating the board can be done in three ways: by passing the board through a reflow oven, by using an infrared lamp to supply the heat, or by soldering each connection individually with a hot air pencil. Flow soldering is the most common way of attaching surface-mount components onto circuit boards. The difficulty faced with flow soldering is melting the solder and the surfaces to be joined together without overheating the electronic components and damaging them.

The application and use of quality control

All manufacturers need to ensure the products they are making are of a good and acceptable quality. To make certain of this, **quality control** systems are set up by manufacturers. These systems involve checking and inspecting parts as they are made and collecting information, and checking items against set standards and tolerances. Inspections take place at identified stages and after the final product has been assembled. The frequency of checks would be decided when planning for manufacture, for example every 50th PCB from production is inspected, or samples to be taken every hour.

Figure 2.8.98 A pick and place machine in operation

Figure 2.8.99 Workers at an automated soldering machine

When manufacturing PCBs using the etching method, there are three main stages in manufacture:
- exposure to ultraviolet light (UV)
- developing
- etching.

The length of time photoresist board should be exposed to light is suggested by the board manufacturer, but it is initially determined by trial and error, and is usually between two to three minutes. Once established, it does not vary and should not affect the quality of the PCB.

The amount of time a photoresist board spends in the developing solution is affected by temperature, correct dilution of the developing chemicals, and the number of boards previously developed. Developing times need to be carefully recorded and quality checked regularly as the quality of the PCB can be greatly affected. Too little time in the developing tank, or the developer being too cold, or too dilute a solution, will mean that the photoresist will not be totally removed from the board, and therefore the circuit will not be etched successfully. Too much time in the developer, the developer getting too hot, or the solution being too concentrated, could mean that all the photoresist will be removed from the board, and therefore no circuit will be left to etch.

Inspection tests need to be carried out frequently when manufacturing PCBs. These can include simple visual inspections after developing and after etching set against data comparison (exposure time, developing solution temperature, developing time, etching chemical temperature, and etching time), and electrical continuity testing of the etched board.

Figure 2.8.100 A PCB being checked for electrical continuity

STRETCH AND CHALLENGE

Produce a flow chart that shows all the stages of the process for manufacturing etched printed circuit boards. Show in the flow chart what quality control checks should be carried out.

KEY POINTS

- Jigs and templates can increase accuracy and speed of production.
- Printed circuit boards can be produced by machining or etching.
- Tolerance is the acceptable difference of a component from its perfect size.
- Pick and place assembly and flow soldering are used in commercial production of printed circuit boards.
- Inspection tests need to be carried out frequently at various stages when etching PCBs.

ACTIVITY

Make a list of reasons of why quality control is important to manufacturing companies.

Check your knowledge and understanding

1. Give an example of a process where a jig could be used.
2. Explain what is meant by the term 'populating a PCB'.
3. Give an example of an electronic component that has its tolerance marked on the outside.
4. Explain what difficulty has to be overcome when flow soldering.
5. Discuss what is meant by quality control and why it is important to manufacturers.

2.9

Surface treatments and finishes

What will I learn?

In this topic you will learn about:
- → the range of finishes available for a variety of materials
- → how to prepare materials before applying a range of finishes
- → how a finish can be used to improve the aesthetic of a material
- → how a finish can be used to improve the performance of a material.

There are a huge range of surface finishes available for all material groups. The choice of finish can be influenced by many factors, but most finishes are applied for one of the following reasons:
- → to increase the durability of a material or product
- → to protect a material from decay
- → to improve the aesthetic of a material
- → to improve the working properties or performance of a material.

Papers and boards

Printing

Paper and board can be printed in a number of different ways. When at home or in school or college it is usually sufficient for us to use an inkjet printer, but when producing thousands of the identical items it is quicker and more cost-effective to use a different type of printing. The most common of these commercial types of printing is offset lithography.

Offset lithography

Offset lithography is based on the principle that oil and water do not mix. The images are put on aluminium plates by exposing them to UV light. The plates are then chemically treated so that the non-image areas are water absorbent. They are then wrapped around a cylinder. The plate is then dampened on the parts of the image where ink is not needed. The cylinder is then passed over a vat of ink and the ink is transferred on the plate. Because oil and water do not mix, the ink sticks to the image (the dry parts of the plate). The image is then transferred on to a blanket cylinder and then on to the paper. That is why this process is called 'offset' lithography – the paper does not come into direct contact with the printing plate.

Offset lithography is used for medium and long print runs of products such as magazines, posters, packaging and books.

In offset lithography, four main colours are used: cyan, magenta, yellow and black. These are often referred to as CMYK. Each colour is printed

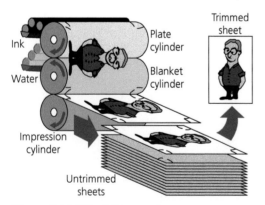

Figure 2.9.1 **Offset lithography**

individually. Before any of the aluminium plates are made, an image must be created for each individual colour.

As in the image in Figure 2.9.2, each colour is printed separately, one on top of the other, to make a full-colour image.

Screen printing

This process uses a stencil over a mesh screen through which ink is pressed. A rubber squeegee forces ink through the open area of the mesh onto the material you wish to print on. This process can be done by hand or it can be fully automated, depending on how many prints you need to complete.

Figure 2.9.2 **Separation of colours**

Screen printing tends to be used on lower volume print runs for things such as posters, display boards, fabrics and wallpaper, but can be also used on a range of materials from glass to wood.

Flexography

Flexography is a type of relief printing, which means that it is done using raised images or letters which are pressed on to the material. It is used for very high volumes of printing, due to its high set up costs, and can be used on a range of materials including plastic, foil, brown paper and newsprint. Flexography uses flexible rubber or plastic printing plates. Raised text or images are etched on the rubber surface of the rollers. Thin, fast-drying ink is then used, which is good on cheaper and non-absorbent materials, but a limited quality can be achieved. It is used to produce repeating patterns in packaging, gift wrap, wallpaper and so on. Flexography normally produces up to 60,000 prints per hour, and due to the use of faster-drying inks, it can be a quicker process than lithography.

Figure 2.9.3 **Flexography**

Flexography produces blocks of colour, as opposed to lithography and screen printing, which produce a series of coloured dots.

Embossing

Embossing is the process of making raised images or letters on paper or card. It involves creating a textured roller or stamp which is pressed over paper or card to make a textured image. Debossing works in the same way but creates a recessed imprint.

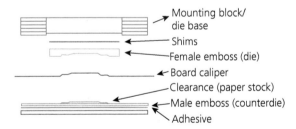

Figure 2.9.4 **Embossing and debossing**

Embossing presses can be automatic or hand operated. The process uses two dies, one male and one female. The dies fit into each other so that when the paper is pressed in between them, the raised die pushes the paper into the recessed die and the paper remains in this shape after the dies are released. To ensure the paper stays in that position permanently, a specific level of pressure is applied to the dies.

Figure 2.9.5 **Embossed paper**

Varnishing

Varnishing is the most common finishing process used on paper and board products. It can be done at the time of printing or after printing. Varnishing provides a glossy or a matt finish to the material, which protects and improves the visual look of the product. It can be applied to the whole material or to set points on the surface, which is called spot varnishing. Spot varnishing improves the visual appearance of an image against an unvarnished area and can sometimes involve adding additional effects, such as glitter. Varnishing the whole surface is sometimes called roller coat or book varnish and is normally applied during the printing process.

Ultra-violet varnish is varnish that dries when passed under UV light. UV varnish achieves very glossy and reflective smooth surfaces, which is similar to if the material were laminated. Their advantage over laminates is that they are completely recyclable. UV coatings are cured very quickly which means that products can move through this part of the process and on to the next without costly delays.

KEY POINTS
- The most common commercial printing process is offset lithography.
- The four colours used in offset lithography are cyan, magenta, yellow and black.
- Screen printing and lithography are done using small dots of ink.
- Flexography is a relief printing process.
- Embossing is the process of creating raised images or text on paper or card.
- UV varnishing is a quick-drying process that improves the durability and appearance of the material.

Check your knowledge and understanding

1 Explain the offset lithography process.
2 Sketch a diagram showing the screen printing process.
3 What is the difference between flexography and lithography?
4 Name two products that have been embossed.
5 Why would UV varnishing be used on the cover of a book?

Timber-based materials
Surface preparation for timber

Before applying a finish to a timber surface, it is important to get the surface smooth and flat. Any imperfections in the surface, such as scratches and dents, are often accentuated by the application of a finish. If you are using a clear finish, such as varnish or oil, it is also very important to remove all pencil and pen markings from the surfaces.

To get a surface smooth and flat on natural timbers and the edges of manufactured boards, a sharp, finely set, smoothing plane is most efficient. On curved surfaces, or on the faces of manufactured boards, glass paper is best. You will need to start with coarse grit glass paper (60 or 80 grit) and work your way through to fine glass paper (400 grit). If the surface has a natural grain pattern on it, it is important to sand in the direction of the grain. To help get a surface flat, wrap the glass paper round a cork or wood block. To sand the end-grain of a piece of natural timber, a disc sander is very effective. You will then need to remove the scratches left by the sander by using fine glass paper wrapped around a block. Dust will need to be cleaned from the surface thoroughly before starting to apply the finish.

Finishes for timber

When choosing a finish for timber, it is important to think of the effect that you want to achieve, the environment in which is going to be used, and what sort of timber it is used on. If you are using natural timber, or plywood with an attractive outer layer, you may want to use a finish which allows you to see the natural grain pattern of the timber. You may want a completely clear finish, or one that changes the colour of the timber while still showing the grain pattern. With these and coloured finishes, you will also need to decide whether you want a matt, silk or shiny finish.

Stains

Stains are applied to enhance or change the colour of the natural timber. Many are variations of the brown-red and black colours of hardwoods, so a softwood can be stained to mimic a hardwood. Sometimes bright colours are used (for example, for children's toys). It is important to apply the stain evenly, with a cloth or brush, and it is worth testing it first on some scrap materials. The stain itself does not give much protection to the timber; it is mainly used to change the colour. It will need an additional clear finish, such as varnish or wax, applied over it, to give protection. Alternatively, you can buy varnishes which have coloured stains in them. These are more difficult to apply evenly, as each layer of varnish will intensify the colour change, so coats need to be applied very evenly.

Figure 2.9.6 Using stain to change the colour of wood

Preservatives

These products are applied to wood to help it repel water and moisture, and in some cases resist insect attack. Traditionally, a brown product called creosote was used, typically on sheds and fences. In recent years, manufacturers have developed an attractive range of preservative colours which are easy to apply with a brush and repel water effectively.

Figure 2.9.7 Modern preservatives have excellent water-repellent qualities.

Figure 2.9.8 Brightly coloured preservatives used on beach huts

Figure 2.9.9 Tanalised wooden decking

Commercially, preservatives can be added to timber through a process known as tanalising. Here, the timber is placed in a sealed chamber where the preservative is pumped in under pressure, forcing the preservative liquid to penetrate the outer cells of the timber. The level of penetration, and therefore level of protection achieved, can be adjusted by altering the pressure and time. You can see in Figure 2.9.9 how the preservative has penetrated the outer layer of the timber decking section.

Timber-based materials

Varnish

Varnish can achieve a similar look to polish, but gives better protection to the surface. This is particularly true of modern polyurethane and acrylic varnishes. Like polish, the varnish needs to be built up in thin layers, with each layer drying thoroughly before sanding. The final coat can have wax applied to it to achieve an even higher-quality finish.

Figure 2.9.10 **Re-oiling some teak garden furniture**

Figure 2.9.11 **Brightly coloured paints**

Oils

Oils are applied to enhance the natural oils already in the wood. Oil is applied with a cloth, and can be built up in layers. The oil soaks into the timber, enhancing its ability to repel moisture without creating a layer on top of the wood that would flake off in time. Teak oil is excellent for oily hardwoods such as teak, and is widely used on yachts. Danish oil (which is mainly linseed oil) is excellent for lighter coloured woods, and is used both indoors and outside. Beech worktops are often treated with Danish oil. Oiled timber products often need recoating periodically.

Paints

Painting gives a solid colour finish to the surface of the wood – you cannot see the wood grain through it if the surface is flat. You will need to build up the paint in layers, starting with a coat of primer to seal the wood, then a layer of undercoat, followed by at least one coat of top coat. The finish may be matt, silk or gloss, giving different levels of reflection of light. The paint surface needs to be rubbed down with fine glass paper between each coat of paint to help ensure a smooth finish. If the wood has knots, these need to be treated with a knotting compound or solution before applying the primer, to stop any sap bleeding through the paint finish.

When choosing paint for wood, oil-based paints are generally tougher and more durable, but take longer to dry. Acrylic paints usually dry quicker and are often non-toxic, making them suitable for a child's toy.

Paint can also be applied from an aerosol spray can, or for larger areas, with a roller. Emulsion paint, designed for house walls, is often more suitable for brush application on large surfaces.

Commercially, spray finishes can be applied to timber using specialised spray equipment that runs off a compressor. The quality of finish that can be achieved commercially is generally of increased quality to the finish achievable by brush or roller. The paint used can be water- or solvent-based. The appropriate PPE that the operator would wear will depend on the type of paint being used. You can see that extraction and respiration equipment are hugely important in a spray paint environment. Developments in technology have meant that timbers can now also be powder-coated (see below). It is common to find that MDF furniture is finished in this way, and kitchen doors.

Figure 2.9.12 **Spray paint operator**

Figure 2.9.13 **MDF painted furniture**

2.9 Surface treatments and finishes

Check your knowledge and understanding

1 Why is it important to sand timber with the grain before applying a finish?
2 Describe the process of preparing a piece of timber before applying a stain.
3 What is the difference between a water-based and an oil-based finish?
4 What is tanalising?
5 What PPE should be worn when spraying a piece of timber in a school environment?

Metal-based materials

Painting

Metals can be painted in the same way as timber can; by applying a primer, undercoat and topcoat, either by brush or by spraying. Like timber, metals need to be well prepared. All grease and dirt need to be removed, and rough surfaces need to be made smooth with a grinder, file and emery cloth. Dents and holes need to be filled with appropriate filler if painting. This is particularly important when spraying, as the thin layers of paint tend to accentuate blemishes. Ninety per cent of the work is in the preparation for spraying. The primer can then be applied, which provides a smooth surface for the following coats to adhere too. Primers can sometimes be red in colour when they have had zinc added. The zinc adds a level of corrosion protection to the metal (usually steel).

Figure 2.9.14 **A car primed ready for its next coat of paint**

Dip coating

This process provides a thin layer of polyethylene plastic over the surface of the metal. The part to be coated is cleaned, and then heated to 200 °C before being dipped in a fluidised (air blown through it) bath of polyethylene powder for a few seconds. The heat makes the powder stick to the metal and fuse together to give a smooth, shiny, colourful surface that protects the metal. Handles of tools are often dip coated to give better grip on the tool.

Figure 2.9.15 **Pliers with dip-coated handles**

Powder coating

Powder coating is a commercial finishing process. The item being coated is electrostatically charged and the paint is applied in a powder form. The powder is attracted to the charged object, where it forms an even layer. The object then passes through an oven where the paint cures and hardens. Powder coating provides a more durable finish than other painting methods and it can be used with a variety of metals.

Figure 2.9.16 **Powder coating**

Galvanising

Galvanising gives excellent protection from rusting to steel parts. The steel is dipped into a bath of molten zinc, giving the surface a bright grey colour. The process is used for many products that are used outdoors, such as steel gates and fencing. The zinc is more reactive than the metal that it is coating, and acts as a sacrificial anode, corroding at a faster rate than the steel and protecting the base metal.

Figure 2.9.17 **Galvanised steel safety barriers beside a road**

Figure 2.9.18 **A galvanising tank**

Figure 2.9.19 **A range of anodised bike components**

Anodising

Anodising is an electrolysis process where aluminium parts are dipped in a chemical bath and an electrical current is passed through the product being anodised. This causes the surface of the aluminium to oxidise, forming a hard surface, which is resistant to wear and scratches. Coloured dyes can be added during the process giving the surfaces an attractive shiny metallic finish.

KEY POINTS
- Preparation of the metal is the most important factor in successfully applying a surface finish.
- Dip coating provides a thermoplastic coating to a metal.
- Galvanising is the process of coating a metal in zinc.
- Ferrous metals cannot be anodised.
- Powder coating provides an even, durable surface finish.

Check your knowledge and understanding

1 What does electrolysis mean?
2 Why do we use a primer?
3 Identify five products in the workshop that have been dip coated.
4 What materials are generally galvanised?
5 What two functions does anodising fulfil?

Polymers

Most plastics are self-coloured; this means that they are made in a range of colours and you would choose the most appropriately coloured parts as you make a project. Most plastics are also resistant to wear and decay, so you do not normally need to apply a finish to plastics.

Polishing

The surfaces are often smooth and highly polished. If you cut a sheet of plastic, either using hand tools or with a laser cutter, you will normally need to make the edge smooth with a file and/or abrasive papers before polishing on a buffing machine.

Flame polishing can also be used to finish the edges of polymers. In this process the edge of the polymer is exposed to a naked flame, which melts the outside surface. This can provide an exceptionally high-quality finish.

Figure 2.9.20 **Polishing mop**

Printing

Decoration and detail can be added to polymer products by various printing techniques. Pad printing or screen printing can be used to transfer a chosen image or design. It is possible to print on flat or curved surfaces; both of these can be seen in the detail added to a child's toy in Figure 2.9.21.

Figure 2.9.21 **This toy is a good example of screen printing**

Figure 2.9.22 **Vinyl decal logo**

Vinyl decals

In applications where screen printing is not a viable process, you will often find that vinyl decals are applied to polymer products. These can either be screen-printed images that are then cut out and applied in a similar way to a sticker, or can be individual text or shapes that have been cut out using a CNC knife cutter.

Check your knowledge and understanding

1 Identify five polymer products that have had a screen-printed design applied.
2 Why do you not always need to apply a finish to a polymer?
3 What safety precautions would need to be considered when flame polishing?
4 When would you use a vinyl decal?

Textile-based materials

Adding colour and pattern to fabrics

Colour and pattern are major aesthetic considerations when consumers select textile products for particular uses. When fabrics come from the loom or knitting machine, they are often a grey or beige colour and need to have colour and pattern added. Before this can take place, the fabrics need to be cleaned to remove natural impurities such as waxes, and oil and dirt picked up from machines during the processing of the fabrics. If these impurities are not removed, the colour will not attach itself evenly to the fabric. Some fabrics need to be bleached to make the fabric evenly white before colour is added.

Dyeing fabrics

Textile dyeing involves the permanent application of a colour to a fibre to give a uniform colour.

Colour can be added at different stages in the manufacture of a textile product.
- Spin dyeing is when colour is put into the spinning solution of synthetic fibres.
- Stock dyeing is when natural fibres are dyed before they are spun into a yarn.
- Yarn dyeing is dyeing yarns before they are made into fabrics.
- Piece dyeing is the dyeing of the woven or knitted fabrics.
- Garment dyeing is when made-up garments are dyed as required to meet consumer demand for different colours.

In order to be successful, the colour must be able to be absorbed by, or react with, the textile fibre to give an even colour to the fabric. The process of dyeing consists of three basic steps:

- immersing the textile into the dye bath
- the dye attaching itself to the textile fibre
- fixing of the dye within the fibre.

The strength with which the dye is held in the fibre is called **colour fastness**. Fabrics may need to have fastness to washing, sunlight and rubbing.

Resist dyeing allows patterns to be made with dyes using a coating, such as wax, or a barrier such as string, to prevent the dye from reaching certain parts of the fabric.

Batik uses wax or a flour paste resist to draw a pattern on the fabric, which is stretched over a frame. When the resist has dried the fabric is dyed. Small cracks that appear in the resist as it dries allow some of the dye through to the fabric, giving it the characteristic crazed batik background.

Tie-dye involves folding, twisting, pleating or crumpling fabric then tying it with string or rubber bands before dyeing it. The folding and tying of the fabric prevents the dye from reaching certain parts. Multi-coloured designs can be made by untying the fabric when it has dried the re-tying and dyeing with a different colour.

Printing fabrics

Printing is a method of applying a coloured pattern to fabric, and has some advantages over dyeing:

- it gives more opportunities for designing
- more colours can be used
- complicated designs can be produced.

The dye used for printing is made into a thick paste so that it can be applied easily.

Screen printing is based on the Japanese method of stencilling. The screen consists of a fine mesh stretched over a frame. Parts of the mesh are blocked off so that a pattern is made. Each colour in the design has a separate screen. There are three main types of screen printing used commercially: rotary, flatbed and carousel.

Roller printing uses engraved copper rollers to print the design. One roller is required for each colour in the design, which runs the full width of the roller, and the circumference of the roller is the same as the pattern repeats. The roller is coated with the printing paste, then rolled over the fabric. Although a quick method, roller printing is very expensive so is only used for very long print runs.

Sublimation printing is similar to ironing transfers on to fabric. The design is printed on to a special paper using computer software. The design is the full width of the cloth to be printed, and is transferred to fabric using a heat press. The temperature of the heat press is high enough to cause the dye to turn into a vapour, which then transfers to the fabric. This method works best on fabrics made from synthetic fibres – in particular, polyester.

Digital printing uses a computer to design the print, which is then printed straight on to the fabric from the computer. The fabric is steamed to fix the design on to the fabric. This method is only suitable for printing small amounts of fabric.

Figure 2.9.23 The batik method uses a wax or flour paste resist to create a pattern.

Figure 2.9.24 A typical spiral tie-dye pattern

Figure 2.9.25 Roller printing

Textile-based materials

Figure 2.9.26 **School uniforms treated with Teflon or Scotchgard stay cleaner for longer so will not need to be washed so often.**

Applied finishes

Fabrics need to be fit for their intended use. Many fibres used to make fabrics have disadvantages, but it is possible to cancel out some of the disadvantages by applying a fabric finish. Some finishes improve the functionality of a fabric, some are used for aesthetics, and others make a fabric safer. An applied finish always costs money, so a manufacturer will need to think about how important it is to put a special finish on to a fabric depending on what it is to be used for.

Stain resistance is a functional chemical finish used on many different textile products. The finish also makes fabrics water repellent and is especially useful on furnishings, school clothing and outdoor garments. Stain resistant finishes use fluorocarbons that are sprayed on to fabrics to stop water and oil-based stains from attaching themselves to the fabric without affecting the colour, feel or weight of the fabric. The finishes are environmentally friendly and will biodegrade over time. Teflon and Scotchgard are the two best-known stain-resistant finishes.

ACTIVITY

Materials needed:
- cotton fabric
- string, elastic bands, bulldog clips, coins, buttons, small pebbles or marbles
- a packet of dye
- salt
- a plastic bucket or other non-metallic container large enough to hold the fabric and water
- rubber gloves.

Method:
1 Wash the fabric to remove any starch or stains.
2 Fold, pleat or crumple the fabric, then tie it tightly with the string or rubber bands in one of the ways shown in Figure 2.9.27. Coins or small pebbles can be inserted into the fabric before it is tied.
3 Wearing rubber gloves, follow the manufacturer's instructions to mix the dye with water in the bucket.
4 Place the fabric in the dye and leave it for about an hour.
5 Remove the fabric and rinse it well in clean water. Remove the string or rubber bands to see the pattern you have created.
6 Repeat the process with another colour if you want a two-colour effect.

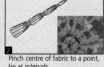
Crumple fabric and tie tightly with elastic bands.

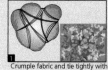
Pinch centre of fabric to a point, tie at intervals.

Concertina-fold fabric, tie at intervals.

Concertina-fold fabric, place pegs or bulldog clips at intervals along folds.

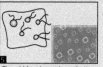
Tie pebbles, rice-grains or buttons into the fabric at intervals.

Figure 2.9.27 **Different designs for tie-dye**

Check your knowledge and understanding

1 Explain what is meant by resist dyeing.
2 What is meant by colour fastness?
3 List two different stages when dye can be applied to a textile product.
4 Name three different methods of printing fabric.
5 Name a stain-resistant finish.

Electronic and mechanical systems

There are a huge range of products that contain electronic circuits. These products contain printed circuit boards that are expected to work perfectly in different environments. Printed circuit boards can be exposed to a variety of different conditions – hot and cold temperatures, moisture, salt and chemicals in the air; this means that boards need protective coatings to prevent them from corroding.

Types of coatings

There are a variety of coating materials available for PCBs. The coating material is chosen after considering the environment the PCB is likely to operate in. A range of questions are asked before the choice of coating is made, such as:
- What temperature range will the circuit operate in?
- Will the circuit be expected to operate in a moist environment?

The most appropriate coating material is then chosen from materials such as silicon, epoxy, acrylic and polyurethane among others.

Coating methods

There are a range of methods by which the coating material can be applied, including brushing, dipping and spraying.

Brushing is used when there are only low numbers of PCBs to coat, as it is difficult to apply and difficult to get an even covering without getting bubbles in it.

Dipping is a process where the PCB is lowered into a **lacquering** solution. It is often used in high-volume production where the coating needs to penetrate every part of the PCB.

Spraying can be carried out by an operator with an aerosol can, and is therefore very popular among those who make low numbers of or individual PCBs. It can also be carried out by machine. These machines can use a variety of nozzles of differing sizes and spray patterns. This method is also used in very high-volume production because it is fast, accurate and can be applied in varying thicknesses to areas of board.

Reducing friction

When one surface of a material moves across another, the resistance to the movement is called **friction**. Friction is an advantage in some mechanisms, such as braking systems, and belts and pulley systems, while in others it can greatly reduce the efficiency of the mechanism and increase wear and tear dramatically. There are different ways of trying to reduce friction in mechanisms.

Making the surfaces as smooth as possible

Rough surfaces rubbing together produce more friction than smooth ones, so one way to reduce friction is to make the surfaces of the materials rubbing together as smooth as possible.

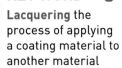

KEY WORD

Lacquering the process of applying a coating material to another material

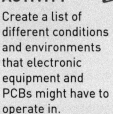

ACTIVITY

Create a list of different conditions and environments that electronic equipment and PCBs might have to operate in.

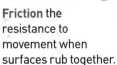

KEY WORD

Friction the resistance to movement when surfaces rub together.

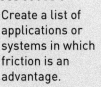

ACTIVITY

Create a list of applications or systems in which friction is an advantage.

Figure 2.9.28 **The smooth polished surface of a hydraulic ram reduces friction.**

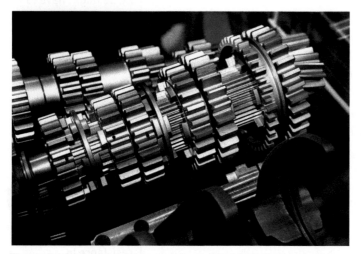

Figure 2.9.29 **A car gearbox requires oil to lubricate the moving parts and reduce friction.**

Lubrication

Lubrication provides a thin layer of liquid between two surfaces. This fills all the small spaces and irregularities and stops the materials from rubbing together. This restricts friction and heat, and in doing so this reduces wear. There are a range of lubricants available, such as oils and grease. These lubricants come in a range of different thicknesses, or viscosity, depending on their use.

Dry lubricants are now available and are made from materials such as graphite and molybdenum disulfide. These lubricants are applied to mechanical components by spraying, dipping or brushing. They offer lubrication at much higher temperatures than liquid and oil-based lubricants can operate at, even at temperatures in excess of 350 °C in some applications. The other advantage they have is that once applied, they do not get thrown off chains and gears like oil-based lubricants do.

Another method of reducing friction is using ball bearings and rollers. Bearings greatly reduce friction between surfaces, as they go between the surfaces and change the movement from sliding to rolling.

STRETCH AND CHALLENGE

Write an article for a design magazine explaining the advantages and disadvantages of the different methods of applying PCB lacquers.

KEY POINTS

- Printed circuit boards need protective coatings to protect them from the environments in which they operate.
- Friction between materials can be reduced by coatings, lubricants and bearings.

Check your knowledge and understanding

1 Give an example of one way of reducing friction.
2 Explain why spraying is the preferred method of applying lacquer onto PCBs in high-volume production.
3 Discuss the advantages of using dry lubricants for certain applications.

PRACTICE QUESTIONS: specialist technical principles

1   Choose one of the products below and circle your choice.

remote-controlled car   bread bin   cushion   wedding invitation   child's spoon

Name the main industrial process used in the manufacture of the product. Use
notes and sketches to explain this process in detail.

Name of process: _____

[4 marks]

2   Choose **one** of the stock forms below. Name the raw material it is
made from.   [1 mark]

	Raw material
Polypropylene sheet	
Cartridge paper	
Steel bar	
Nylon fabric	
Plywood board	

3   Describe two ways that materials can be reinforced to resist
certain forces. Give examples in your answer.   [2 x 2 marks]

4 Look at the following working drawing. Fill in the chart to show the measurements of A and B.

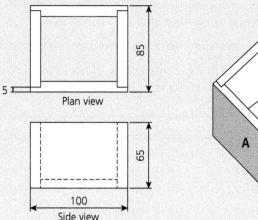

85

5

Plan view

65

100

Side view

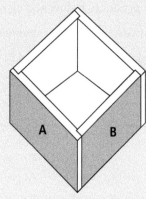

A

B

	Width (in mm)	Length (in mm)
A		
B		

[4 marks]

5 Explain why stock forms are used in the manufacture of products? [4 marks]

6 Name a suitable stock form for one of the following:

cartridge paper   plywood   wool   brass   [1 mark]

7 Products are often supplied flat to be assembled by the user. Explain why this is beneficial to the manufacturer, retailer and consumer. [8 marks]

8 Explain how the use of automation and the use of robotics has changed the way products are made. [8 marks]

9 Products are often protected or enhanced using surface treatments or finishing techniques. For a material of your choice, name a possible surface treatment or finishing technique. [1mark]

Material: _____

Surface treatment or finishing technique: _____

10 Finishes can be used to make a product more durable or extend the life of the product. Discuss this point giving examples in your answer. [6 marks]

11 Explain how rivers and seas can be affected by the manufacture and use of products. [10 marks]

3 Designing and making principles

Designing is a critical activity. Good design can improve people's lives, bring enjoyment and pleasure, and allow us to do things that we never even thought were possible. The possibilities are endless. Unfortunately, poor design achieves the opposite; it can make life difficult and add to people's frustration. In order to create a good design, it is important to understand as much as possible about the problem you are trying to solve or the context that you dealing with.

A large part of designing is concerned with experimenting and trying ideas; the other part is how you communicate those ideas to others. Once the design is refined to a point where the prototype can be made, people can decide if the design has been successful or if it needs improving. You will need to consider the environment that the eventual product or system will exist in, and how well it fits and meets the needs or wants of its intended users.

Making a prototype as accurately as possible will ensure that it is the design that is being tested and not just the making.

This section includes the following topics:

3.1   Investigation, primary and secondary data

3.2   Environmental, social and economic challenge

3.3   The work of others

3.4   Design strategies

3.5   Communication of design ideas

3.6   Prototype development

3.7   Selection of materials and components

3.8   Tolerances

3.9   Materials management

3.10   Specialist tools and equipment

3.11   Specialist techniques and processes

At the end of this section you will find practice questions relating to designing and making principles.

Investigation, primary and secondary data

What will I learn?

In this topic you will learn about:

→ how to use primary and secondary data to understand client and/or user needs

→ how to write a design brief and produce a design and manufacturing specification

→ how to carry out investigations in order to identify problems and needs.

When designing new products, it is important to consider a wide range of information and all the possibilities available to you at the beginning of the project. As you move from a problem to a finished product, you will need to make decisions about every aspect so that your product reaches the right people, performs as you want it to and meets the needs of the client.

Use primary and secondary data to understand client and/or user needs

Research is a continual process of learning more about a subject, so that your decisions are more relevant and can have a greater impact on the people who use your product. When completing a project, you will need to carry out research to gather data that helps you to better understand the needs of the intended user or client for your product.

There are two kinds of research, and you will need to make use of both of these throughout your project:

● **Primary research** is where you find new information yourself. Typically this will include methods such as interviews, questionnaires, analysis of existing products, materials tests and observations.

● **Secondary research** is where you use someone else's information (for example, information from books in a library, magazines or newspaper articles, or the internet). There is nothing wrong with using this type of research provided you make it clear that the information belongs to someone else, that it isn't your own work, and you clearly reference where the information has come from. How you select the information, the conclusions that you draw from it and how you use it will determine whether the project is a success or not, so think carefully about how valuable the information is.

Market research

Market research is carried out by manufacturers in order to gain an understanding of the person or people that they think will use their products. Before a new product is developed it is important to know what people expect from it, how they will react when they use it and whether it will be successful once it is launched. Market research can make use of both primary and secondary sources to gather information about users' needs.

> **KEY WORDS**
>
> **Primary research** any type of research where you collect new information yourself (for example, through interviews, surveys or observations).
>
> **Secondary research** gathering existing data that has already been produced (for example, using books, newspapers, magazines or the internet).

When completing your project, you will need to conduct your own research that relates to the problem and design brief that you have set out to solve. Typical questions that need to be answered in the market research stage are:

- Who is your product aimed at?
- Who are your clients or potential clients?
- What do your clients want from your product?
- How old are they?
- Are they male or female?
- Where do they live?
- How will price change their feelings about the product?
- What kind of lifestyle do they have?
- What products do they use at the moment?
- What are the styling features of the product they currently use?

Successful market research provides a full and accurate picture of how the potential buyers and clients of your product are behaving, as well as how your competitors are reacting to the demands and needs of the same group of people.

Interviews

Interviews and questionnaires are very similar and are examples of primary research methods that gather information from real people. They use questions to find out what people are thinking – interviews usually involve a face-to-face meeting, with a single person answering questions (although the interview could also take place on the phone or via a videoconference), whereas questionnaires consist of a set of written questions for people to answer.

In both interviews and questionnaires you can ask open or closed questions:

- An **open question** is typically used in an interview and allows someone to give any answer they like. For example: 'Looking at the table lamps in the photographs, describe the features and functions you find most interesting.' The answers you get to this question could be wide and varied. A response like this is difficult to put into numbers, and therefore it is harder to compare different people's responses. You will need to spend some time finding the relevant parts of the answer and highlighting them. In some cases the person you ask may not be as precise as you would like, and so you may need to analyse the whole interview and draw conclusions at the end.
- A **closed question** is one where the answers are chosen from a selection that you have provided. For example: 'Do you use a table lamp at home?' The answer to this can only be yes or no. You may recognise these as multiple-choice questions. Closed questions are easier to convert to numbers, so you can very quickly work out how many respondents gave a particular answer. A disadvantage is that the interviewee is limited to the answers you have generated.

When completing a project, it is often useful to use an interview in which you ask just one person questions. This could be someone who is an expert in some way – they could represent your target market or be someone who knows a lot about the kind of product that you are developing. An interview is an appropriate way to gather this type of information as you will be able to ask open questions that allow the person to answer in full, offer important details and explain themselves. When you present this information in a portfolio, you do not need to show every word that was spoken in the interview; you only need to give brief details of the interview itself, such as time, date, who is being interviewed and why they have been chosen, along with important quotes that support your conclusions.

Sometimes it will be necessary to find out what lots of people think, and in this case a questionnaire is a much more effective method. The people who you ask to respond to a questionnaire may not be experts, or even be

Figure 3.1.1 Questionnaires that use closed questions allow you to categorise responses and display results in tables, graphs and charts.

KEY WORDS

Human factors considerations that are concerned with people.

Ergonomics the relationship between people and products, and how they use and interact with them.

aware that they need your new product, but by asking lots of people simple questions you can find out a great deal about what your new product needs to include. If you use a questionnaire you can standardise the way that questions are asked and categorise the responses. The information may not have the depth that is found in an interview, but it can be useful to confirm what a range of people think about the same thing. The results of a questionnaire can be displayed as graphs to demonstrate how each answer compares to the others.

Human factors

Human factors are the considerations a designer must take into account that are concerned with people. The relationship between people and the products they use and interact with is referred to as **ergonomics**. It is concerned with how the human body fits and operates items and equipment, and should be considered when designing all sorts of products, from furniture to the interfaces used on machines. Ergonomic design can prevent repetitive strain injury and other physical problems occurring.

Figure 3.1.2 Driving controls need to be within reach and safe to use for a wide range of people.

Human factors can be categorised as physiological, psychological and sociological factors.

Physiological factors

Physiological factors are related to a person's physical attributes (for example, how the body moves, stamina, strength and coordination). When you are designing a new product you will need to think about the person that will use the product and which of their physical attributes needs to be considered. For example, when designing handles for furniture, the designer would need to think about the strength that people can generate in their hands to allow them to open and close doors or drawers.

The way that people react to physical stimuli is an important method of gaining feedback about how successful a product is, and will help you to make decisions about how the product is used and any changes you may need to make. A well-designed product will be intuitive to use.

An example of where physical attributes are thoroughly considered is the interior layout of a car: a large range of controls need to be kept within reach during driving, while others may well be hidden to ensure a safe experience.

Figure 3.1.3 Red envelopes are often given to people with money inside in China to celebrate the new year.

Psychological factors

The way that people react to new products and experiences also changes how designers plan and make them. Whether people can reach all the right buttons and use a product comfortably and easily (physiological factors) are very important considerations, but the associations that people make between products and aspects, such as colour, sound, smell and appearance are just as important, even if they are more difficult to discover or measure. These are **psychological factors**.

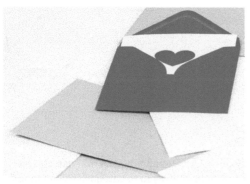

Figure 3.1.4 Red envelopes are given in the UK to hint that the card inside may be romantic.

The use of colour, for example, is very specific to cultures and traditions. White is often associated with purity and cleanliness, and so it is often used in products for kitchens and bathrooms. However, in some countries it is can also associated with death. Red is used in the UK for Valentine's Day and to signify romantic relationships, whereas in China it is used to show good luck, and Chinese New Year celebrations make full use of red items. Taking products from one country to another can sometimes have unintended consequences if these factors are not fully considered by designers.

Sociological factors

The effects that new products have on other people within a group need to be considered alongside the needs of the end-user. These are called **sociological factors**. For example, a new pair of personal headphones may well deliver incredible sound quality for the user, but also produce a significant amount of noise pollution, so that passers-by are forced to listen to music or phone conversations that are supposed to be private.

Ethical questions can also change how people use and purchase products. Clothing production, for example, is often a great source of moral questions. An example of a product that may pose ethical considerations is a down jacket – it is very warm in winter because of the insulating properties of the feathers used from geese and eider ducks; however, the way that these feathers are taken from the birds and the conditions that the birds are kept in could be considered to be cruel by some people, and therefore raises ethical questions. Concerns over the way in which a product has been produced can affect how people react to the product – it may prevent people from buying a product, and therefore designers need to be aware of what is likely to be acceptable to people.

Figure 3.1.5 Would you be happy to know that the birds whose feathers were used to make your winter jacket were mistreated or killed?

Focus groups

A **focus group** involves gathering lots of people together for a discussion to find out what they think. It is a useful way of conducting primary research and helping you to understand the problem you are trying to solve. Large companies use this method to test out new ideas and you can use it too in a project if you have access to a group of people who are likely to use your new product.

At the research stage you are likely to be asking questions, and as the answers will be varied and spoken in quick succession you may need to record the conversation rather than try to write it all down (you will need to ask permission to do this). From this you could draw conclusions about what people want so that you can include these aspects in your design specification.

As you progress through the project and at the end of the project, you may use a focus group to test out your prototype. People in the focus group might use it, offer feedback and change the way that it develops. You will also be able to take photos (again, make sure you ask permission) and see people using your product. This is very satisfying as well as being a valuable source of new ideas to help improve your product.

Figure 3.1.6 Using people to give feedback and test ideas is an important part of product development.

Product analysis and evaluation

Another useful method of understanding client and/or user needs is to learn from existing products with similar features, or that are designed for the same market.

When you examine a product you will need to have a clear question that you are attempting to answer. Looking at a few products and writing down general details is unlikely to provide you with any useful information.

Product analysis can be used alongside other methods of research. For example, to find out if teachers are likely to know how to use a digital device in the classroom, a questionnaire combined with a product analysis could be used. You could ask ten teachers what devices they currently own that fit into this category and list the results. You could then analyse the products that they have mentioned to find out how complex they are and what they can do. One method of comparing the products would be to rank the features that you are interested in to find out if there are any similarities or important differences. An example is shown in Table 3.1.1, where products are awarded a mark out of ten.

Use primary and secondary data to understand client and/or user needs

Ranked out of 10 (10 is high)	Cost	Usability
iPhone	10	9
Samsung Galaxy S7	9	10
Acer laptop	4	6
iPad	7	9
Microsoft Surface Pro	8	7
Macbook Pro	10	8

Table 3.1.1 Using methods such as ranking will allow you to quickly gain the information you need to move forward with your project.

When analysing a product, we can also make assumptions based on knowledge that we already have. To be able to do this you will need to know a lot about the product already, such as who uses it, how it is made and how successful it has been. For example, if you were to analyse a device that helps the elderly open a screw lid on a glass jar, key questions to ask might be:

- What do you already know about the product by looking at it or using it?
- What don't you know about the product?
- Why have the designers and manufacturers used these materials or components?
- What is the history of the product? How long is it used for and what happens to it once it has been used?
- What would come next? If this company were to tackle your problem, what might they come up with?

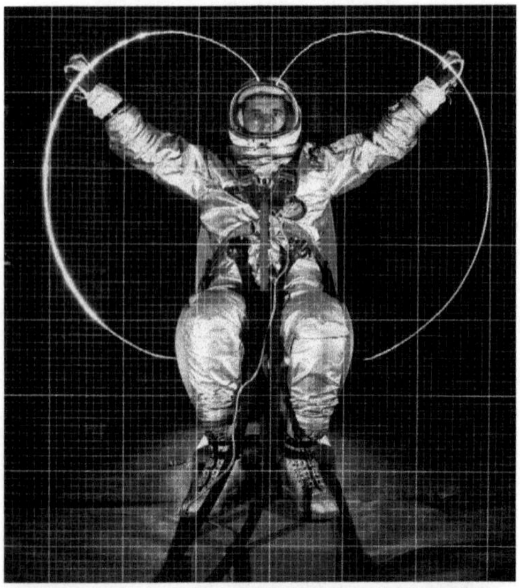

The use of anthropometric data and percentiles

The size of people has been measured and studied a great deal by designers; this information is called **anthropometrics**. It is a long-term record of the sizes of humans (for example, height, weight, waist size, hand span) and is used to relate items to the human body, making it very useful to anyone who intends to produce something that will be used by a large number of people, such as a hair dryer (consider the handle size, the placing of the switches on the handle, and the length of the electrical cable).

When completing a project, you are likely to need to find and use anthropometric data. By using this information to change a design of anything that people will use, you are adding an ergonomic element (making the design relevant to people) to your project.

When you find anthropometric data, it will be presented in one of two ways: as a list or table, or as a graph or visual representation of the information. A typical graph that is used to show how the sizes of people change throughout a population is a **bell curve** or **percentile graph**. This type of graph shows how common a value is across a range of people. For example, if you were to measure the height of

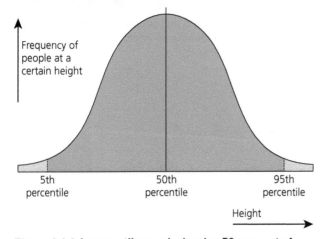

Figure 3.1.7 Anthropometric data is gathered by measuring lots of people in a range of positions. It is useful when considering what size something should be.

Figure 3.1.8 A percentile graph showing 50 per cent of people are of average height or taller, and 50 per cent are of average height or shorter.

100 students in the same year at school you would see the most common size at the top of the curve and the rarest sizes, both big and small, at the ends.

Figure 3.1.8 shows that 50 per cent of people are of average height or taller, and 50 per cent are of average height or smaller. The graph tails off to each end because fewer people are extremely short or extremely tall. To the left of the average, there is a point known as the **5th percentile**, because 5 per cent of the people (or 1 person in 20) is shorter than this particular height. The same distance to the right is a point known as the **95th percentile**, where only 1 person in 20 is taller than this height.

When designing a product it can be important to think about whether to use the 5th, 50th or 95th-percentile anthropometric data. For example, a designer of a car dashboard needs to ensure that all but the smallest drivers can reach the controls.

How to write a design brief and produce a design and manufacturing specification

How to write a design brief

The design activity comes about after identifying a need or problem that has to be solved after studying the context. The brief is a statement that sets out what is to be designed. It can contain some details regarding what the product must do, its appearance and other design constraints.

You can write a design brief in a number of ways, but a simple start to the first sentence is:

'Design and make ...'.

After this you need to include some details. Most of these additional points will come from the design opportunity – the context of the project, and analysis of any research that you may have carried out at the very beginning of the project. You should not include anything in the design brief that you have invented or included because of your opinion or personal interests. The design brief is the first step to showing your clear understanding of the task ahead and the process that you will take to solve the problems.

An example design brief is shown in Figure 3.1.9.

The context
Addressing the needs of the elderly.

Design opportunity
Many elderly people are arthritic and stiff, frail and have some degree of memory, hearing and sight loss, and they can also have trouble balancing due to the consequences of ageing. Therefore, they may need help with bending, reaching, carrying, lifting, opening letters and food jars, turning keys in doors, standing for long periods of time, as well as reminders for appointments and medications.

Design brief
Design and make a device to assist elderly arthritic people who have trouble turning keys in doors. Consider the range of keys in use, the ways to increase the leverage and ease of grip for the elderly user, as well as how the device will engage with existing keys. Consider colour and texture.

Figure 3.1.9 **An example design brief**

Analysis of the design brief

When you are faced with a design brief your first job is to analyse it. The analysis can take the form of asking and finding answers to as many relevant questions as possible that can be posed by studying the brief. The questions will be different for different contexts and briefs. Some example questions are listed below for the example design brief shown in Figure 3.1.9.

- What is the function of the product?
- What does it need to do?
- How often will it be used?
- What can it be made from?
- What 'ergonomic' factors need to be taken into account?
- What designs already exist? How successful are they? Could they be improved?
- Who can be interviewed that understands the problem? What questions should be asked?
- Where can research material be collected to help with the design?

How to produce a design specification

The design specification expands upon the design brief by setting out more detailed design constraints. It generally takes the form of a list of design requirements after further thought and consideration has been given to the design brief, and research carried out. It states what the design must be, must do, and must have, and is referred to when designing and evaluating design ideas. The starting point for producing your design specification may be to look at all the information you have found after analysing your design brief, and identifying the key issues that will guide you when designing. It is important that the design specification covers the key areas that relate to the product you are designing, and it should be composed of simple clear statements that may include information relating to the following factors:

- function
- performance criteria
- safety
- environments it may operate and be kept in
- measurable targets – maximum or minimum sizes, weights, etc.
- ergonomics
- aesthetics
- material availability.

How to produce a manufacturing specification

The manufacturing specification should contain the information necessary to make the product. It should be produced after the final idea has been developed, and it should explain how the product is to be manufactured. It needs to contain all the relevant information so that someone else could use the specification to manufacture the product, such as:

- materials to be used, including quantity and sizes
- components to be used to assemble the product
- working drawings showing dimensions
- tools and equipment to be used
- health and safety considerations
- instructions regarding the sequence of making
- acceptable tolerances.

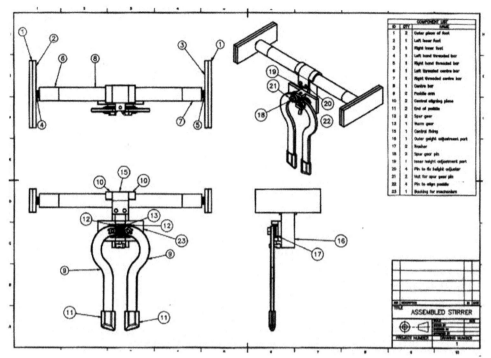

Figure 3.1.10 **An example of a detailed working drawing**

Carry out investigations in order to identify problems and needs

Why a designer considers alterations to a brief and modifying the brief

As designers work through projects and make attempts to solve the problems that have been identified, new problems may be found that might change how the design brief is understood. In one sense, this process of finding things out by developing solutions is an example of primary research because it is new information that has been found out by doing something yourself. This research can then alter the design brief in ways that were not thought of at the start of the project.

Using the example design brief from Figure 3.1.9, after looking at existing solutions and considering what elderly users and other people need from the device, it is possible that new questions might arise that change the way that the design brief needs to be written. New technologies may be found that could unlock doors without the need for a traditional key, or a range of devices might need to be developed for different shaped key bows (the bow is the part left protruding from the lock that the user turns). Therefore a change to the design brief may be necessary – this may come about by carrying out further research, materials testing, prototype testing, design development and evaluation, including of client feedback. If this is necessary, it is an acceptable part of the design process.

If further research and testing finds aspects that are critical to the success of the project, and without which the project would not work, changes to the design brief should be made. Some research might lead you in a direction but offer some scope for change. An alteration to the design brief, provided it is justified by relevant research, is a useful addition to any design project.

KEY POINTS

- Market research is carried out to gain an understanding of the target market for a product.
- Interviews, questionnaires and focus groups are examples of primary research methods that use questions to find out what people are thinking.
- A designer must take into account human factors when designing products, including physiological, psychological and sociological factors.
- Anthropometrics is a record of human measurements that is very useful to anyone who intends to produce something that will be used by a large number of people.
- A design brief describes the problem or situation that needs a design solution.
- When you have completed initial research, you will have answers to many of the questions that you set yourself at the beginning of your project. These answers will influence ideas for your project. You will not necessarily have done all the research at the beginning of the project, and you may not have all the answers. Further, ongoing research may be needed during the design process.
- A design specification is a list of design criteria; it is used to determine the success of your ideas.
- A manufacturing specification contains all the information necessary to make the product.
- As designers work through projects and make attempts to solve the problems, new problems may be found which change how the design brief is understood.
- If further research and testing finds aspects that are critical to the success of the project, and without which the project would not work, changes to the design brief should be made.

Check your knowledge and understanding

1 State three examples of sources of secondary research.
2 Explain the difference between anthropometrics and ergonomics.
3 Describe how would you present primary research data that you had gathered from a questionnaire.
4 What two key documents are needed at the start of a project?
5 Why might a designer need to alter a design brief when working through a project?

3.2 Environmental, social and economic challenge

What will I learn?

In this topic you will learn about:

→ how the following might present opportunities and constraints that influence the processes of designing and making:
- deforestation
- possible increase in carbon dioxide levels leading to potential global warming
- the need for fair trade.

When designing and making in any scale of production, it is important to consider the effects of your product and its manufacture on the environment and the community. This is a shared responsibility between the designer, maker, retailer and consumer. Careful design and effective management of the production process can substantially reduce the impact that a product has on the environment.

Deforestation

Deforestation occurs for two reasons:

- Trees are cut down to use for timber- and paper-based products, and the trees are not replaced/replanted
- Trees are cut down to create grazing space for animals, or to create space for growing crops.

The cutting down of trees is not always managed in a sustainable way, with hardwoods (which take a long time to grow) being cut down and not effectively replaced.

Designers and manufacturers have a responsibility to source materials from sustainably managed forests, where new trees are planted to replace those cut down.

Designers might choose to use a Forest Stewardship Council (**FSC**) certified material. The FSC is an organisation that reviews how materials for products made from trees are provided and gives approval to sustainable practice. It allows consumers to recognise the logo and know that the wood, paper, board and so on used in the product comes from well-managed forests. These forests need to go through a number of checks before accreditation. These include the need to:

- comply with all laws
- protect indigenous communities and their right to be on areas of land
- be positive for the local communities
- maintain ecosystems and the environment.

Designers may consider fast-growing softwoods, rather than slow-growing hardwoods, for their products. They may also want to ensure that only FSC materials are used. Companies can make an impact by introducing 'zero deforestation' policies that clean up their supply chains.

Figure 3.2.1 **The FSC logo**

Companies should set ambitious targets to maximise the use of recycled wood, pulp, paper and fibre in their products.

Global warming

Global warming is the increase in the average temperature of the Earth's atmosphere and oceans, which has been seen increasingly over recent years. This change in temperature is believed to be a permanent change in the Earth's climate and has an impact on the whole planet. Global warming occurs when gases such as carbon dioxide (CO_2) collect in the atmosphere. These gases absorb sunlight that has reflected off the Earth's surface. These pollutants trap the heat that would ordinarily pass through the atmosphere and escape into space, and the result is that the Earth becomes gradually hotter. This is known as the greenhouse effect.

There is some discussion about whether global warming is real, but climate scientists looking at the data and facts agree that the planet is warming. Carbon dioxide and other greenhouse gases are released because of a number of factors, but most of these are created by humans. They include the burning of fossil fuels and the use of vehicles in transportation, land clearing and agriculture.

Figure 3.2.2 Thomas Brisebras's Preserve

The planet is undergoing changes attributed to global warming, and it is anticipated that it will continue to do so at a greater rate. These changes may include the rising of sea levels due to the melting of the polar ice caps and a change in weather patterns.

The products that we design and make have an impact on the amount of CO_2 released into the Earth's atmosphere. Some of these products can be improved by choosing different materials, or using alternative energy to power machinery. It is possible to design products that use less material, or consume less power, or can be recycled at the end of their use. All of these things can improve the effect the product has on the environment in terms of the amount of harmful gases released.

Designers often have to attempt to improve their products to make them have less of an impact on the environment. For example, car manufacturers have targets set by the government that they have to meet in terms of the emissions from their vehicles. Some designers make new products solely for the purpose of illustrating energy use or encouraging consumers to consider sustainability.

The designer of the product in Figure 3.2.2, Thomas Brisebras, has designed an hourglass which shows the energy consumption in a home each day. When the hourglass empties, the electricity that runs through the house turns off automatically, indicating the daily maximum.

Fair trade

Fair trade is a trading partnership that works towards fair prices and better working conditions for farmers and workers who produce goods all around the world. More than 1.5 million farmers and workers benefit from being part of Fairtrade. The idea is that these people gain a fair price for their goods and that they are protected from exploitation. It is about supporting the development of communities and giving them a better chance of protecting the environment in which they live and work. Typical fair trade items include food, flowers, gold and cotton products.

You may have seen the **FAIRTRADE Mark** on products. This Mark means that the ingredients or materials in the products have been given approval and meet the Fairtrade's social, economic and environmental Standards. The Standards include a Fairtrade Premium, which is invested in business or community projects such as better facilities for health care, education and transport. Mostly, Fairtrade helps smaller-scale farmers and workers, who are often the most vulnerable in comparison to larger businesses.

As a designer you may wish to use materials from Fairtrade licensees. This is a difficult decision to make as Fairtrade products are often more expensive due to the need for farmers and workers to gain extra profit and investment from buyers. The producers are all based abroad, so the ethical benefit could be offset by transportation costs, CO_2 used in transport, and so on. These issues need to be considered, as Fairtrade can not only be an advantage ethically, but can also make the product more appealing to an increasingly ethically conscious consumer.

Figure 3.2.3 **The Fairtrade mark**

STRETCH AND CHALLENGE

Research three products that you feel have successfully considered one of the three topics discussed above; deforestation, global warming or fair trade. Discuss how the designer has addressed these issues.

KEY POINTS

- Designers have the responsibility for choosing their materials and processes carefully, so as to have the least impact on the environment and vulnerable communities.
- Deforestation can be avoided with the correct management of forests. Designers can choose to use FSC materials, which are wood, paper or board that has been taken from sustainable and well-managed forests.
- Fair trade products have been made by communities who have been given a fair price for their goods. This offers some protection to these communities against exploitation.

Check your knowledge and understanding

1 Explain the term 'deforestation'.
2 What are the main causes of global warming?
3 What does fair trade do for farmers and workers?
4 What are designers and makers trying to do to address problems of deforestation, increased carbon dioxide and unfair trading?

3.3 The work of others

What will I learn?

In this topic you will learn about:
→ the work of well-known designers and companies
→ how their work can help us with our own designs.

Design movements

Design movements are particular styles of design popular with a group of people. We often look back at these design movements today to try to gain inspiration for our own ideas. Looking at the work of others will help you to develop a better understanding of design history and will help you to make informed choices about your own designs.

While you are not expected to have a detailed knowledge of these design movements for you GCSE, they will help you to understand the work of designers and companies that follow on pages 243–250. Many of these designers belong to or have been inspired by these design movements.

Arts and Crafts Movement: 1853–1907

The Arts and Crafts Movement was set up by a small number of artists and designers, including William Morris. These designers were concerned about the effects of industrialisation and the use of machinery on designing and making. They did not like products made by newly developed machinery and wanted to use traditional methods of craftsmanship to make beautiful products for all. The key features of Arts and Crafts products are:
● simple forms with little ornamentation
● natural materials
● flowers and natural forms
● celtic patterns
● handmade products.

Figure 3.3.1 **An Arts and Crafts product**

Art Nouveau: 1880–1910

Art Nouveau was a highly decorative style which combined traditional craft skills with new materials and machined surfaces. Art Nouveau did not deliberately avoid the use of machines, as the Arts and Crafts Movement did. It shared the same belief in fine craftsmanship and quality, but was happy with mass production. It was a deliberate attempt to create a new style that did not try to replicate historical designs, but looked to the future. Unfortunately, the Art Nouveau-styled products were out of reach for most people, as they were very expensive. This was due to many being made as one-off products by highly skilled workers who would take a great deal of time over one individual product.

Figure 3.3.2 **Art Nouveau ironwork**

The most obvious characteristic of Art Nouveau is its organic, undulating lines, often depicted as flower stalks and vines, insect wings, and other delicate, natural objects. This was often shown through the use of elaborate wrought-iron scrolls.

Art Nouveau architecture made use of many new technological innovations, including the use of exposed iron and large, irregularly shaped pieces of glass. In graphic design particularly, designers used organic and geometric shapes to create flowing lines in designs often featuring women and flowers. These are seen, for example, in the wrought – iron train stations.

The key features of Art Nouveau are:
- sinuous, curvy lines
- the typical whiplash line
- stylised flowers and other natural forms
- the female form, with long, flowing hair.

Art Deco: 1908–1935

Art Deco is characterised by the use of geometric shapes and patterns which are usually strongly symmetrical. The use of mass production meant that this design movement was not simply for the wealthy, but for all. One of the most iconic examples of Art Deco is the Chrysler Building in New York. You can clearly see from the images of this building the detailed repetition of geometric shapes.

At this time, travel became popular and so the style often involved materials that had rarely been used before in the Western world, such as tortoiseshell, ivory and animal skins. In 1922, Tutankhamun's tomb was discovered, and it became fashionable to use symbols such as Egyptian pyramids and sphinxes in design. Art Deco designs were often bold and contrasting in colours and materials. New materials such as Bakelite were used to contrast with more traditional materials, such as silver, to create striking products. Art Deco designers were famous for using the 'stepped profile', where lines copy each other a set distance apart. Lines were always perfectly set away from each other and often in parallel. The stepped profile is the epitome of the Art Deco style, but zigzags, chevrons and lightning bolts are also commonly found.

The key features of Art Deco are:
- simple natural motifs – shells, sunrises, flowers
- geometric and angular shapes
- the Hollywood style of mirrors, chrome, glass, shiny fabrics
- stylised images of aeroplanes, cars, and skyscrapers
- theatrical contrasts – highly polished wood and glossy black lacquer mixed with satin and furs.

Modernism

De Stijl: 1917–1931

De Stijl was founded by two Dutch abstract artists, Piet Mondrian and Theo van Doesburg. Their art was strictly geometric and based on horizontal and vertical lines. Mondrian used pure, bright colours which were bold and striking.

The artists focussed on simple, bold primary colours and rejected fussy decoration. De Stijl artists applied their style to a range of media in the arts and in the designing of functional products.

Figure 3.3.3 Art Nouveau poster

Figure 3.3.4 Chrysler Building

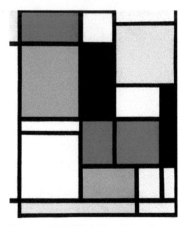

Figure 3.3.5 Piet Mondrian focussed on simple, bold primary colours and rejected fussy decoration.

Design movements

De Stijl simply means 'the style' in Dutch and was promoted through a journal of the same name and written by the artists. They planned to use art to try to heal the scars of the First World War, and saw art as a way of escaping from the memories of a troubled time.

The key features of De Stijl are:
- strong focus on graphic design
- contrasting primary colours
- no fussy decoration.

Bauhaus: 1919–1933

The Bauhaus was a school of art and design founded by Walter Gropius in 1919. The idea of the school was a collaboration of all arts, including architecture, photography, graphic and industrial design. Unlike other design styles which have come in and out of fashion with the general public, the Bauhaus style has been a constant inspiration to a range of designers and artists throughout different time periods.

The designers involved in the Bauhaus were keen to embrace new technology and the opportunities mass production gave them to produce practical, functional products for a range of consumers. Before this period, ordinary objects had never been deemed beautiful, and the design of such items was given little thought. Bauhaus encouraged designers to produce objects that ordinary people could afford and that were mass produced, yet beautiful.

Figure 3.3.6 Barcelona Chair by Mies van der Rohe

The ethos of the design school was 'form follows function' and designers believed that the beauty of a product was in its functionality rather than in excessive decoration. It focussed on geometric shapes and products were simplistic and balanced.

Bauhaus designers often used bold, contrasting colours and materials. Tubular stainless steel or chrome was often paired with bold block colours of black or red. Designs were often considered futuristic, using new materials, techniques and exposing aspects of the designs, such as joining methods, which would previously been covered with ornamentation. Graphic designers used bold contrasting colour schemes and experimented with typography. Designers set out to grab attention with striking letters in lower-case or upper-case fonts, often standing lettering vertically or diagonally.

The key features of Bauhaus are:
- new materials and manufacturing techniques
- products are often mass produced
- simplistic designs
- clean geometric shapes
- affordable products
- less ornamentation.

Figure 3.3.7 Joost Shmidt Poster

Postmodernism: 1970–1990

Postmodernism was one of the most controversial of design movements, with design groups such as Memphis and Alessi taking a new pleasure in the designing and making of products in a way intended to spark interest and opinion.

Postmodernism shattered preconceived ideas about what design should look like. It was often amusing, certainly controversial, and less about the simplicity of the modernism movement. Postmodernism's designers based their products on complexity and contradiction.

Memphis: 1981–1988

Memphis was an Italian design movement of designers and architects who wanted to shock the design world with unusual products which did not need to conform to other ideas of the time. Ettore Sottsass, one of the founding members of the group, called Memphis design the 'New International Style'.

The designers felt that designs at the time lacked humour and were dull and uninteresting. They felt that designs were too slick and often boring and dark in colour. The Memphis designers created the opposite style of design to this. They embraced colour, creating shockingly bright and bizarre objects in comparison.

The Memphis design group often painted or laminated their work to create bright, shiny colours.

Objects such as the **Beverly cabinet** used the clashing colours of yellow and green with snakeskin-prints. The **Carlton bookshelf** challenged people's ideas of shape and colours with its angled shelves and bright hues.

The key features of Memphis are:
- bright striking colours
- unexpected, shocking combinations
- laminated wood
- angular lines.

Figure 3.3.8 Ettore Sottsass with one of his great creations

Past and present designers

Harry Beck

Henry Charles Beck, known as Harry Beck, was an engineering draughtsman at the London Underground Signals Office. In 1931, he created a new map to show the London Underground system. Beck could see that the network had increased rapidly and had become too big to represent geographically, so he worked on a solution to this problem.

Beck said: 'Looking at an old map of the Underground railways, it occurred to me that it might be possible to tidy it up by straightening the lines, experimenting with diagonals and evening out the distance between stations'.

Beck's solution was based on electrical wiring diagrams and created a much simpler design, which is now a common system for transport maps all around the world. The Underground's publicity department were worried about his initial proposal feeling that it was too different from what people were used to seeing. After a few modifications, they agreed to create a small trial version of 500 copies. Soon after, and due to the success of this version, the map was published fully in 1933 when 700,000 copies were created. It was so successful that it was printed again after only one month.

Marcel Breuer

Marcel Breuer was a student and teacher at the Bauhaus school between 1920 and 1928. He used new technologies and new materials to design and make iconic objects. One of most famous pieces was the Wassily Chair, designed for Wassily Kandinsky in 1925. This piece was made from polished, bent, tubular steel, which later became chrome plated. It reflected the style of the Bauhaus design movement in that the object was deemed beautiful because it did not decorate or over embellish, but was functional, sleek, geometric and clean in appearance. It is still copied and sold today.

Breuer used tubular steel frequently at this time due to its affordability. He used it to create strong products which could be mass or batch produced, and found that he could use this material to create hammock-like structures for seating which allowed for a lack of springs. He designed a range of furniture, including stools and cupboards, which despite selling at the time, became even more popular years later.

In 1935, Breuer was concerned about his safety due to political unrest and moved to London. Here he worked as an architect and product designer, but his style changed to use a different type of material. At this stage, and inspired by Alvar Aalto, he used plywood rather than his previous metal pieces. This new work was often a reflection of his earlier designs but in a new material.

Coco Chanel

Gabrielle Bonheur 'Coco' Chanel was a French fashion designer who built her own fortune after a troubled childhood. She grew up in an orphanage after her mother died and her father abandoned her. She was known for great shifts in fashion for women – at a time when women wore corsets, she developed new styles that were more comfortable and normally worn only by men. These included collared shirts with ties and boater hats. In 1925, she introduced the Chanel suit with a collarless jacket and well-fitted skirt, which are well-known designs today. Chanel also created legendary designs such as the 'little black dress', which originally would not be seen as appropriate due to the colour black being known as a colour associated with mourning.

Chanel was the first fashion designer to create her own fragrance. This is now a popular way for fashion designers to diversify, with many lending their names to perfumes or makeup. Chanel's signature scent, Chanel No. 5, has become an iconic product that is still very popular today.

Norman Foster

Norman Foster was born in Manchester and later studied at Yale University where he gained a Master's degree. It was in America, at Yale, that he met his future business partner, Richard Rogers, who worked with Foster, Su Brumwell and Wendy Cheeseman to create a series of houses in the UK.

After this time, Foster Associates was created and Norman Foster created a number of key pieces of architecture commissioned by multinational companies such as IBM, and later pieces such as Stansted Airport in Essex.

Foster's buildings are of a high-tech contemporary style with great attention to detail, consideration for the environment and innovative use of materials and technology. He uses modern materials such as steel and glass to create his contemporary feel, and creates large open spaces in the centre of buildings to allow for greater communication in built spaces such as offices.

Figure 3.3.9 Singapore's Expo Station

Figure 3.3.9 shows one of Norman Foster's designs. The building is modern and futuristic in design, using a combination of titanium steel and glass. There is a mirrored stainless-steel ceiling which reflects the passengers and trains as they pass. Little artificial light is needed as the structure reflects light onto the platform. The titanium panels reflect the sun away from the building creating a cooler environment in the hot Singapore.

Sir Alec Issigonis

After studying engineering, Alec Issigonis joined Morris Motors in 1936 as a suspension designer. There he developed the Morris Minor, which remained in production from 1948 to 1971. This was the first completely British car to reach over 1 million sales. Original examples of this car are still collectors' items today.

Figure 3.3.10 **The Mini**

Issigonis then went on to develop the Mini in 1959. This was a reliable car which was particularly small and compact yet had enough space for four passengers. It was an inexpensive and fuel-efficient car, which made it more accessible to a larger group of consumers. By the time of Alec Issigonis' death, more than 5 million had been sold, making it the best-selling British car in history.

Alexander McQueen

Alexander McQueen was a London-based, English fashion designer who worked for the Givenchy fashion line and Gucci before starting his own line. He said that working for other people stifled his creativity, and he developed his own style when working on his own.

McQueen was a controversial designer who pushed the boundaries and created very theatrical fashion shows. He was known for creating visual impact and sparking conversations. Examples of this include casting models as chess pieces or a model being spray painted by robots.

Alexander McQueen gained four British Designer of the Year awards and was a highly successful fashion designer.

Figure 3.3.11 **Alexander McQueen was a controversial designer.**

William Morris

William Morris was one of the founders of the Arts and Crafts Movement. He is one of the most famous designers of this time. He set up a company called Morris, Marshall, Faulkner & Co. in 1861 with fellow artists, which produced a wide range of products. Morris is famously known for his highly decorative wallpaper, which can still be bought today. His patterns are inspired by his knowledge of natural forms, such as plants and birds. Morris took the natural forms that he found while in the English countryside and used them to decorate the inside of our homes.

Morris also designed a range of other products, including furniture, textiles and jewellery. His textiles pieces were often similar to his wallpaper designs, having a strong focus on influences from nature. His furniture designs were functional, but hand-crafted and ornate. The original Morris Chair, pictured in Figure 3.3.13, had dark-stained woodwork, turned spindles and heavily decorated upholstery.

Morris famously said, 'Have nothing in your house that you do not know to be useful, or believe to be beautiful.'

Figure 3.3.12 **A wallpaper design by William Morris**

Figure 3.3.13 **The Morris Chair**

Figure 3.3.15 **'Dragonfly' Tiffany lamp**

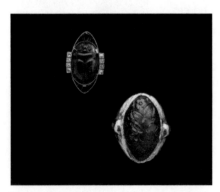

Figure 3.3.16 **Raymond Templier rings**

Mary Quant

London designer Mary Quant wanted fashion to be affordable for the younger generation, and opened her own retail outlet on King's Road. She was unhappy with the range of items available in her shop and decided to start to create her own clothes to sell there. The result was the white plastic knee-high boots and tight sweaters we associate with 1960s fashion. After the success of her first shop, a new outlet opened in Knightsbridge, and soon Quant was mass-producing and exporting to the USA to keep up with demand.

Her famous mini skirt was created in the mid 1960s, although Quant insists it was not her who invented this but the girls who came into her shops requesting skirts of shorter and shorter length. Quant, at this time, expanded her range further offering patterned tights and a cosmetics range.

Louis Comfort Tiffany

Louis Comfort Tiffany was an American painter, craftsman and designer who became particularly famous for his work with glass. By the 1890s, he was a leading

Figure 3.3.14 **Mary Quant**

glass producer, experimenting with interesting ways of colouring the material. He became famous for 'Favrile' glass, an iridescent and freely formed material sometimes combined with bronze alloys. Having started the Tiffany Glass and Decorating Company, Tiffany was commissioned to decorate part of the White House in Washington, D.C. Here he created a huge stained-glass screen in the entrance hall. Tiffany's most famous designs were his stained-glass lamps, which are widely replicated today. These lamps have brightly coloured and ornate shades, often using moonstones to enhance aspects of the design. Their stems are also highly decorated, often with the female form, and are reflective of the curves and shapes found in the natural world.

Raymond Templier

Raymond Templier was born in 1891 and came from a family of jewellers who lived and worked in Paris. Templier studied at the École Nationale Supérieure des Arts Décoratifs in Paris until 1912, and there he developed his own individual style. When Templier entered the family firm in 1919 he was less interested in the traditional jewellery being made by the company at the time. His designs were much more Cubist, using geometric patterns and Art Deco stylings. He often used lacquer or enamel with white gold or silver, rather than the heavy use of diamonds. He was also known for using dark stones, such as onyx, to contrast white metals such as platinum. His designs were geometric (using straight lines and shapes) and bold with geometric stones and contrasting materials.

Gerrit Rietveld

Rietveld designed for mass production. His designs were simple and ultimately suitable for machine manufacture. He originally designed his famous Red and Blue Chair in 1917, but only changed the chair's colours to red and blue after becoming influenced by the De Stijl movement in 1919. His design became an icon of the De Stijl movement and reflective of the industrial design that emerged due to this design style.

In the design of this chair, Rietveld transformed the traditional armchair from a bulky, heavy design to a geometric work of art. He did not want to create an object that focussed on the material but one that could 'stand freely and brightly on its own two feet'.

Figure 3.3.17 Red and Blue Chair

Figure 3.3.18 Zig Zag chair

Rietveld stopped working with De Stijl in 1928, and became associated with a more functionalist style of design. At this time he focussed on architecture and was concerned with social housing, inexpensive production methods and new materials. He was ahead of his time in the use of certain materials, such as prefabricated concrete slabs, but worked primarily for independent employers, so was unable to impact on social housing until much later. When he did begin to develop social housing, his aim was to create more affordable 'standard dwellings' that used standardised materials and construction more suited to an assembly line. This allowed for cheaper buildings that could be built on small timescales.

Rietveld designed the Zig Zag Chair in 1934, which was a much more minimalist design and free of the bold contrasting colours of his time with De Stijl.

Charles Rennie Mackintosh

Mackintosh was born in Glasgow in 1868 and attended the Glasgow School of Art. Here he met Herbert MacNair and the Macdonald sisters, with whom he shared a similar style, and they became known as 'The Four'.

They liked the long sinuous lines of Art Nouveau, and focussed on animals and vegetation in their designs. They produced a wide range of products from posters to furniture. Some of Mackintosh's best known work includes his chairs and the design of the Glasgow School of Art.

Figure 3.3.19 Charles Rennie Mackintosh chair

Aldo Rossi

Aldo Rossi was an Italian architect and product designer. He believed that rather than disrupt the atmosphere of an area such as a city, it was important to design buildings that respect their environment and build on the common architecture of the area.

Figure 3.3.20 San Cataldo Cemetery

One of Rossi's first buildings was the Cemetery of San Cataldo in Modena, Italy. Rossi's design for the cemetery building was a geometric cube with square windows in perfect symmetry. It was very severe in appearance and lacked any kind of ornamentation.

This building mirrored many elements of the style of local factories.

Aldo Rossi also designed smaller household products in connection with the Italian design group Alessi. He developed a range of coffee-related products, such as espresso makers, which he is well known for. These products were sleek and sophisticated in form. He used materials such as stainless steel and copper to create this modern look.

Figure 3.3.21 Flying Carpet Armchair by Ettore Sottsass

Figure 3.3.22 Vivienne Westwood

Ettore Sottsass

Ettore Sottsass was the most influential designer of the Memphis design movement. He was born in 1917, and died in 2007, after a long career in design. His work ranged from furniture to jewellery and was wide and varied, from his work with Memphis to his later work with Alessi. Ettore Sottsass was an interesting and unusual character, and did not like to conform to society's expectations. He wished for his designs to be the same. For example, he believed that objects should not just fulfil a function, but suggest a new way of functioning: a chair might suggest 'a new way to sit'.

Philippe Starck

Philippe Starck is a famous present-day designer. He has worked with a number of design groups and other designers, including Alessi, and his designs are typical of postmodernism. He enjoys pushing the boundaries of design with unusual form and ideas, and has been known to push political ideas in the design of products. Starck's most famous design was the Juicy Salif Lemon Squeezer. This design has been widely discussed due to its unusual form and its controversial style. It has been said that the form of this product has meant the sacrifice of function: the product looks striking but isn't particularly efficient at getting juice from lemons. However, this product is still widely sold today.

Vivienne Westwood

Vivienne Westwood was born in Derbyshire in 1941. She worked as a teacher but made jewellery in her spare time. She met Malcolm McLaren (an art student and future manager of the Sex Pistols) and began to develop her creative skills and passion for artistic pursuits. In 1971, McLaren opened a shop in London to sell Vivienne Westwood designs and the outlet became an iconic part of the punk movement.

She was instrumental in the fashion decisions of the Sex Pistols, and influential in all fashion at that time. As this fashion began to change, so did Westwood, creating different designs including frilly shirts and tweed suits that many people copied.

Westwood had a controversial style that was different to the typical style of the time, and an outspoken nature that made her a topic of conversation among the British public. For many years, Westwood lived in a small and modest South London flat despite having made her fortune, and travelled to work on her bike every day.

Past and present companies

Braun

Braun is a well-known German company designing and manufacturing small appliances. It was started by Max Braun, who was a mechanical engineer who started a small engineering shop. He achieved his original success producing components for radios. After eight years, he was making entire radios, and soon after, Braun's company became one of Germany's biggest radio manufacturers. The company grew and diversified, producing a wide range of products including cameras, electric shavers and coffee makers.

The Braun brand is seen as a combination of functionality and technology. During the pop art era, this was combined with an attention to colour. For nearly 30 years, Dieter Rams worked as head of design for Braun. Many of his designs are now displayed at the Museum of Modern Art in New York.

Figure 3.3.23 A Dieter Rams' radio

Dyson

Dyson was started by James Dyson after he became frustrated with the workings of his vacuum cleaner and wanted to develop a product using new, improved technology. Dyson worked for five years and produced 5,127 prototypes before he created his first bagless vacuum cleaner. This product used 'cyclone technology' to capture small particles of dust without the use of a bag. The cyclone idea came from a sawmill that used cyclone technology.

Other inventions include a washing machine which uses two rotating drums (this product was discontinued after a lack of success). Dyson also developed a hand dryer which uses a thin sheet of moving air to force the water away from your hands.

Apple

Today, Apple is an extremely well-known company producing consumer electronics and computer software. Apple has produced many famous and successful products, such as the Apple iPhone, the iPod, and the Mac personal computer. Apple was founded by Steve Jobs, Steve Wozniak and Ronald Wayne in April 1976, but the company has developed and changed over the years, and Jonathon Ive has become another well-known name associated with it.

Ive worked with Steve Jobs to improve Apple after a low point in the company's history. Now Ive is the chief design officer, and has led Apple's design team since 1996. He has over 5,000 patents and has received a number of design awards for his work with Apple. Ive has designed many of Apple's most famous products, including the iMac, iPod, iPod Touch, iPhone, iPad and Apple Watch.

Figure 3.3.24 James Dyson and his bagless vacuum cleaner

Apple products are sleek and stylish, and the technology they use is fundamental to their successful design.

Figure 3.3.25 **Michael Graves kettle**

Alessi

Alessi is an Italian design company designing products from chairs to kitchenware. Alessi are known for creating striking and iconic designs for everyday products and their products are known as 'designer' in the modern use of the term. The company was founded in 1921 by Giovanni Alessi and remained in the Alessi family for many years, from Carlo and Ettore Alessi, to more recently Aldo Alessi. Alberto led the company through particularly successful collaborations with designers iconic to this time, such as Achille Castiglioni, Richard Sapper, Alessandro Mendini and Ettore Sottsass. Alberto Alessi led an exciting stage of postmodern design, stating, 'A true design work must move people, convey emotions, bring back memories, surprise, go against …'.

Alessi designs are often statement objects intended to start conversations and allow people to rethink what they believe products should look like. Often these designs place more importance on personality and interest than on functionality. This is true of the Michael Graves kettle, which is intended to bring pleasure to the user by whistling when boiling. This adds a humour to the product and has proved very successful, with this design being a bestseller for many years.

The key features of Alessi are:
- products with personality
- creative, thought-provoking designs
- often using bright colours and metal
- mass-produced items.

KEY POINTS

- Design movements are particular styles popular with a group of people. Within these design movements there were key influential designers.
- Harry Beck is famous for the design of the London Underground map.
- Marcel Breuer was part of the school of design named 'Bauhaus'.
- Norman Foster is a successful architect.
- Sir Alec Issigonis designed the Morris Minor and the Mini.
- William Morris is famous for his furniture and wallpaper designs
- Alexander McQueen, Coco Chanel, Vivienne Westwood and Mary Quant were all fashion designers of different times and styles
- Louis Comfort Tiffany designed the famous Tiffany lamps, which are still manufactured and widely copied today.
- Raymond Templier is famous for his influence in the Art Deco style.
- Gerrit Rietveld designed products for the group De Stijl.
- Charles Rennie Mackintosh is a famous Scottish designer who produced a wide range of products, from posters to furniture.
- Aldo Rossi, Ettore Sottsass and Philippe Starck all worked at some point for the Italian design group, Alessi.
- Braun, Dyson, Apple and Alessi are successful companies that still design products today.

Check your knowledge and understanding

1. Give an example of a product which may have been inspired by an alternative design movement.
2. Give an example of a designer and explain the key features of their products.
3. Name a designer that has influenced you and explain what you like about their designs.
4. Explain the key features of the Alessi design group.
5. Explain the key influential changes in the fashion world due to the work of Coco Chanel.

3.4

Design strategies

What will I learn?

In this topic you will learn about:

→ coming up with ideas that are imaginative and creative
→ collaborating with other people to broaden your ideas and develop them
→ understanding the needs and wants of users
→ applying a systems approach in complex design situations
→ the use of iteration to improve a product or system
→ techniques that prevent designers becoming fixated on a single idea
→ exploring, developing and evaluating your ideas.

Very rarely do designers get ideas from looking at a sheet of plain paper. And yet, isn't that what you sometimes feel you are expected to do? So, what can you do to be creative, use your imagination and be inspired to create innovative and creative designs such as an energy-saving lamp? There are several things you can do that will make it easy for you to come up with lots of exciting ideas from the start of your project and through its development. This topic will explain some different approaches which you might use on their own, or together, to give you the best outcome.

Design strategies

It is worth remembering that not all successful designs come from professional product designers. Many are from young people who see an opportunity and rise to that challenge! You can see examples of this on television. Perhaps your prototype could be a potential invention for a design-based television programme, like law student Rachel Lowe's was. Rachel needed a part-time job and so she became a taxi driver. One evening while working on her shift, she came up with the idea for the *Destination* board game. It went on to become a best-selling game at Hamleys, the famous London toy store.

Rachel Lowe later negotiated a deal with Warner Bros. to produce *Destination Hogwarts*, set in Hogwarts, the school in which J.K. Rowling set her famous Harry Potter stories. The players are students at the school, amassing house points on their journeys around it.

Rachel had a lucky spark of inspiration that ended in success. To do the same in your project, you need to start planning the designing so that you have lots of resources to help you do this.

Collaboration: working together

Large successful companies such as Rolls Royce and Airbus have designers who work in teams to solve problems, in the belief that this is not a task that is performed well in isolation. This allows them to share understanding of what is required when analysing the task, and to come up with a range of possibilities that would not perhaps have arisen when working alone.

> **ACTIVITY**
>
> Investigate products that are innovative and creative. Discuss with your group what you like about them and why you think they are successful.
>
> You may find the following website a useful starting point: www.innovate-design.co.uk/dragons-den

When you are in the initial stages of a project it is sometimes a good idea to work with others in a group, especially if you are working on a similar project or context. For example, rather than working on your own, your group could find a range of images that could help to influence your design and use these to create an image board or mind map for everyone to use. By bringing together everyone's thoughts you will end up with a much wider set of resources to help you. You could do this using the internet or magazines, or even by taking photos yourself to show examples of products, images, textures and patterns. Designers often use these image boards to show to a client to ensure they are meeting the brief.

It is important that you go on to analyse this yourself, but even then you could work together initially. Analysing a task often feels more difficult than analysing an existing product, but it is a critical part of any design project and one that is all too often skipped in favour of other seemingly more interesting parts of the project.

If you go shopping for a new pair of trainers, you may well 'analyse' a few before you buy – we call this choice. Obviously you don't write this all down – you do it in your head, often subconsciously. You ask yourself a lot of questions: What colour or design do I want? How much money do I have? Will they suit me? Are they comfortable? Are they fashionable? Do I want leather or fabric? And so on. When you have to analyse a design task it feels more difficult because you are not sure what questions to ask. This can be much easier if you work together. When you were starting to learn how to design, your teacher set the design brief. You may have analysed your task using a simple mind map, and you were limited to a fixed range of materials, skills and time. At this stage in your design education, you are expected to write your own brief, and although a mind map will still be a good starting point you will need to look much deeper into the range of problems and possibilities you will encounter! Analysis is about asking questions. You could work collaboratively to set the questions and then answer them independently to fit your own specific design brief. Sometimes we use a set of reminders to help us formulate these questions – Access FM and CAFEQUE are common examples. Remember you don't need to answer all of these questions before you start a design. Some will be answered through your development work and this is good as it is one of the most important parts of a project!

ACTIVITY

Work together with others to produce an image board.

Aethestics	What the product will look like; its colour, finish, etc.
Cost	The price someone will pay for the finished product. Also the total cost of all resources required to make the product.
Customer	Who is going to buy the product and why (the target market)
Environment	The effects the product will have (environmental – the 6 Rs, social, ethical, moral)
Size	The size the product must be (measured in mm)
Safety	To be considered both during and after manufacture (i.e. for the people making the product, and those who buy it)
Function	What the product does; what it includes, holds, etc.
Materials	What the product will be made from.

Figure 3.4.1 **Access FM**

Cost	The price someone will pay for the finished product. Also the total cost of all resources required to make the product.
Aethestics	What the product will look like; it's colour, finish, etc.
Function	What the product does; what it includes, holds, etc.
Ergonomics	How easy/comfortable the product is to use.
Quality	How well the product is made/built; the quality of the materials used to make it.
User	Who is going to buy the product and why (the target market).
Environment	The effects the product will have (environmental – the 6 Rs, social, ethical, moral)

Figure 3.4.2 **CAFEQUE**

Figure 3.4.3 **Lighting department in Ikea**

Research into a task can also be done collaboratively. When we go shopping, to the cinema, or even play computer games, it can be much more fun to do this in a group. It's the same when you are working on your investigations: you could arrange to visit a local store to look at existing products. This is better than looking on the internet or at printed pictures as you can handle the products and see how they are put together, what the materials really look and feel like and how they compare. The best bit though is that you can talk about them with your peers and share opinions, and you will be more likely to remember those questions you thought about in your task analysis.

You are likely to be working with the same group of people throughout a project, and so once you begin your design work, you will all feel confident to share concepts and give each other feedback to help with the evaluation of your initial ideas.

User-centred design

Trying to decide how to approach a task can take some thought and you can use a variety of different design opportunities to lead you in the right direction.

Figure 3.4.4 **The Dyson supersonic hairdryer**

Solving a problem

Sometimes people invent a product because they have come across a specific problem that they want to solve. This might be an improvement to an existing design or a completely new idea.

For example, the first product invented by James Dyson was the ball wheelbarrow. When doing his garden he found that the traditional wheel got stuck in the mud and did not turn easily. He worked out that a ball at the front of the barrow would give more surface area to reduce sticking and would also help to turn in different directions. Another Dyson product, the supersonic hairdryer, is also a product that has been reinvented. It's different, perhaps more efficient and effective, but in principle does the same job.

Equally, the one-handed sticky tape dispenser is another example of a reinvention on a smaller scale. Look at the difference between these two solutions to the problem.

ACTIVITY

Select a common personal or household product, such as a bag, torch or water bottle. Now use your designing skills to reinvent this product to be used in a different way (for example, for a sporting activity, for the beach or camping, or by children or the elderly). Consider the ways in which the product needs to be adapted and which materials you might use to make it fit for its new purpose.

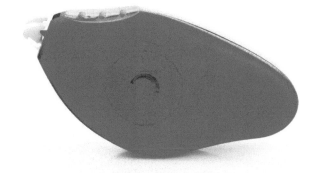

Figure 3.4.5 A cast sticky tape dispenser from the 1950s

Figure 3.4.6 A plastic one-handed sticky tape dispenser

Sometimes there are problems to be solved that don't already have a solution. For example, the little feet sock pairer is a plastic clip that holds your socks in pairs when in the washing machine or tumble dryer to prevent the problem of 'odd socks'. It isn't going to change the world, but it does solve a problem.

Client-based approach

Sometimes the best way can indeed be to solve a problem for other people. They can help you in your project every step of the way, from analysing your task, though to evaluating your ideas and testing your solution. In business, industrial designers are frequently asked to design products for large chain stores. They are often looking for a new look to an existing product or a new product to compete with other stores items, especially at specific times of year such as Christmas and Valentine's Day.

When using a client for your project you should ensure that they would be available to support you throughout the project. They should agree to your design specification, give feedback on your initial ideas and be able to test and give feedback in your final evaluation.

Figure 3.4.7 Children trying out designs

Designing through customer feedback

Product evolution often comes about through customer feedback and the demand for alternative solutions to a problem. Potential customers may be asked to suggest improvements to existing products. Focus groups are often central to this when testing a new design or an existing product.

A good example of a product that has developed in this way is the digital camera. Cameras used to contain a roll of film that had to be developed, and this took time. Feedback from Kodak customers led to the development of the first digital camera in 1975. This used a cassette tape and took 23 seconds to record. The designer, Steven Sasson, believed that one day the technology would exist to enable customers to own a product without any mechanical moving parts and that could process pictures immediately.

Think about the times you have used a product and wished it worked differently, or times your family have been frustrated when performing a task. Designers are always striving to make peoples' lives easier, quicker, less stressful and more efficient.

Design strategies

Figure 3.4.8 This camera from the 1970s uses a one-use-only roll of film storing up to 36 pictures.

Figure 3.4.9 This camera uses the latest digital technology and can store thousands of pictures on a rewritable SD card.

Most design companies no longer conduct their own market research: there is a wide choice of different market research companies available to do this for them. Some use traditional methods of data collection, such as questionnaires and 'cold calling', and others tend to move towards more innovative methods that require up-to-date knowledge and expertise in this field. When you use your phone, tablet or computer to search online your searches are recorded. One benefit to the consumer is that the advertisements you see are personalised – meaning you are more likely to respond to the advert. The benefit to the market research companies is a huge data bank of consumers' needs and wants, which can be added to other research and passed on to companies requiring their services.

A systems approach

Systems are usually described as being a group of interconnected parts or components that do something. They can be very complex so we need a way of dealing with them. This is called the systems approach.

What the system is trying to achieve is known as the system goal. In the case of a central heating system it would be to keep the house at a comfortable temperature. The component parts would be the boiler, radiators, pipes, pump and controls such as timers and thermostats.

Systems thinking means looking at the whole of the problem or operation, rather than the individual parts in isolation. This makes it very useful when analysing a problem and identifying possible solutions.

Systems can be divided into two types – hard systems that are machine or hardware dominated, and soft systems where the actions of humans decide what happens.

When producing design solutions, hard systems are much easier to model, as they will have set behaviours, for example, switches will always be either on or off, unless they are special types. Lamps will light, but how bright they are may depend on voltage or current flow; both of these can be calculated and added to the model.

How the system behaves when modelled is very important and allows us to accurately predict how successful the system is likely to be when constructed. There are lots of computer applications that allow us to model electronic, programmable and mechanical systems.

Describing a system can be more difficult than you first imagine. An example might be a car; it comprises a lot of interconnected parts, but without a driver and fuel it doesn't do anything. Once there are lots of cars, the system needs to include rules, such as which side

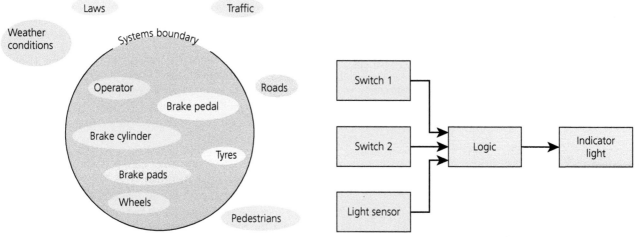

Figure 3.4.10 **Systems boundary diagram**

Figure 3.4.11 **Simple systems block diagram**

of the road to drive on, what happens at junctions and what the maximum speed is. To overcome this, we set up a system boundary to help decide what is essential. Things outside of the boundary are said to be in the **systems environment**; normally we have no control over things in the environment. So, although we can design a car and decide about technical issues, such as which fuel it should use, we cannot change the laws that govern how it is to be driven.

Complex systems can be divided into a number of systems, so a car will have an engine divided into sub-systems for controlling fuel supply, cooling and a transmission system for getting the power from the engine to the wheels. There will also be systems for braking and steering. Within each of these there will be sub-systems, such as pumps, which, although fairly complex, can be regarded as simple black boxes when designing.

We use systems block diagrams to show in simple form how the system operates.

These follow the usual rule with inputs on the left and outputs on the right, so you can follow the sequence of events in the same way as you read. More complex systems follow the same rules, although there may be many connections shown.

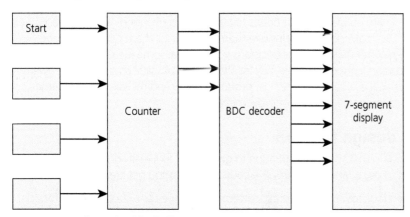

Figure 3.4.12 **Complex block diagram**

The example in Figure 3.4.12 shows a circuit being designed for timing. It is already looking fairly complex, so a simpler solution might be to use a microcontroller (Figure 3.4.13).

Design strategies

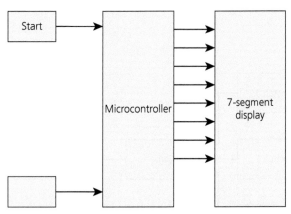

Figure 3.4.13 **A simpler solution using a microcontroller**

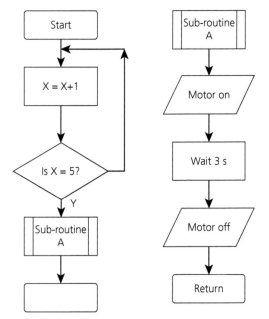

Figure 3.4.14 **Example of a flow chart**

Figure 3.4.15 **Geometric patterns are often used in textile designs.**

Most systems rely on feedback, this can take two forms: **negative feedback**, which is useful for keeping things constant, and **positive feedback**, which tends to magnify any trend.

An example of negative feedback is learning to ride a bike. The target is to ride in a straight line, but as you start to veer to the left you turn slightly to the right. However, normally you correct too much and start to go the right, so have to correct by steering left again, and so it goes on until you learn how to cope with this overshooting of the target.

When an automatic system, such as a central heating system, behaves in this way we call it **hunting**. What happens is the thermostat switches on the system, so the temperature rises, it reaches the desired temperature, so switches off, then as it cools down, the thermostat calls for heat again, and so it continues.

Often we show what happens in a system by producing a **flow chart**. Flow charts use standard symbols so they are easily understood by different users.

In the example in Figure 3.4.14, the system is controlling the length of time a motor is on. There is feedback from the decisions box. Flow charts are also used in some control applications, with the flow chart being converted to code before being downloaded to the microcontroller.

Iterative design

Iterative design is a method of designing based upon prototyping, testing, analysing and refining the product. By testing the most recent iteration of a design, changes and refinements are made. You will see this happen often in updates for your phone's operating system, apps and computer games.

This will happen as a result of your thoughts, which will influence your sketching and your interactions with others. The process starts with an idea of what needs to be solved or changed. You will need to make sketches, annotate them and discuss them with your client, teacher or peers. After making some decisions you will trial some ideas to make sure your conclusions are correct, and this may lead to further physical changes in the final prototype. A critical evaluation of the final product by you and the client or consumer may lead to further ideas for future improvements.

Avoiding design fixation

One way of finding that spark of inspiration to get you off to a really good start in your design work is to use a range of alternative sources to help you get creative.

Designers often look towards geometric patterns and shapes as a starting point for design. This can be seen in many examples of surface patterns, especially in textiles for both fashion and home furnishing. In the 1960s, Mary Quant used a lot of these patterns in her clothing range, and today Marc Jacobs continues to use these shapes as a source of inspiration, not just in clothes but also in other fashion products and accessories.

Geometry is also used in three-dimensional design, and some of the best designers have used this in their work. Taking the most basic of shapes and transforming them into a seemingly simple and effective design may not be easy, but it is certainly effective. The Braun geometrical kettle designed by Emi Schenkelbach uses triangle, circle and square shapes that are merged carefully without any obvious joins. It looks clean and simple – inspired by three basic shapes you first learned as a young child.

Other mathematical patterns have also been used successfully in designing products. Italian mathematician Leonardo of Pisa, known as Fibonacci, introduced his mathematical sequence in 1202, and this is still used today by designers to help find the best look and proportions for products.

The Fibonacci sequence is characterised by the fact that every number after the first two is the sum of the two preceding ones: 0, 1, 1, 2, 3, 5, 8, 13, 21, 34, 55, 89, 144.

Golden Ratio

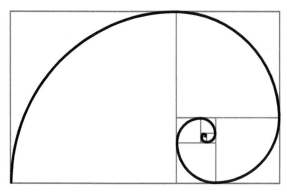

Figure 3.4.16 **Fibonacci squares**

Figure 3.4.17 **The pattern of seeds on a pine cone uses the golden ratio**

ACTIVITY
Using an example of a Fibonacci pattern repeated in nature, sketch a simple design for a pendant, charm or earrings.

Designing from natural forms: biomimicry

Biomimicry is an innovative approach to design that strives to copy nature's time-tested patterns and structures. The idea is that nature has already solved many of the problems we encounter, and that plants, animals, and even micro-organisms are naturally proficient engineers.

Observation of the habits of a beetle in Namibia allowed South Korean designer Pak Kitae to use nature-inspired technology to design the Dew Bank Bottle.

Figure 3.4.18 **Namibian beetle**

Figure 3.4.19 **Dew Bank Bottle**

Design strategies

We have a wealth of beautiful shapes and patterns in nature that we can use to inspire our designs. These mass-produced designs are very fashionable and popular in people's homes and are frequently used by professional designers, such as Cath Kidston.

Designers of handmade products also often use nature for inspiration, as it not only helps their designs to be creative, but also demonstrates their passion for their local area. Lovebird Jewellery designer, Kirsty Dyer, uses images from the Devon coast and countryside as a basis for her work, which portrays stylised versions of natural forms such as feathers, seagulls and local flowers.

You can use images you find in nature to help with your designs, but you might also find it useful to adapt the patterns and shapes of natural forms. The following piece of work was produced from an original observational drawing. After the initial sketch, an interesting part of the pattern was chosen and mirrored to produce a symmetrical pattern. After choosing a colour scheme for the design it was repeated several times to produce the final design. This process was made easier by the use of CAD, but could be done just as effectively using traditional tracing techniques.

Throughout history, designers and design movements have been a source of inspiration for new designs. Some early Dyson models were available in a special limited edition De Stijl colour scheme, in homage to the Dutch design movement of the same name. All these are coloured a combination of purple, red and yellow.

ACTIVITY

Look at a recent Dyson model. Die Stijl is Dutch for 'the style'. Can you see how the Dyson company have used this to influence their designs?

Figure 3.4.20 **Cath Kidston shop and shopping bag**

Figure 3.4.21 **A repeat pattern based on a drawing from nature**

Cultural influences

In 1851, London held the Great Exhibition in the Crystal Palace in Hyde Park. It was the first of many international exhibitions that celebrated examples of cultural and industrial product design. This was probably the first time people could see all these innovations in one place, and it holds a significant place in the history of product design.

Over the last 100 years it has become easier to travel, easier to connect through the media and the internet, and so we have become much more aware of the vast wealth of design around the world.

We have also learnt how important it is for society to invest in products that help developing countries to be self-sufficient and live more comfortably and safely. The wind-up radio, designed by Trevor Baylis, helped people to stay in contact with the rest of the world in places where there was no power or connectivity.

Figure 3.4.22 **Wind-up radio, designed by Trevor Baylis**

Exploring and developing design ideas

These design strategies should help to provide a starting point for your designs and also help you later on in developing your work. You will also need to investigate the range of technologies available to you: materials, systems, components, tools and equipment, including their purpose and how they meet personal and social needs. You should consider how society and environmental sustainability factors influence design.

Sketching

Before you start sketching, ensure that you have the best tools available to help you produce a high-quality piece of work. This doesn't mean it has to be a perfect drawing, but should show different aspects of your design clearly: keep the drawing simple, clear and accurate, and your notes simple and precise. Take time to decide which aspects of your design need improving and which need further investigation. Drawing in three-dimensions can help you to show what you expect your design to look like, although some parts of the design may be better shown in a two-dimensional sketch. (Some good examples of sketching can be found in Topic 3.5.)

ACTIVITY
1 Practise your sketching skills using an example of an existing product. Imagine you want someone to make this for you and work out what information they would need to produce it. Sketch the product from different angles – use two-dimensional and three-dimensional sketches and focus upon details of the manufacture.
2 Make a sketch of an idea of your own. Use this drawing to help you explain your idea to someone else. Ask them to re-sketch it and notice if they need to ask you any further questions. You may be able to use their sketch and ideas to help you develop your design.

Figure 3.4.23 **Card model cut on a laser from corrugated card**

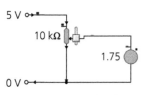

Figure 3.4.24 **Modelling part of a simple circuit**

Modelling

You can use simple modelling techniques to quickly test your thoughts, enable you to understand what the product will look like in three-dimensional and help you visualise the correct proportions and sizes. There are several ways you can do this – for example using card, foam, fabric or wire.

You might decide to make a full size mock-up in a cheaper material. If you are making part of your design from sheet material such as plywood or acrylic, thin corrugated card is one possibility – it can be cut using CAD/CAM, is much quicker to cut and it will give you the general idea of how the parts will look and will also help you work out how it will go together.

For textiles products you can use a cheaper material such as calico to make a mock-up known as a 'toile'.

Construction kits can be used to model mechanical systems, and contain a range of mechanisms that you can try out without having to manufacture them all yourself.

Electronic systems can be modelled using a virtual circuit in a simulation application. This makes it easy to try alternative ideas without the need to obtain the actual components. Components such as resistors or capacitors can have their values changed very easily, and the effect this has on the circuit can be tested.

Testing

Testing should be an ongoing activity throughout your designing and manufacturing. You will already have had experience in analysing products and should keep this in mind when producing your own designs. You could refer back to this analysis and review how your ideas compare.

Consider how you will judge which of your ideas is the best to take forward – you could use a comparison table or test out your ideas on a client. If you are designing your product to fit a person or for someone to use effectively you should also test to ensure that the product is ergonomic.

Materials and construction methods may need to be tested for strength and durability, appropriateness and aesthetics. It is important to ensure quality and safety, but also, conversely, designs should not be over-engineered as this may not be time or cost effective.

In systems, circuits are often 'breadboarded' on prototyping boards. It is called breadboarding because early circuits were often wired up on wooden boards. This means the circuit can be tested using real components.

Many products start out well, but using inappropriate or badly applied finishes spoils the end result. Make sure you test these on a sample material before you go ahead. Consider how long these take to apply and set or dry.

> ### ACTIVITY
> On a piece of softwood try out a range of different finishes, such as wood dye, paint, sanding sealer, varnish or wax. Experiment with different application methods, including a brush, sponge, roller, cloth and spray.

Figure 3.4.25 **Using a toile to check fit**

Figure 3.4.26 **Finishing techniques, such as enamelling, require experience, and it is worth testing these out.**

Figure 3.4.27 **Breadboarding**

Figure 3.4.28 **Parent and child testing a product**

Evaluation of work to improve outcomes

As you will be testing aspects of your designs throughout the process, so you will also need to evaluate the results, make improvements and refine your design. The need for your final summative evaluation should be recognised from the outset, as this will help you to make informed choices along the way. Your design specification provides an initial checklist and makes a good starting point for your evaluation. Rather than writing it and forgetting about it until the end of the project, try to make it a work in progress that you can come back to, add to if necessary and review at various stages. Designers rarely achieve perfection, and you will be able to show how you have evaluated your product at every stage of its development.

You will no doubt have analysed a similar existing product in the early stages of your project and this will have helped you produce your own criteria for your specification. Ensure that you have enough time in your summative evaluation to compare your prototype to other products. In reality these are what consumers will be looking at when deciding what to buy. How is your design more appealing to the customer and why? Don't be afraid to be critical at this stage either. You could only improve the design further by being honest with yourself!

If possible, try out your design on a client or other appropriate user. You will again need to plan this beforehand and sometimes it can take time to organise. This is not something you can do easily in a classroom; field-testing is an essential part of design testing and evaluation. Children, for example, not teenagers, should evaluate children's toys! Bear in mind that the consumer here is the parent or carer so their opinion is valid along with the reaction of the child.

KEY POINTS

- Negative feedback is used in a system to hold an output at a fixed level, whereas positive feedback is used to make sure that something happens by magnifying a small change.
- Hunting is when a system is trying to achieve something but keeps overshooting the target and tries to correct but overshoots again.
- Flow charts are used to show or plan sequences.
- Complex systems can be split into sub-systems, which simplifies both designing and testing.
- When designing systems, look at the opportunities that incorporating feedback would offer.
- Modelling, whether actual or virtual, allows the system to be tested before committing to manufacture

Check your knowledge and understanding

1 Give an example of a design you think is innovative and provide two reasons for your choice.
2 Explain one way in which a company may get their team to work together in analysing a task.
3 How might a collaborative approach to design improve your initial ideas?
4 Using a specific product as an example, explain what led to developments in this product over time.
5 Give three advantages of using a client during the design process.
6 Give three examples of ways in which designers can produce ideas that are unusual and creative.
7 What is the difference between a hard and a soft system?
8 What effect does positive feedback have on a system?
9 Give two advantages to using a virtual rather than actual prototyping system.

3.5

Communication of design ideas

What will I learn?

In this topic you will learn about:

→ freehand, isometric and perspective sketching to convey design intent
→ enhancing sketching using a range of advanced techniques
→ rendering techniques to improve communication
→ incorporating a range of techniques across a design ideas sheet
→ systems and schematic diagrams.

There are many different skills and techniques that can be used to effectively communicate information to a reader, viewer or client. These can range from a quick thumbnail, detailed and exploded sketches through to presentation drawings, CAD models, 3D-printed models, toiles to working prototypes. The purpose of all of these techniques is to communicate and convey the designer's **intent** for a product, considering the research that was undertaken and the criteria that have been generated.

Some of the skills are faster to learn than others, and you do not need to be good at every one of these to be a successful designer. Find what you are good at and work to your strengths to communicate your design intent effectively.

Freehand sketching, isometric and perspective

The best designers who use **freehand sketching** are those who are confident in the lines they sketch with, even if some of them are wrong. Freehand sketching is done without the use of rulers or drawing aids and is a way that a designer can quickly express thoughts and ideas. Confident sketches display long flowing and fluid lines, shapes and forms. Confidence in communication through the use of sketching starts with understanding the basics and practising them, and repeating them, until they are embedded.

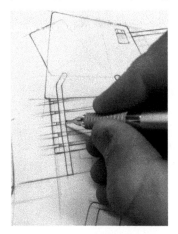

Figure 3.5.1 A mechanical pencil always remains sharp and gives a consistent line quality.

Sketching confidence can even be down to how you hold the pencil when you start to sketch or the type of pencil chosen. A hard (ideally H or 2H pencil), sharp pencil is fine to use, although a **mechanical pencil** is recommended as it is well balanced, it always remains sharp and gives a consistent line quality, and it has a grip where the pencil should be held to allow for a range of sketching techniques to be used. Blue pencil lead is often used by professionals so they can see all construction lines and details.

To develop confidence in sketching it is a good idea to start with freehand two-dimensional sketching of some basic three-dimensional forms. If you then apply some simple rendering techniques to these sketches, they will look much improved and communicate more information. You can then combine these shapes together to create more complex forms and design ideas.

Freehand sketching in 2D

Freehand two-dimensional sketching develops confidence in using construction lines, particularly a centre line, to produce a series of basic shapes. These sketches will then be used to demonstrate how they can form the underpinning construction of more complex two-dimensional sketches. A simple way to start is to sketch in two the five different three-dimensional forms shown in Figure 3.5.2. These could be considered the 'building blocks' of freehand sketching in 2D, and once you have sketched these shapes you can then combine and refine them to create additional shapes.

Figure 3.5.2 **Basic three-dimensional forms for freehand two-dimensional sketching**

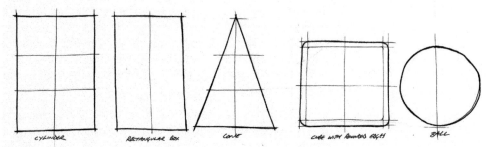

Figure 3.5.3 **Freehand two-dimensional sketches of three-dimensional forms – without rendering**

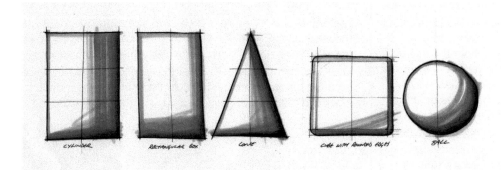

Figure 3.5.4 **Freehand two-dimensional sketches of three-dimensional forms – with rendering**

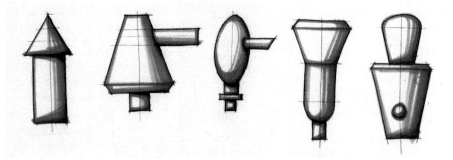

Figure 3.5.5 **Freehand two-dimensional sketches of combined two-dimensional shapes – rendered**

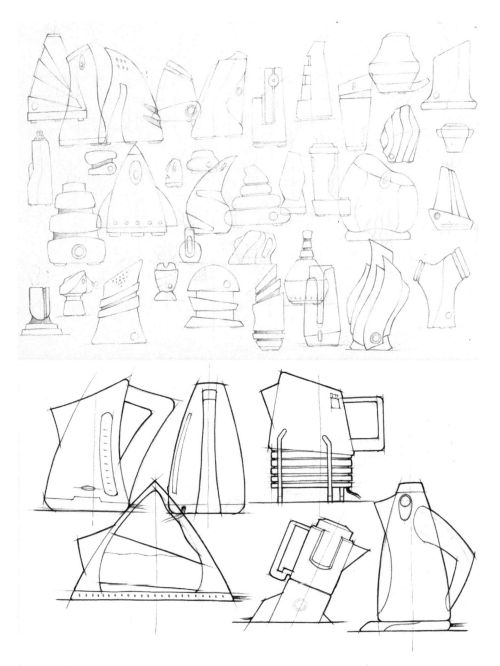

Figure 3.5.6 and Figure 3.5.7 Freehand two-dimensional sketches 'styled' into design ideas for products

Figure 3.5.8 Enhanced freehand sketches of ideas for fashion bags

Freehand sketching, isometric and perspective

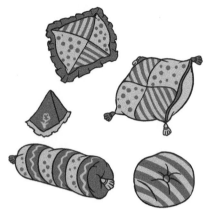

Figure 3.5.9 Showing colour, pattern and texture in drawings of textile products

When presenting ideas for textile products, showing the pattern and texture of fabric will enhance the drawings.

Enhancing two-dimensional sketches

You can make your design sketches look even better through a few enhancement techniques, such as adding line weight, using thick and thin fine-line pen work, and colour-rendering techniques. To enable you to effectively enhance your design sketches it is recommended to use the following equipment: 0.7 mm mechanical pencil (with blue leads), 0.5 mm fine black pen, 0.8 mm fine black pen, thick outline pen, and a very soft black pencil. If you wish to marker render your work then you will also need A3 marker paper and some marker pens.

Figure 3.5.9 Showing colour, pattern and texture in drawings of textile products

Figure 3.5.10 Enhanced two-dimensional sketches

Weight of line

Good design sketching often begins with a construct or framework – an initial set of sketched lines laid down so lightly that they are barely visible. This is often referred to as 'crating'. These initial sketched lines allow for mistakes and experimentation in order to get things right. They establish a footprint for the form of the product and then fade into the background once final darker and bolder lines are added. Adding a faint centre line or 'parting' line (where two halves of a product might meet) adds additional detail and realism to design sketches and can help with the rendering of the form. You can then pick out areas of a sketch to add more 'weight' to the line (for instance, by pressing harder with a sharp pencil, which will make the sketch look more dramatic). While there is no set rule on adding line weight (it is more by look and feel), it is recommended to add more line weight to **concave** curved edges.

**Figure 3.5.11 Freehand two-dimensional sketch sheet demonstrating weight of line (2D) –
note that some lines are darker than others. Compare this to Figure 3.5.6 to see the
difference the weight of line has made.**

Thick and thin lines

The outline is a continuous and connected line that defines the outer boundary of an object.
It is the boldest and thickest line of all. Firstly, sketch out your design idea(s), add some weight
of line to enhance the sketch and define the outline, then using a 0.5 mm black fine line pen
enhance the relevant lines in your sketch. After a sketch has been fully defined the outline
can be darkened and thickened further using an 'outline' black pen. This could be a 1.0 mm
black pen or a thicker pen. It is often heaviest or thickest at the bottom of the object where
it contacts the surface it is sitting on, so you could go over this line a couple of times. This is
also the furthest point from a light source so it is usually in shadow. Always avoid thick lines
on the inside of your design sketches, as you will lose the definition of the outline and the
quality of detail. Figure 3.5.11 shows this technique applied to make some design ideas stand
out more than others. The technique highlights preferred ideas above others and so can help
communicate decisions, or design intent.

Techniques such as thick and thin lines are often used to make an idea stand out more clearly
on a page. It is best employed on a three-dimensional sketch; however, it can also work
effectively on a two-dimensional sketch. The trick with thick and thin lines is to ensure that the
outer edge of a design idea is a thicker black line and the inner lines are thin. This enhances
the sketch and 'lifts' it off the page. This can be seen in Figure 3.5.12.

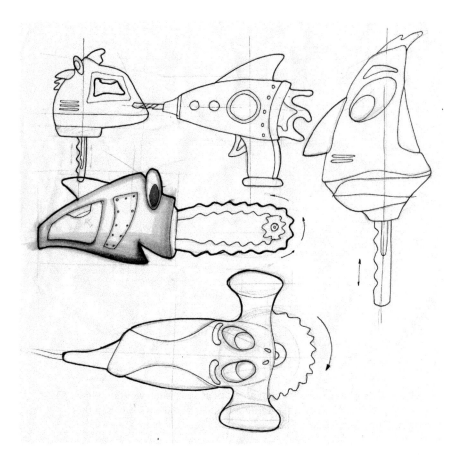

Figure 3.5.12 Enhancement of a freehand sketched idea using thick and thin lines

ACTIVITY: STYLING TWO-DIMENSIONAL FREEHAND SKETCHING

Using the basic shapes and forms shown in Figure 3.5.3, style them into something recognisable.

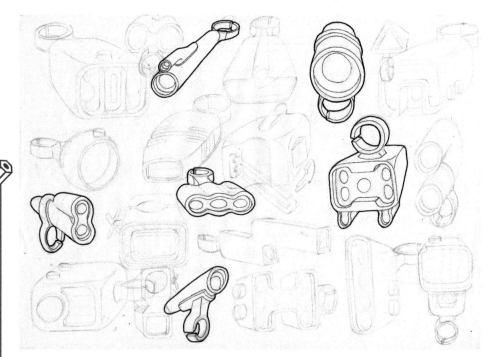

Figure 3.5.13 Thumbnail sketch sheet. The ideas highlighted are then taken forward for development and refinement.

Rendering techniques

Rendering can be defined as the addition of colour, or texture, to enhance a sketch to better communicate design intent. Coloured pencils are the most easily accessible way to render a sketch, and can be very effective for adding tone. The use of marker rendering pens along with marker paper can simply enhance sketched design ideas. Marker pens can be used to add a colour 'wash' (Figure 3.5.15), to 'colour in' (Figure 3.5.16), and render (Figure 3.5.17). Detailing can also be added through the use of black pencil or a soft white pencil (Figure 3.5.18) over the top of the marker rendering. The detailing gives the sketch greater depth as it makes the shadow areas darker and can pick out highlights. A simple rule is 'all edges are always dark'. This typically enhances your sketch if the dark edges are blended, or faded, inwards as it helps give the two-dimensional sketch an impression of a three-dimensional one. These techniques are shown in Figures 3.5.15, 3.5.16, 3.5.17 and 3.5.18.

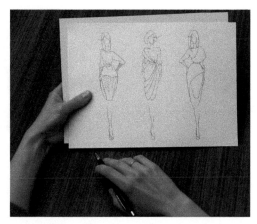

Figure 3.5.14 Sketching design ideas for fashion garments

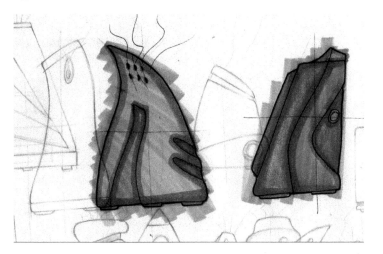

Figure 3.5.15 Colour wash

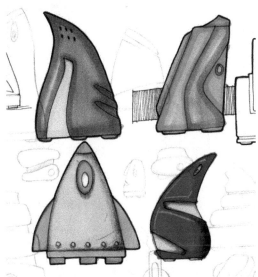

Figure 3.5.16 Colouring in with markers. This is similar to a colour wash but the marker pen remains inside the outline.

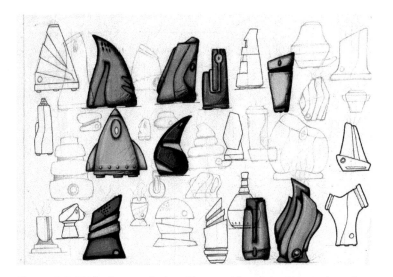

Figure 3.5.17 Marker rendering. The most important part of marker rendering is the use of light – an area where marker is not applied to signify light hitting the object.

Freehand sketching, isometric and perspective

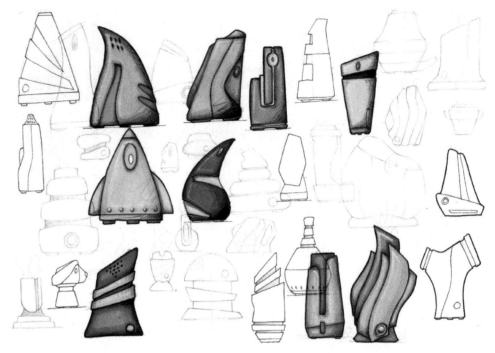

Figure 3.5.18 **Adding detail using a black soft pencil**

Freehand sketching in 3D

A recommended way to begin sketching in 3D is to start with an idea that you already have sketched in 2D, as this will enable you to **translate** the shape into a three-dimensional form, rather than trying to think of the idea and details and then sketch it in 3D.

The example in Figure 3.5.19 shows a two-dimensional sketched idea turned into a three-dimensional sketched idea.

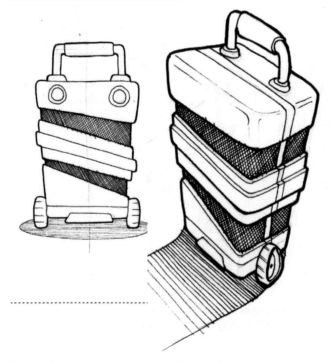

Figure 3.5.19 **Three-dimensional sketching**

3.5 Communication of design ideas

Isometric drawing

Isometric drawing, which means 'equal measure', is a way of presenting a design sketch in three dimensions. A 30-degree angle is applied to the sides, which gives the appearance of three dimensions. This form of three-dimensional drawing is a more constructed, technical way of presenting an idea in three-dimensions and is used to show more information than just a two-dimensional sketch. The use of isometric drawing equipment or isometric grid paper makes it easier to sketch in this way as the angles can be seen quite easily. However, objects do not look real as there is no perspective applied, but it can represent the three-dimensional form. **Isometric sketching** differs slightly from this technique – an isometric grid is placed underneath a sheet of plain white paper and is used as a guide.

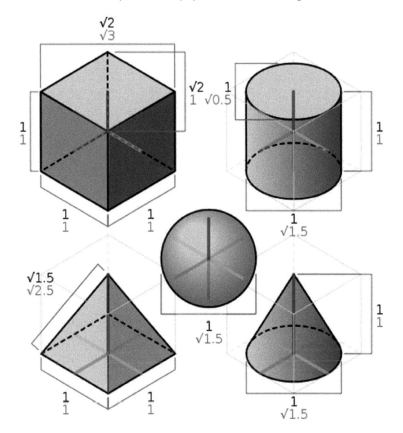

Figure 3.5.20 **A number of basic forms in isometric projection**

Figure 3.5.21 **A sketch for a sweet dispenser sketched in isometric projection**

Perspective drawing

To make objects appear more realistic, perspective drawings can be used. Receding lines are used to show how things appear to get smaller as they get further away. If these receding lines are extended, they will meet at points called **vanishing points** which are placed on the **horizon line**.

One-point perspective drawing

One-point perspective is way of representing three-dimensional objects and space using intersecting lines drawn vertically and horizontally that radiate from one point on a horizon line.

One-point perspective drawing shows how things appear to get smaller as they get further away, converging towards a single 'vanishing point' on the horizon line.

Drawing in one-point perspective is usually appropriate when the subject is viewed 'front-on', such as when looking directly at the face of a cube or the wall of building, or when looking directly down something long, like a road or railway lines. It is a popular drawing method with architects and illustrators, especially when drawing room interiors.

Figure 3.5.22 **A one-point perspective station**

Two-point perspective drawing

This form of drawing is a more realistic way to present forms in three dimensions; however, it can take some practice to get it right. Two-point perspective has two vanishing points on either side of the page on a horizon line. A vertical line is usually drawn to represent the leading/front corner of the form you are sketching and then most lines connect to either of the two vanishing points.

If you want to draw a box in two-point perspective you should start by adding two vanishing points at either side of the page and a vertical line roughly around the middle of the page. If you then put two points on this vertical line and connect them to the vanishing points, you will end up with two V shapes. Add sides to gain depth and then connect the top intersection corners to opposite vanishing points to realise the box in two-point perspective, as shown below.

Figure 3.5.23 **Two-point perspective three-dimensional forms**

Figure 3.5.24 **Two-point perspective three-dimensional forms – lots of boxes**

A box sketched in two-point perspective can be used as a framework to help guide a three-dimensional sketch by dividing the box into sections before adding details. This can be seen in Figure 3.5.25, where the box has been modified to present a design idea for an old-fashioned style lamp.

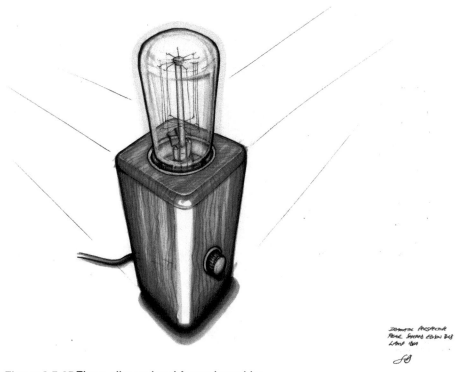

Figure 3.5.25 **Three-dimensional forms lamp idea**

Use of shadows

Once you can sketch in three dimensions you can add further techniques to enhance the sketch. A shadow can help give your sketch a degree of realism and 'sit' it onto a surface that will give it weight, size, and form. Your shadow does not need to be accurate at this very early stage – it is more just a suggestion. Adding a shadow gives the sketch a context rather than the sketch (object) 'floating' in space.

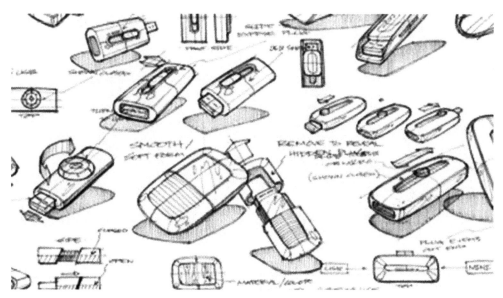

Figure 3.5.26 **Sketch demonstrating use of shadows**

Freehand sketching, isometric and perspective

Use of linking boxes

A background box can be used to join together similarities of ideas and can help the overall look of a design sheet (which is a selection of design sketches that a designer produces in response to a set brief or problem). Linking boxes are used to 'link' similar ideas together or when an idea has been sketched two-dimensionally and/or three-dimensionally to demonstrate the form. You can add boxes with a black fine line pen (thin) or a complementary (light) colour. They should be added behind the design ideas sketches. It is really important that the lines all go in the same direction (that is, all horizontal, all vertical, or all diagonal). The box should not detract from the product sketches; it should complement the sketch and enhance the page.

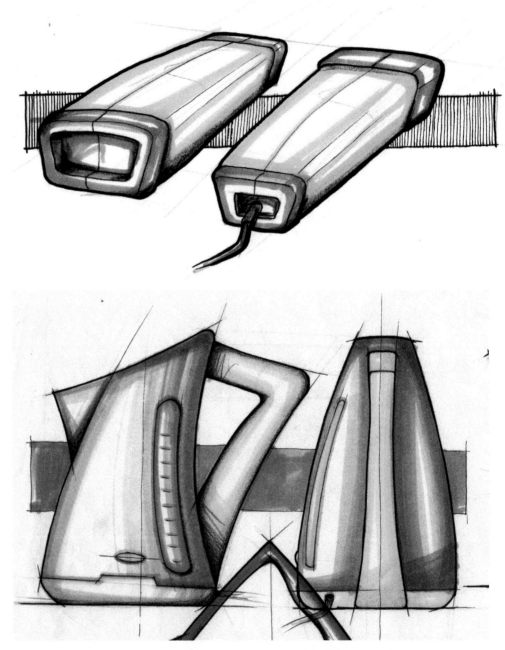

Figure 3.5.27 and 3.5.28 **Sketch ideas demonstrating the use of linking boxes**

Organic forms

Sketching an organic form is difficult, but is made easier if you use a centre line and work with a two-point perspective. Any underpinning framework, such as two-point perspective lines, will help. A good way to do this type of sketch is first to imagine what the form should look like. It is then a good idea to sketch the form in two dimensions from different viewpoints to demonstrate what you are trying to communicate. Then the use of a centre line and perspective lines can help 'guide' the form and the early 'construct' (framework) sketching. Figure 3.5.29 demonstrates an organic sketch using a two-point perspective underpinning construct.

The further use of grid lines, to show curved or contoured surfaces, can help aid the form to look more three-dimensional.

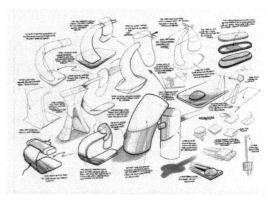

Figure 3.5.29 **Sketch demonstrating organic three-dimensional forms**

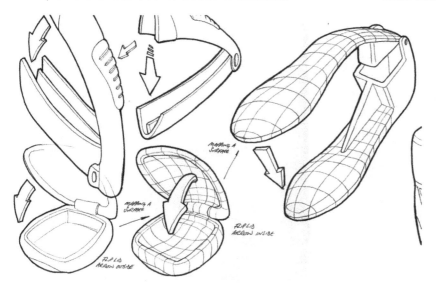

Figure 3.5.30 **Sketch demonstrating organic three-dimensional form with grid lines to show curved or contoured surfaces**

Aid to three-dimensional sketching in two-point perspective

A technique that you can use to sketch in two-point perspective is to get a photograph and use that to sketch around, as a photographed object will naturally be in two-point perspective. Alternatively, you could set up your own photograph using some foam blocks. If you are sketching a design for a handheld product then get a photograph of a hand holding an object and then use this to help you sketch your own three-dimensional idea in two-point perspective. Figures 3.5.31 and 3.5.32 demonstrate this technique.

Figure 3.5.31 **A hand holding a smartphone**

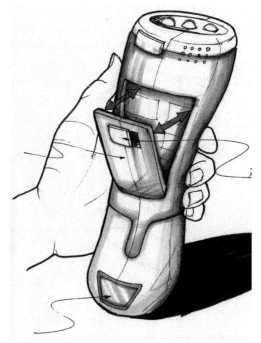

Figure 3.5.32 **Using the image in Figure 3.5.31 to add the relevant receding lines and a centre line to sketch in two-point perspective**

Freehand sketching, isometric and perspective

Putting it all together

The design idea shown in Figure 3.5.33 is for a portable bluetooth speaker system with a detachable handle, sketched in two-point perspective. Note that this design idea starts as a two-dimensional sketch (on the left of the image) which identifies the shape that the form will follow. The three-dimensional sketch translates the two-dimensional idea into a three-dimensional form. The use of shadows helps to 'sit it' on a surface. A linking box, in a complimentary colour, joins the similarities together. Colour rendering has been applied to enhance the sketch. The addition of an **exploded view**, where the parts have been separated, then demonstrates further detail of how the idea is intended to work. When you put all of these techniques together it helps to better communicate your design intent.

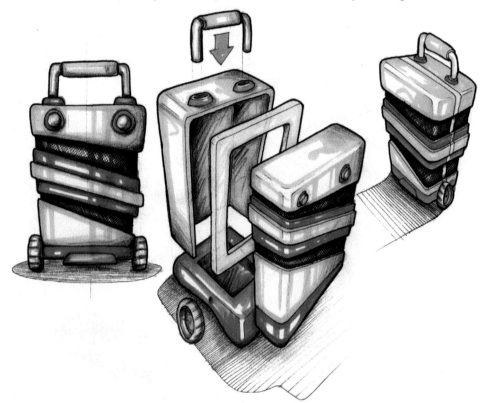

Figure 3.5.33 Example of putting it together

Design ideas and sketches: a good way to start

Starting initial design ideas, even thumbnail sketches, can sometimes be a very daunting task. One way in which designers approach this starting point is by using an inspiration board. Think about the things you like. For instance, what shapes, forms, colours, textures, and so on do you like? Look at nature. Look at mathematical patterns. Look at existing products and conceptual products that you like. Look at what your user group likes. Look at architecture. Use your inspiration board to inspire you when you design – lay it out in front of you. Pick out some of the shapes, forms, colours, patterns and details to help you start to come up with initial sketches. Remember, after generating an initial sketch you have the opportunity to develop it thoroughly, so don't think that every sketch has to be perfect or a completely developed solution.

Drawing the human body can be challenging for many people. You can trace a garment shape from an example on an inspiration board or use a fashion template to help show ideas for clothing. You can place the fashion outline under the drawing paper and trace over it to get the basic shape right. This will give you confidence in presenting your ideas. Fashion templates often show an exaggerated form of the human body to make the clothes look more attractive.

An inspiration board is an excellent source of images to us as a starting point for a repeat pattern to be used on fabric. Use a cardboard 'window' to select part of an image that you can develop into a pattern. You can scan the basic image into a software program and manipulate it to make a repeat pattern.

For a project, you will be expected to generate a range of feasible design ideas, based upon the criteria in your design specification. A good way to start this is by creating a **thumbnail** sheet of design sketches with lots of ideas.

All sorts of sources can be used for inspiration – your design specification and research, of course – but you may also decide to base some aspects of your ideas on a theme such as:
- natural forms – look around you at nature, animals, trees and plants, clouds, etc.
- the work of other designers or a design movement, such as Starck, Bauhaus, Arad, Newson, etc.
- influences from music, films, media, fashion or current trends
- new technological developments, such as using new materials or processes, new smaller circuit boards, electronic developments, etc.

Figure 3.5.34 Fashion illustrations are often based on an exaggerated human shape.

You should also show evidence of the influence of your research on your design ideas. As you experiment with first ideas and gradually refine your thinking, you should start to think about the possible materials or processes you could use and research them, to work out if your ideas are feasible. You may also start to play around with combinations of ideas or work on the fine detail of some of your ideas. Making use of the information you acquire during your research phase should enable you to develop and refine alternative ideas to the stage where you can select one or two that are the most promising to be developed.

To start getting a flow of ideas it is strongly recommended that you begin by creating a series of quick thumbnail sketches of your potential ideas. These should be quick thoughts and ideas (with some details), and you could completely fill at least one A3 sheet with lots of ideas (aim for between 10 and 20). Anything is possible at this stage, however ridiculous some of your ideas may be, and you just need to get any thoughts from out of your head and on to paper. Don't worry if some of them are not good. You do not need to render them or present them in an advanced way (although you can if you wish!) as the sheet should be used as the first sheet of ideas as part of a project. You should ideally fine line pen some of the ideas (but not all if you do not want). For speed it is recommended that you produce these ideas mainly two-dimensionally (although some of the examples shown in Figures 3.5.35, 3.5.36 and 3.5.37 also have quite a few three-dimensional sketches on them). You can then use this sheet(s) to pick out some of the better ideas to sketch in more detail for your actual design ideas submission.

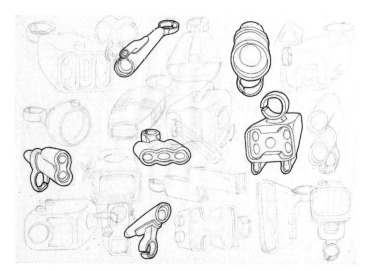

Figure 3.5.35 Thumbnail sketches for a bike light

Design ideas and sketches: a good way to start

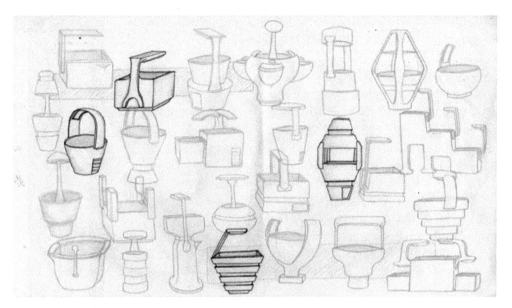

Figure 3.5.36 **Thumbnail sketches**

When communicating and presenting design ideas, you should try to use a range of media, skills and techniques to help you, although you should try to use those which you can demonstrate to best effect. Try not to perfectly 'lay out' your designs on a page – sketch naturally, and if an idea goes over the top of another idea, don't worry about it. Design ideas are supposed to be quick sketches, not perfect, precious final designs.

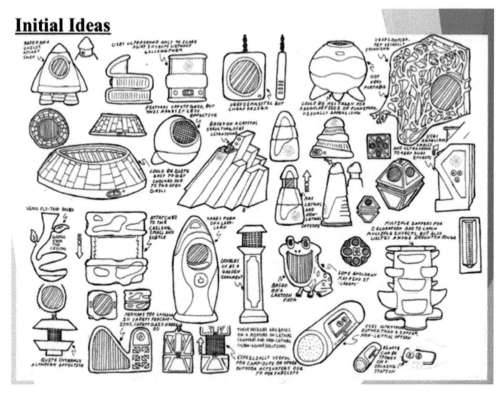

Figure 3.5.37 **Student example of communicating design ideas**

Figure 3.5.38 **A fashion designer will often include swatches of fabric or component samples when presenting ideas.**

Systems and schematic diagrams

To aid the planning of a project a systems diagram is highly recommended. It will act as a guide and let you show your thinking regarding your design ideas and whether or not they may solve the initial problem. It also allows you to think systematically and logically about the problem and how you may be able to solve it. A systems diagram is similar to a flow chart, although it may appear much simpler.

A simple systems diagram can be seen in Figure 3.5.39. The key to a good systems diagram is communicating where every process can be divided into input, process and output.

In the input area, write about all the problems or questions you are trying to solve. Look back to your design problem, your initial research or your specification and write a summary under a simple diagram or picture that shows all the inputs you can think of.

In the process area, describe the type of electronics, components or mechanisms that you could use in your project. Do not be too specific about the circuits or mechanisms that you might use. Explain how the circuits or mechanisms control the way your product will work.

In the output area explain what you think will be the outcome of your product.

Schematic diagrams are representations of systems using graphical symbols rather than realistic pictures or drawings. Schematic diagrams are used regularly in engineering. A schematic diagram in electronics uses symbols to represent components such as resistors, transistors and capacitors, showing how connections between them are routed while omitting the physical detail of the components. A schematic diagram for a central heating system uses symbols to represent valves, pumps and other vessels and shows the interconnecting paths of the pipework.

In fashion design, schematic drawings are called 'flats' as no body is drawn. Flats clarify the technical details of a garment and are less ambiguous than many fashion design sketches. All the details of the garment are shown without shading to avoid any ambiguity. Fashion flats are usually hand drawn, but computerised drawing programs give greater accuracy.

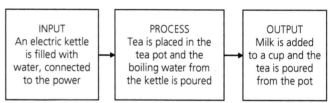

INPUT	PROCESS	OUTPUT
An electric kettle is filled with water, connected to the power	Tea is placed in the tea pot and the boiling water from the kettle is poured	Milk is added to a cup and the tea is poured from the pot

Figure 3.5.39 **Example of a systems diagram for making a cup of tea**

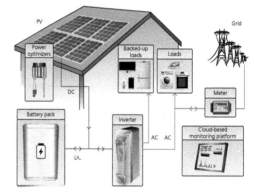

Figure 3.5.40 **Example of a schematic diagram for solar panels**

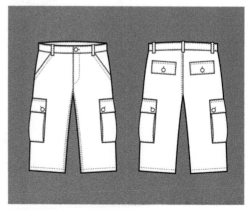

Figure 3.5.41 **A schematic drawing or 'flat' for trousers**

Annotated drawings

Annotation is a comment used to **describe**, **explain** and **discuss** a design idea to improve your communication of your design intent. It can express a lot and is considered very important in the communication of an idea. Often, a sketch does not show all of your design intent, and this is where discussing your idea, on paper, can help your communication to the reader (client). Through annotation you can draw particular attention to a part of a design idea.

When using annotation to describe your design idea, think about its shape and form – does it resemble anything familiar (for example, a known shape such as an ellipse or a cone), another product, a famous building, something from nature?

If you are explaining an idea, you could mention what materials you think it could be made from, what type of finish it might have, how the materials might be joined together, what scale or size it is, how you intend it to work, and why you have designed certain features, such as rounded corners, holes or grooves.

Imagine discussing your design idea with a friend, a family member or your user group. What would you say? Don't assume that someone looking at your sketched work understands fully what you are trying to communicate, so annotation is also key to help explain your design intent.

To annotate a sketch, first comment upon the aesthetics (what you see) – describe the shape or form. Then comment upon material selection and explain the reasoning for your choice. Further comments could focus on size, weight, finish, features and joining methods. Figure 3.5.42 demonstrates the addition of annotation to a sketched design idea.

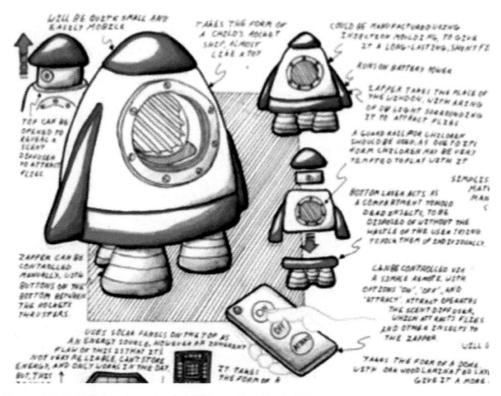

Figure 3.5.42 **Addition of annotation to a sketched design idea**

Exploded diagrams

An exploded diagram can be used to demonstrate more detail about the intent of the designer. These details could be the construction of the product, how a product might fit or join together, how any components may fit on the inside of a product, how a product can be strengthened, and generally give and show more information about the design intent.

To complete an exploded diagram you need to separate the parts of your design to show inner details, construction and joining methods. A good technique to create an exploded view is to quickly trace your sketch and then place it underneath your paper to create an underlay. Sketch one part of your work and then move the underlay a small amount in one direction and sketch the adjoining part. Repeat this as many times as you feel necessary to demonstrate all details. Figure 3.5.43 shows a sweet dispenser as an exploded diagram.

Figure 3.5.43 An exploded diagram of a sweet dispenser

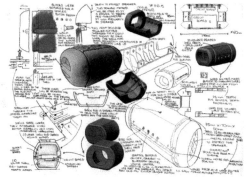 

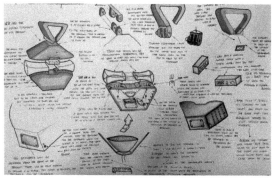

Figures 3.5.44, 3.5.45 and 3.5.46 Student examples of adding exploded diagrams to their design ideas sketches

Working drawings

A working drawing is usually a scale drawing of your chosen design idea, and this serves as a guide for the construction or manufacture of your product. It is a technical drawing and usually consists of a front, side and plan view of a final design idea or proposal. The most common format to present this working drawing is as a third-angle orthographic drawing. To construct a third-angle orthographic drawing for the L-shaped object in Figure 3.5.47, the plan view of the L-shape is drawn as a view from above. The front view is drawn as if standing in front of the L-shape. The side view is drawn as if standing at the side.

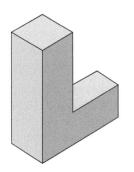 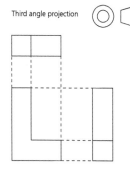

Third angle projection

Figure 3.5.47 Isometric and third angle orthographic drawings of an L-shaped object

Dimensions are added so that any person using the working drawing could potentially manufacture the design proposal. Usually there are at least six dimensions, but you can add as

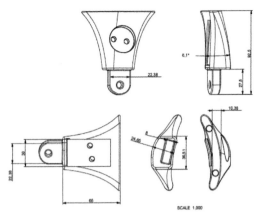

Figure 3.5.48 **Orthographic drawing with dimensions**

The video above shows a child around the age of my target market joining in to use a coffee grinding machine. An important aspect of my product will be the involvement of the child in the process. In the video it is clear that the child is excited to be involved in the process even in the smallest ways. The empathy shown between the child and the product is very important in order for it to be successful. In the video the child gets involved by watch the lights show up counting down to when the coffee if finished. Simple lights flashing like this are effective in showing when the machine is finished in a simple way. This could be something that I could implement into my design to show the smoothie is finished blending. This video presents the fun and enjoyment aspect of a product and how it is important to involve the child in the activity. Although a young child couldn't really do this activity completely by themselves the large button at the top of the product and the simple light system to show when the coffee beans are complete are the kind of simple features I would like to include in my product.

Figure 3.5.49 **Slide containing a visual recording to understand client needs**

many as you feel are required in order for someone else to understand your final proposal and to be able to make it.

The working drawing should be precise and drawn to a scale, so if constructed by hand, rather than on a computer, drawing equipment such as a ruler and set square should be used. If the drawing is half the size of the final proposal then the scale is 1:2. If the drawing is a third of the size of the final proposal then the scale is 1:3. You should use a fine black pen for the final outline. The dimensions are added above and along the dimension lines, with thin arrow heads. Figure 3.5.48 shows a third-angle orthographic drawing with critical dimensions added.

Audio and visual recordings

To support your design work there are opportunities to use audio or visual recordings. This could be interviewing a client or user group to find out their needs. It could be to gain feedback on your design ideas or developments before deciding on a final proposal. You could use audio or video to record you explaining a design idea. You could use video to test a model in an appropriate place or position, to demonstrate comfort, ergonomics and so on, or to experience a 'day in the life', where you put yourself into the position of the client to try to experience first-hand their needs (for instance by wearing some thick glasses or foggy goggles to experience what life may be like for a visually impaired client). You could use also use video to record the analysis or disassembly of a product to explain as you take it apart. Figure 3.5.49 shows a slide containing a video of a family routine involving a toddler to learn more about the interaction with machines.

Modelling

To help you develop a sketched design idea you should also use modelling to help you realise your design in three-dimensions and make decisions as to the direction your design idea will take. As you develop a model it is really useful to test your design with your user group to see how they respond to your idea and ask your client what they think of your design.

A three-dimensional model is excellent for working out ergonomics, form and practical details, such as how different parts of the product will be joined together. Modelling materials should be used, rather than a resistant material, as these should be easier and quicker to cut, shape, form and join. Popular materials for modelling are corrugated card, blue styrofoam, green foam, modelling board and MDF. All of these materials are readily available and relatively cheap. All of the materials can be very quickly formed and shaped to realise a design three-dimensionally. A model can be made to scale (either bigger or smaller: twice the size, half the size, etc.) or made approximately to the same size as the final model.

You should select a design idea, or elements of different ideas, that you feel best solve the initial problem and satisfies your user group. Many designers like to translate the idea into 3D so that they can start to realise the form and to test the idea. The advantage of modelling is that it produces a three-dimensional model where all the sides, the top, back and bottom have to be considered. A model can be held and tested for ergonomics by a user, and then this information can be fed back into developing the design. Models can then be photographed and used to communicate more information as you can annotate the photos or even sketch over the top of them.

You should look to develop all aspects of a selected design. If, for example, you were designing a radio, you might consider some of the following: the battery compartment, knobs, dials, switches, speakers, speaker configuration, screen, layout of buttons, stand, positions, aerial, sizes, split lines, how parts fit or join together, lugs, snap fittings, how components are held in place and so on. At this point you could use some digital photos of details that you may have taken and include them on a development sheet as areas to be addressed, looking at the finer details and developing them to show your design intent. Development sheets are critical for your project as this is where you further refine your chosen idea to test, evaluate, modify, check, refine and develop the idea towards a final proposal. This will also allow for additional development research to be carried out where needed (for instance, anthropometrics data (sizes) for dials, knobs, switches, etc). Figure 3.5.50 shows an example of the development of a radio where further details and refinements are carried out.

If you were developing a new garment or textile-based product then you could produce a toile, which is an early version of the finished garment made up of cheap material so that you can test, refine, modify and perfect the design before using more expensive materials. A toile shows how the garment actually sits on a human body, how it drapes, if it is possible to put it on and take it off. You could trial the use of electronics within the toile to see how they could be positioned; you could experiment with additional materials, fabrics, stitching, patterns, construction or smart materials to test how they may perform to enhance the final garment.

For an electronics project, the circuit can be modelled to make sure that it meets the requirements of the brief and the specification. You could test this first using circuit software and then make a physical test model. This allows you to test that the circuit works with similar components, or alternative components, to those intended for the final proposal. The more common forms of physical electronics modelling use either a prototype board, also known as a breadboard, or a stripboard, also known as veroboard. A breadboard is a board covered with small sockets into which components can be plugged and requires no soldering. They are a temporary construction form of modelling as the components are not permanently attached, so that you can easily move them around or replace them in order to test and evaluate possible modifications or improvements to the circuit design. A stripboard is a board coated with straight copper tracks and pre-drilled holes at regular intervals, and components need to be soldered into place. This is often used as a last test model before final manufacture.

If, for example, you were developing a portable speaker, you could model the battery compartment or power source, at two or three times the scale (2:1 or 3:1), to show how that area is to be developed in your design or how a mechanism works (sliding, snap fit, rotating, etc.), and why it will be done in a certain way. You could also look at the type of texture or finish you wish to achieve. You could use CAD to help you develop your design in as a three-dimensional model by creating simple shapes or forms and then use these to draw or design over. CAD can also help you demonstrate how parts could fit together if you want.

Your development should show a clear progression of your idea and you should annotate your designs honestly – you must discriminate, analyse and evaluate all areas of your design, stating what is good and bad, what works and what does not, and why.

In industry, modelling is a key process because it enables manufacturers to test and modify products and processes. Even though modelling may

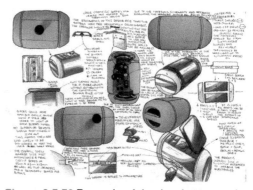

Figure 3.5.50 Example of the development of a radio where further details and refinements are carried out

Figure 3.5.51 Example of a the development of a garment toile where further details and refinements are carried out

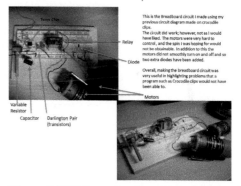

Figure 3.5.52 Example of an electronic circuit where testing has been used

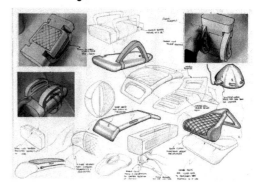

Figure 3.5.53 Example of a portable speaker where further details and refinements are carried out

involve making a number of development models, in the long run, it saves time and reduces manufacturing costs because products and processes are tried and tested before manufacture. This avoids costly mistakes such as the product not functioning properly or the product not meeting customer requirements.

Figure 3.5.54 **Examples of models made by Dyson**

As a product develops it becomes more accurate until it becomes a 'prototype product'. This is a detailed three-dimensional model made from suitable materials to test the product before manufacture. By making a detailed three-dimensional model it could help you plan:

- the most appropriate assembly processes – what order you are going to make things in and why
- how long different processes might take
- the materials, components, equipment and tools you need
- the order in which to assemble your product
- how easy the product will be to manufacture in the time available – if anything needs simplifying
- where and how you will check the quality of your product – quality assurance and quality control checks against your working drawings, production plan and design specification.

Card modelling

Corrugated card is a commonly used modelling material because it is widely available in a range of thicknesses. To make an even cheaper model, you could use a recycled box. The card can be cut using scissors, a scroll saw or a craft knife. It can be joined very easily using PVA glue, masking tape, sticky tape or double-sided tape.

Card modelling using CAD and a laser cutter

If you are confident in your skills in CAD and can produce virtual models quickly, then you could use CAD along with a laser cutter to produce a model of your design to realise the form.

If you model a part in CAD that you would like to be printed in a three-dimensional form or prototype (made from corrugated card) then you can transfer it into the relevant slices using a free piece of software that you can download to manufacture using a laser cutter (see Figures 3.5.56–3.6.59).

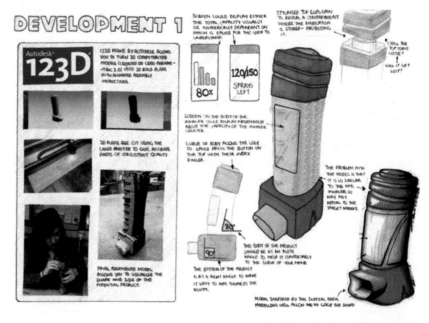

Figure 3.5.55 **Card modelling using a laser cutter further developed through sketches**

Figure 3.5.56 Step 1: import your CAD model and choose how you slice the model.

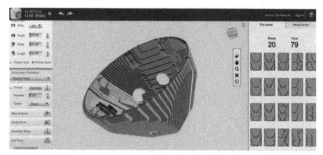

Figure 3.5.57 Step 2: edit the number of layers.

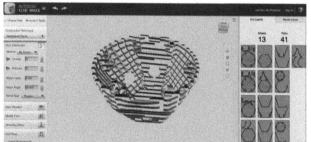

Figure 3.5.58 Step 3: output the layers as an STL file.

Figure 3.5.59 Step 4: photograph the model for use on design sheets.

Audio and visual recordings

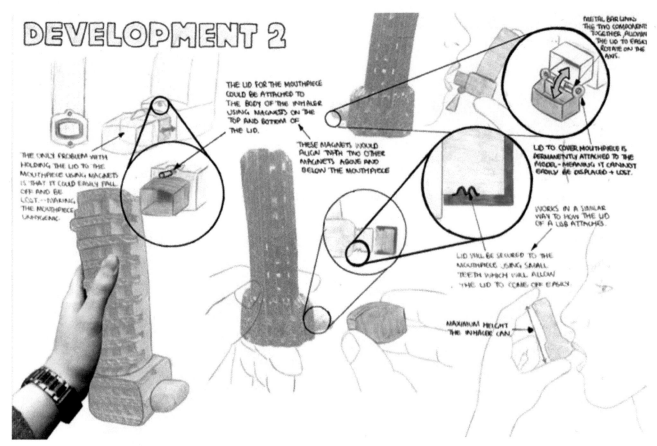

DEVELOPMENT 2

THE ONLY PROBLEM WITH HOLDING THE LID TO THE MOUTHPIECE USING MAGNETS IS THAT IT COULD EASILY FALL OFF AND BE LOST - MAKING THE MOUTHPIECE UNHYGENIC.

THE LID FOR THE MOUTHPIECE COULD BE ATTACHED TO THE BODY OF THE INHALER USING MAGNETS ON THE TOP AND BOTTOM OF THE LID.

THESE MAGNETS WOULD ALIGN WITH TWO OTHER MAGNETS ABOVE AND BELOW THE MOUTHPIECE

METAL BAR LINKS THE TWO COMPONENT TOGETHER ALLOWIN THE LID TO EASILY ROTATE ON THE AXIS.

LID TO COVER MOUTHPIECE IS PERMANENTLY ATTACHED TO THE MODEL - MEANING IT CANNOT EASILY BE DISPLACED + LOST.

WORKS IN A SIMILAR WAY TO HOW THE LID OF A USB ATTACHES.

LID WILL BE SECURED TO THE MOUTHPIECE USING SMALL TEETH WHICH WILL ALLOW THE LID TO COME OFF EASILY.

MAXIMUM HEIGHT THE INHALER CAN.

Figure 3.5.60 **Printed card model cut using a laser cutter**

Virtual modelling

Digital media and new technologies have developed enormously over the last few years, and designers at all levels have been very quick to use these methods for both designing and making. There are many different three-dimensional modelling software choices for designers widely available. These software systems allow photorealistic three-dimensional models to be produced and rendered. CAD can be used to develop design ideas and can be invaluable in visualising what a product will look like when it is made. It is excellent for checking the proportions of a product, trying out different colours and materials, placing the visualisation into a scene or context, and being able to quickly modify the design to re-render (see, for example, the use of virtual prototype by fashion designers on page 293).

Styrofoam or green foam modelling

Cutting and shaping high-density polystyrene foam, or styrofoam as it is more commonly known, is a fast way of creating solid models to realise the size and shape of your design idea. Styrofoam is available in a limited range of colours, the most common being blue. It is supplied in large flat sheets of different thicknesses, ranging from 15 millimetres to 100 millimetres. It can be easily cut using a range of hand tools, such as a coping saw, finishing materials such as abrasive paper, and some power tools such as a scroll saw. Final shaping and finishing can be carried out through using files and abrasive paper, such as 'wet and dry' paper. Green foam is more dense, and as such, allows for greater details to be added. It is more commonly used for final test models.

Foam-cored board can also be used. This consists of a polystyrene foam with a facing of paper on either side. This lightweight or easy-to-cut material is often used for architectural scale models and prototypes for small objects.

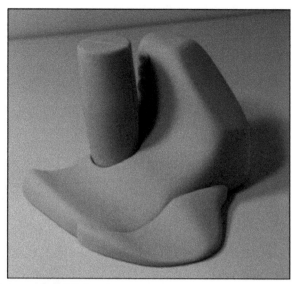

Figure 3.5.61, 3.5.62 and 3.5.63 Test mechanism models using a laser cutter, MDF and green foam

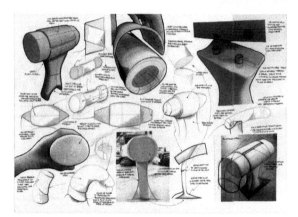

Figure 3.5.64

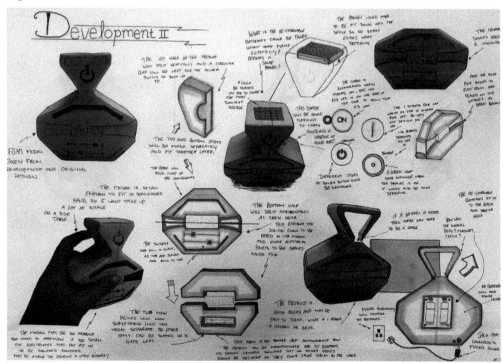

Figure 3.5.65

Audio and visual recordings

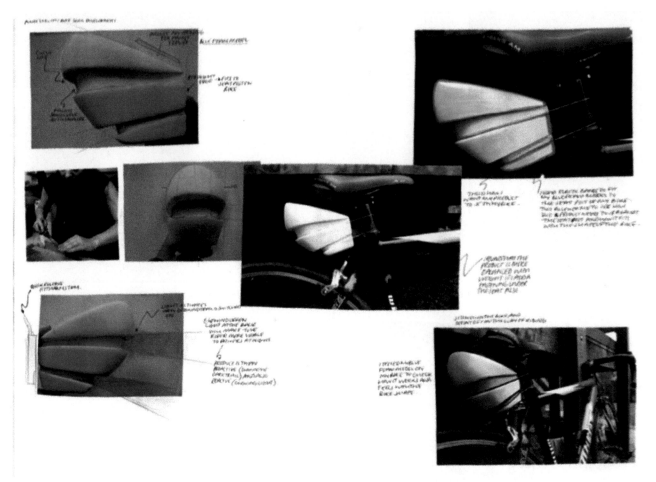

Figure 3.5.64, 3.5.65 and 3.5.66 **Development examples showing integrated modelling**

Check your knowledge and understanding

1 What techniques can be used to make a design idea stand out on a page of ideas?
2 What is the preferred method of freehand sketching in 3D?
3 What is a construct of framework for sketching in 2D or 3D and how is it used?
4 What is meant by the term 'weight of line'?
5 What is annotation used for?

3.6 Prototype development

What will I learn?

In this topic you will learn about:
→ how to design and develop prototypes in response to client wants and needs
→ how the development of prototypes:
- satisfies the requirements of the brief
- responds to client wants and needs
- demonstrates innovation
- is functional
- considers aesthetics
- is potentially marketable
→ how to evaluate prototypes and be able to:
- reflect critically, responding to feedback when evaluating your own prototypes
- suggest modifications to improve them through inception and manufacture
- assess if prototypes are fit for purpose.

A prototype is a model of a product used to evaluate a design, its performance and its ability to be manufactured. When working on a project, you will probably make one or more models when developing your product to test out your ideas and evaluate your design. This lets you see what might and might not work and allows you to make changes to your design so that it meets all of the points on your design specification. It also allows your client to see what your product might look look like, so that they can decide if it is suitable for their wants and needs. The prototype lets you try out some manufacturing processes so you have a better idea of how it might be made.

Designing and developing prototypes

One of the functions of a prototype is to test out ideas and concepts, so a first model will often be quite basic and without much detail. It may be used to give the client an early idea of the designer's thinking. These early models are usually changed and developed, often many times, until a final design is decided upon.

Prototypes and models are usually made out of cheaper materials, especially if the final material that is going to be used is expensive or difficult to work with. Sometimes it is necessary to use the actual materials that will be used for the final product, so that realistic testing can take place, for example, when testing an electronics product.

Some materials commonly used for modelling first ideas include:
- paper and card, which are inexpensive and easily available
- MDF, which is easy to work with and can be used for intricate shapes
- expanded foam such as polystyrene, which is not good for detail, but can be used to get an overall idea of the shape, form and size of a product
- foam board, which is lightweight and used to make simple models
- plastic, which is easily shaped and used to make representational models

- clay, which is easily shaped and useful for testing ergonomic shapes
- calico, which is an inexpensive cotton fabric used when modelling textile products
- a breadboard, which is used to test prototypes of electronic circuits.

Figure 3.6.1 **Expanded foam is useful for testing concepts.**

Figure 3.6.2 **Foam board is used for simple models.**

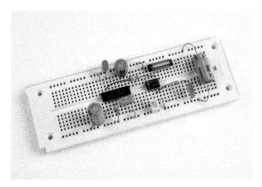

Figure 3.6.3 **A breadboard is used to prototype electronic circuits.**

Satisfying the requirements of the brief

A designer will be given a design brief by the client. This outlines the important features of the product, such as size and function, and also some features that are desirable but not essential.

A prototype can help the designer to work out the best ways of including all the features required and make modifications until the specification points have been addressed. Sometimes a prototype of a small section of the product is made to check how it fits in with the whole product, or to develop it further without having to model all of the product.

Responding to client wants and needs

In commercial manufacture, CAD and CAM programs are often used to try out ideas. Making a physical prototype is expensive in terms of the materials used and the time taken, especially when modifications have to be made to the design and further models produced. Many products are now developed using CAD to make virtual prototypes which allow designers to see how the product will perform without having to make a physical model. The client can also test the virtual product to see if it performs as expected, and ask for changes if the model is not quite right. This speeds up the time taken to develop and manufacture a product. The client is able to start manufacturing and selling the product earlier than if many physical models had to be made and tested.

Prototypes are often part of a manufacturing specification and the agreement between the manufacturer and the client, as they show the manufacturer exactly what the product must look like and perform.

Demonstrating innovation

Prototypes are used for other purposes in commercial manufacture. Physical prototypes can be used to test a concept to see if a product is likely to have appeal to the target market and its viability for manufacture on a larger scale. Such prototypes are often for very new and innovative products that are ahead of trends and fashions and may or may not be produced on a larger scale in the future. Some very new or different products that may be ahead of their time are often only produced as prototypes and do not go into bulk production.

Figure 3.6.4 **A three-dimensions virtual reality bicycle allows the client to review the product.**

Functionality

Being able to test a product to see how it performs before it is manufactured in large numbers is important if materials, money and time are to be used efficiently. A prototype product may be made from a cheaper material and have less detail than the intended design, but can allow the designer to evaluate how it works and what may need to be changed. For example, an electronic circuit may need to fit into a small space, such as a phone, so a physical model will allow the designer to check this out. Certain ergonomic features will be required in a design for a new hairdryer, and these can be trialled in a full-size prototype. Fashion designers make models to check how a garment fits and drapes, and if it can be worn easily.

Figure 3.6.5 **A concept car used to test the likely market**

Figure 3.6.6 **The ergonomic features of a hairdryer can be tested by making a full-size model.**

Figure 3.6.7 **Fashion designers often make physical prototypes to test out their ideas.**

Aesthetics

While many prototypes are made to test out the functionality of the intended product, aesthetics play a large part in determining the success of a new product. Fashion and trends often dictate the colours used, the pattern and the style lines. Using CAD programs and virtual prototyping allows designers to experiment with colour schemes and to superimpose patterns on to the basic product. This technique is especially useful in fashion design, so that garments can be viewed in different colourways and made in different fabrics. A virtual fashion show can be viewed by clients all over the globe and allows them to see how virtual prototypes of garments will drape and move on a real person.

Figure 3.6.8 **Virtual prototypes allow fashion designers to see how garments drape and move on a real person.**

Designing and developing prototypes

Marketability

Most prototypes are not intended to be sold, but rapid prototyping does allow small numbers of a product to be manufactured economically, as a large production line does not need to be set up.

Rapid prototyping uses CNC machines to make a model of a part of a product in a very short time using a three-dimensional printing program. This system can produce very complex shapes. It can also be used to manufacture parts in small numbers for specialised orders, which is usually cheaper than other manufacturing methods.

Rapid prototyping may also be used for customisation of mass-produced goods to meet the requirements of individual clients.

Figure 3.6.9 **A 3D printer printing a cube**

Evaluation of prototypes

If a prototype is to be used effectively it needs to be tested and used in the way the designer intends to check that it is fit for the intended purpose. This gives valuable feedback about what works, what the product looks and feels like, and whether the client is happy with it or requires further modifications to be made.

When testing your own models and prototypes, it is important that you find out what your client and likely users of your product think about it, not just other students in your class. You might also check it against a similar product that is already on the market, by analysing the similarities and differences to see how your product compares. This will help you to suggest improvements that make your design better. Making modifications to your models and prototypes as a result of testing and feedback from others is an important part of the design process and can help make sure that your final outcome is as good as it can be.

In commercial manufacture, a rigorous testing schedule will involve the client, members of the target market group and experts in the design, manufacture and use of the type of product. Sometimes a focus group will be set up to seek opinions of the potential users or the product might be shown at a sales exhibition to judge its likely appeal to the intended users.

The prototype product will be tested to see that it is functional and fit for the intended purpose, safe to use and manufacture, that it can be manufactured efficiently with materials that are easily available and at the right price for the target selling price. Comments and suggestions for improvement will be taken into account when making any modifications before the design is finalised.

The final prototype will be used to work out costings for materials and labour, and the likely profits to be made.

KEY POINTS
- Prototypes are used to test and modify design ideas.
- Prototypes can be physical or virtual models.
- Prototypes often form part of the agreement between a manufacturer and the client.
- Prototypes can be used to show innovative designs and check their viability.
- Rapid prototyping allows small numbers of specialised products to be made economically.
- A rigorous testing and evaluation scheme is important to decide if the product design will be successful.

Check your knowledge and understanding

1 Name three different materials commonly used for making prototypes.
2 Give two reasons why virtual prototyping is used in commercial manufacture.
3 List two commercial uses for rapid prototyping.
4 Give two ways that a virtual prototype can help a client to decide if a new design is right.
5 Identify four ways that a prototype can be evaluated.
6 Give three reasons for making a prototype of a new design.

Selection of materials and components

What will I learn?

In this topic you will learn about:
→ how to select and use appropriate materials and components considering functional need, cost and availability.

During the development of your design you will need to consider what materials and components will be suitable for the manufacture of your prototype. There are three main factors to consider; functional needs, cost and availability.

Papers and boards

Functional need

When designing products it is important to consider a range of materials and assess their suitability for the function of the product. Different properties of the alternative materials need to be considered in order to determine which product will last longest, be strong enough, or resist moisture and so on. An example of this might be in food packaging, where a board might be chosen to resist moisture or create a seal so that the product cannot be contaminated. Different materials may be tested to ensure that they do not absorb the moisture content of the food over long periods of time, or tested to see if the print is successful and looks of a high enough quality.

Figure 3.7.1 Card drinks carton

Cost

Cost is another important factor in the decisions designers and manufacturers make. In order for a product to be commercially successful there must be enough of a profit margin. There are many issues involved in creating a commercially viable product, but one is consideration of the material cost so that it matches the quality of the final outcome. Materials such as solid white board, therefore, are only used in products which require a high-quality outcome, such as hardback books. These products need to be sold at a higher price to recover the original outlay of cost. Solid white board would be an unlikely choice for a packaging manufacturer – despite this material being able to perform the necessary functional role, it would be too expensive, and therefore inappropriate for the product.

Availability

Materials are often chosen due to availability. This is very true of small-scale production, where a material may be available and the designer therefore designs the product with the material in mind, rather than looking for a material after the design has been completed. When you consider your own projects you may have had to use the most appropriate material that is available in your school or college.

KEY POINTS
- When selecting materials for manufacture, functionality, cost and availability are all important issues.

Check your knowledge and understanding

Consider the following products. Name a potential material and explain why you have chosen this material as a suitable choice:

- cereal box
- perfume box
- poster
- drinks carton.

Timber-based materials

Functional need

There are many opportunities for using wood when designing and making prototypes. Wood is typically used when making items of furniture and for providing structure to a design. The properties of wood vary considerably. Balsa, for instance, is light in weight, soft and relatively weak, but is very easy to work, which makes it ideal for use as a modelling material. Teak is a strong hardwood and contains a natural oil that makes it an excellent choice for use in garden furniture. Oak is a very strong material that has a very decorative grain known as 'figuring', which makes it suitable for manufacturing high-quality furniture.

Figure 3.7.2 **Wooden patio furniture**

Cost

Financial cost

The price of wood can vary considerably. Rough-sawn softwoods are relatively inexpensive, whereas hardwoods such teak, oak and ash can be quite expensive. Balsa is expensive, but that reflects the amount of cutting to size and transportation required, and the small amounts it is usually purchased in.

Manufactured boards such a MDF (medium density fibreboard) are inexpensive, while thicker sections of marine plywood are much dearer.

Environmental cost

Natural timber is considered to be the most environmentally friendly material as it is renewable, reusable, recyclable and has less negative impact on the environment when being processed from a tree to its stock form.

Manufactured boards are less environmentally friendly, as they have undergone additional processes that involve using extra energy and adhesives. Some manufactured boards, such as MDF (medium density fibre board), also produce fine particles of dust when being machined or sanded. This fine dust can be harmful if inhaled and has been linked with cancer.

KEY POINTS

- Wood is an ideal material for producing furniture.
- Different types of wood have very different properties.
- Manufactured boards are available in much larger sizes than natural timber.

Availability

Wood is a readily available material that comes in a wide variety of stock forms. Solid wood is sold in a number of different states:

- rough sawn – timber that is straight from the saw, it will have a rough surface with lots of splints
- PSE (planed square edge) – the timber will have been planed smooth on two adjacent surfaces
- PAR (planed all round) – timber that has been planed on all four sides
- manufactured boards are available in thicknesses from 1 millimetres to up to 40 millimetres, and most commonly in a standard sheet size of 2,440 millimetres by 1,220 millimetres.

Check your knowledge and understanding

1 State the stock size of manufactured board.
2 Give an example of a modelling timber.

Figure 3.7.3 **Metal patio furniture**

KEY POINTS

- Metal is generally used when strength is important.
- Metals can be permanently and non-permanently joined.

Metal-based materials
Functional need

Metals are generally used when strength is a high priority. Low-carbon steel is a strong metal that can be joined in a wide variety of ways, and is the most common metal used for manufacturing products such as car bodies and the casing for white goods (fridges, freezers, ovens, washing machines and driers). Aluminium is light in weight and is naturally resistant to corrosion, which makes it suitable for use in manufacturing aeroplanes, stepladders and bike frames. Precious metals, such as gold and silver, are durable, attractive and are used in jewellery. Metals can be permanently and non-permanently joined in many different ways and cast to form intricate shapes.

Cost
Financial cost

Low-carbon steel is a relatively inexpensive material, whereas precious metals, such as gold and silver, are very expensive.

Environmental cost

Metals can only be considered to be environmental friendly if they are recycled at the end of their life. Sourcing and processing metals uses large amounts of energy and releases significant amounts of harmful gases into the atmosphere.

Availability

Most metals are readily available and come in a variety of stock forms.

Polymers

Functional need

There are many reasons why you could consider using polymers in the manufacture of your prototype.

- One of the main reasons for selecting thermosetting polymers is due to their ability to be moulded with the use of heat. This allows you to produce a wide variety of interesting shapes with relative ease.
- Many polymers have a natural high-gloss finish, which means there is no need to apply a finishing coat.
- They can be obtained in a wide variety of colours, and therefore do not need to be painted or stained.
- They do not corrode, therefore they do not need to be protected from water.
- They are a good thermal insulator.
- They are a good electrical insulator.
- They are unaffected by many chemicals. However they can be affected by solvents.
- Thermoset polymers are unaffected by heat once they have been moulded.

Figure 3.7.4 **Polymer patio furniture**

Cost

Financial cost

Polymers vary in cost, but are considered to be expensive when compared to most metal- and timber-based materials. However, due to their suitability for volume production, the cost of polymer-based products is generally low. This is because processes such as injection and blow moulding are mechanised processes that require relatively low energy and a small workforce.

Environmental cost

Similar to metals, polymers can only be considered to be environmentally friendly if they are recycled at the end of their life. Sourcing and processing polymers uses large amounts of energy and releases significant amounts of harmful gases into the atmosphere. Some thermoplastics and most thermoset polymers are not able to be recycled.

Availability

Most polymers are readily available in a wide variety of stock forms.

KEY POINTS

- Polymers have many properties that make them different to wood and metal.
- Polymers have a high environmental cost

Textile-based materials

Functional need

Fibres and the fabric construction methods used give different properties to textile materials, and these need to be considered in relation to the product to be made and the priorities identified as being important. The need for strength and resistance to abrasion, absorbency, the ability to insulate, ease of care, crease resistance and ability to stretch will be important in different products.

Figure 3.7.5 Fashion designers select fabric for clothing

For example, a school shirt needs to be hardwearing, easy to care for and maintain a smart appearance, so a plain-woven fabric made with a blend of polyester and cotton fibres is an ideal choice. If the priority is to keep the wearer warm, then a brushed fabric such as polyester fleece, or a knitted wool fabric would be suitable.

The choice of components can add to the efficient functioning of a product. Zips are often easier to fasten than small buttons or press studs and seal the entire opening, Velcro is good for products that need to be fastened and unfastened quickly and easily, such as children's clothes or stage costumes, but it is a bulky fastening.

Cost

Fabrics and components vary in cost from very expensive to very low cost. Fabrics made entirely from natural fibres, such as wool and cotton, are generally more expensive than those made entirely from synthetic fibres or blends with natural fibres. Lycra is an expensive fibre so will add to the cost of a fabric, but it is only used in very small amounts and it gives useful qualities to fabrics. There are different levels of quality in fibres that can affect the cost, for example, woollen fabrics made from recycled wool will be cheaper than those made from new wool. Additional finishes and decorative techniques, such as printing, embroidery and crease resistance, will also make fabrics more expensive. The width of the fabric will affect cost, as wider fabrics tend to cost more – but less needs to be bought. Some fabric patterns, such as checks, may mean that the fabric cannot be used economically, so extra will need to be bought to allow for pattern matching.

The quantity of different components needed, including thread, will be an important factor when working out cost, and must be factored in when calculating the overall cost of making the product.

Availability

Fabrics and components are readily available in many different outlets. Fabrics are made in a limited number of standard widths and come in a huge range of colours, prints, textures and weights. There is a wide range of components, including fastenings, threads, bindings, ribbons, edgings and trims that can be chosen to match the colour of the fabric or provide a contrasting effect.

Check your knowledge and understanding

1 List the functional properties that are important in fabric used for a cushion for a garden seat.
2 List five different components that can be used when making garments.

Electronic and mechanical systems

When choosing materials and components for electronic and mechanical systems, a decision needs to be made whether to design and construct a circuit or purchase a pre-made module.

Functional need

You need to have a clear specification regarding the function of an electronic and mechanical system before considering manufacturing a prototype. This is because there are a number of pre-assembled programmable project boards available that are suitable for almost any outcome, and this option needs to be considered against the cost and time of manufacturing your own systems from scratch. Electronic input and output components are available from suppliers; working parameters such as voltage and current ratings need to be considered when sourcing and using these components. A range of gears, axles and bearings are also available from suppliers along with ready-assembled gearbox modules and other pre-assembled modules that might suit your requirements.

Figure 3.7.6 **A pre-assembled programmable project board**

Cost

The cost of a pre-assembled programmable project board, input and output components, and mechanical gearing and gearbox modules may be high because of the specialist equipment and skill needed to produce them. However, these parts are all reuseable in future projects. Most manufacturers produce components on a very large scale, which can reduce the costs of each component.

Availability

There are a number of specialist suppliers of electronic and mechanical components that can supply individual components and pre-assembled modules for projects. Many components are manufactured globally and imported – often from Asia, with microchips often manufactured in the USA and other parts from Europe. Currently these are widely available but increasing complexity in the technology used to produce them may reduce the number manufactured in the future.

KEY POINTS

- When prototyping electronic and mechanical systems, it is important to have a clear specification in order to make decisions about components or kits required.

Check your knowledge and understanding

Explain why you need a manufacturing specification when sourcing electronic and mechanical components.

3.8 Tolerances

What will I learn?

In this topic you will learn about:

→ how a range of materials are cut, shaped and formed to designated tolerances
→ why tolerances are applied during making activities.

When you measure or cut a piece of material, it is likely that there will be a certain degree of error in either the marking out, the cutting and in some cases both. In many applications, size is critical, and dimensions or materials must be highly accurate. In others, a rough size or nominal size would be acceptable for the application. The range in which this is acceptable is called the tolerance.

Tolerance is expressed as an allowable variation of a dimension. This can either be how much larger a finished piece can be, or equally how much smaller. It is presented as a figure that follows the dimension.

For example, a tolerance expressed as 25 millimetres −1 would mean that the finished dimension cannot exceed 25 millimetres, but anything between 24 millimetres to 25 millimetres would be acceptable. The same dimension presented as 25 millimetres ±1, would mean that any variation between 24 millimetres and 26 millimetres would be acceptable.

Tolerances tend to be more important in situations where components are designed to fit together accurately. The smaller the tolerance the more accurate a product, but with accuracy comes an increase in the manufacturing costs.

You have already been introduced to tolerances in Topic 2.8 Specialist techniques. Tolerances are important factors to consider when manufacturing, but why are they important?

Papers and boards

In the case of paper and card products, length and width measurements of an item are often the main use of tolerances, but it could also be other elements, such as weight or the distance between one object and another.

When working with paper and board products, tolerances are necessary for a range of different products of different sizes. Some products will require a smaller tolerance, meaning that machinery has to create this product to a better level of accuracy and may therefore be more precise and potentially more expensive. An example of this is in the production of money, which has to be produced using machinery which will allow products of a specific size and thickness. This may be used as an indicator of fraudulent notes, and so the tolerance of genuine banknotes needs to be precise. Larger products, such as posters, may not need to be of the same level of accuracy and may therefore be manufactured using less precise machinery and with less frequent tolerance checks throughout manufacture.

Timber-based materials

Timber is one of the most challenging materials to manufacture to accurate tolerances due to the fact that, as a natural material, it is prone to move, split, bow and warp.

In many applications where timber is used as the main material, accurate tolerances are not as vital as with some other materials. If you were manufacturing a large dining table that was 1,500 millimetres long, it would not really matter, or indeed be noticed, if the tabletop's length was actually one or two millimetres longer or shorter than the stated 1,500 millimetres.

If a bespoke piece of furniture was being produced, such as a fitted kitchen or wardrobe, then the accuracy of the dimensions would need to be manufactured with a smaller tolerance, as if the width of the units was too big they would not fit into the intended space.

Manufactured boards are more stable than hardwoods and softwoods, meaning that they are less prone to warp or split. This means that they are easy to machine to accurate dimensions. Hardwoods and softwoods can also be machined accurately, but are less likely to maintain their machined tolerance due to their dimensional instability.

Metal-based materials

Metal is a good material to machine and manufacture to accurate tolerances, especially when produced with the assistance of computer-controlled machinery.

Products with rotating parts, such as wheels or axles that make use of bearings, demand a high level of accuracy in order to be efficient. A skateboard axle is 8-millimetres diameter, therefore the bearing that runs on the axle must be larger than 8 millimetres in order to slide over the axle. If it is too large then the wheel will not run smoothly and the use of a bearing would be redundant.

Metal products that are manufactured by casting are hard to produce to a specific tolerance, as the castings can shrink a little as the molten metal cools, making calculating the finished size tricky. They can be machined once cooled using a centre lathe or milling machine where accurate tolerances can be achieved.

Polymers

Polymers can be manufactured using industrial manufacturing techniques, such as injection moulding, to very accurate tolerances.

Lego bricks are an excellent example of a product where accurate tolerances are vital to the performance of the product. Lego bricks are designed to fit together securely, but also to be able to be removed and rebuilt. The actual tolerance that Lego work to is a closely guarded secret, but if the bricks were too large then they wouldn't be able to fit together, too small and the structures wouldn't stay connected.

Textile-based materials

Accurate tolerances when working with textile materials are less important and often more difficult to achieve than with other materials. The woven and knitted structures have quite a lot of give in them when compared to other materials, such as metals and polymers, so it is difficult to work to very fine tolerances. Seams may not be exactly the same width all the way along, especially in commercial manufacture where many different machinists may be working on the same batch of products.

When bulk manufacturing clothing there are often quite large and acceptable tolerances, between 5 millimetres and 20 millimetres, in the size of the finished garment. Since the size and shape of the human body varies so much, even when people are a standard clothing size 8 or 16, it does not matter that the tolerances are so large, and it will not even be noticed that individual garments are slightly bigger or smaller. However, if a client were to have a made-to-measure individual

Figure 3.8.1 **A smaller tolerance is needed for the union flag appliqué design than for the overall cushion size and the other bedding items.**

garment made it would be expected that the tolerances would be smaller to give a precise fit. Furnishing products such as duvet covers and cushions are also manufactured to quite large tolerances as an exact size is not crucial.

Certain textile techniques require smaller tolerances, for example, patchwork and appliqué, otherwise the different shapes would not fit together accurately and that would spoil the aesthetics of the product being made.

Electronic and mechanical systems

When making any electronic or mechanical system, leaving things open to more than one interpretation could lead to the system not working. In electronics it is important to specify the value of the components, for example, using a larger resistance than specified could result in a larger time delay than you require in a time-delay circuit. In this scenario you may want to be specific about not only the value of the resistor the circuit requires, but also its tolerance. For example to get the time delay you require the resistor value may have to be within ± 1 per cent of the stated value of the resistor. For most circuits you will manufacture the standard tolerance of ± 5 per cent, or ± 10 per cent will be sufficient – the circuit diagram you are working from will state if a lower tolerance is required.

Figure 3.8.2 **Polymer gear wheels**

If building a simple gear train where two different-sized gears are required to mesh together, the holes you drill for the axles may have to be within 0.5-millimetre accuracy. If they were both outside this tolerance then they might mesh too tightly together to turn freely, or at the other end of the tolerance might not mesh at all or just enough that they turn and jam together.

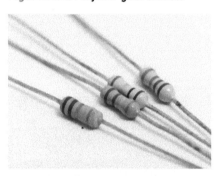

Figure 3.8.3 **Resistors**

KEY POINTS
- Tolerances are the acceptable range of size a product or part can be.
- Tolerances are shown as + or – the acceptable dimension.
- Tolerances are important where components fit together.
- The smaller the tolerance the more accurate the product.
- More accurate tolerances will make a more consistent product, but may also increase manufacturing costs.
- Tolerances are applied when making electronic and mechanical systems in order to ensure the systems function as intended.

Check your knowledge and understanding

1 What are tolerances?
2 Why do we use tolerances?
3 Give an example of a typical tolerance for a paper or board product.
4 Why is it difficult to maintain an accurate tolerance with timber once machined?
5 Give an example of where an accurate tolerance is not necessary.
6 You are cutting a piece of material 270 × 285 mm where each dimension has a tolerance of ± 2. Which of the following is within tolerance?
   a 271.5 × 287.5   b 285 × 270   c 268.5 × 284.9   d 265 × 284
7 Explain what the consequences could be if tolerances were not applied when making electronic and mechanical systems.

3.9 Materials management

What will I learn?

In this topic you will learn about:

→ the importance of planning the cutting and shaping of material to minimise waste, for example, nesting of shapes and parts to be cut from material stock forms

→ how additional material may be required for a seam allowance or a joint overlap

→ the value of careful measurement and marking out to create an accurate and quality prototype

→ the use of data points and coordinates, including the use of reference points, lines and surfaces, templates, jigs and/or patterns.

Efficient use of material is a hugely important factor to consider when producing both one-off prototypes and large-scale batches of a commercial product. It is important to remember that when calculating cost, any waste material generated must still be accounted for. The financial and environmental implications of efficient material management become all the more important as the scale of production increases.

When cutting out shapes in any material, the shapes need to be organised to be as close as possible to each other without reducing the quality of the cut. This is called nesting. How much space you will need to leave between shapes will depend on the equipment used to make the cut. For example, a material cut on the laser will need much less space between each shape than a material cut on a CNC router. On the router, a cutter of 6-millimetres diameter, for example, may be used. This will create a 6-millimetre channel between parts, and so if parts are too close, the cut lines will cross each other. The same shapes can be cut on the laser cutter with less space in between, because there is a much smaller amount of waste – usually less than one millimetre, depending on the quality and maintenance of the laser cutter and the distance from the material.

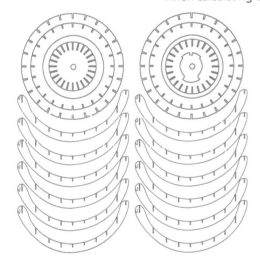

Figure 3.9.1 Nesting of shapes on the laser cutter

Figure 3.9.1 shows carefully nested parts for a laser cutter. Maximum use of the material means that only just enough space is left between parts.

Papers and boards

Cut materials efficiently and to minimise waste

When manufacturing paper and board products, a laser cutter may be used to create shapes, particularly at the prototype stage. Once the product is ready for final manufacture in quantity, it is more likely that shapes will be cut using die cutters. These also require little space between parts, as the blades in the die cutter are fine, and so shapes can be nested together closely. In printed paper and card products manufacturers often print over the cut line to ensure that there are not any white spaces and to ensure that slight misalignment of the cutters does not have as much impact on the final product.

Using appropriate marking-out methods

When making paper and board products a range of marking-out methods are appropriate. Some of these are done by hand and others with the help of computer-aided design. When marking out is completed by hand, tools such as steel rules, T-squares, compass cutters and so on can be used (see Topic 2.8 for more information). When computer-aided design is used for marking out, these drawings are particularly accurate and can be repeated multiple times to ensure consistency. Often drawing programs use coordinates to plot lines accurately. A two-dimensional drawing uses an *x* and *y* axis as though the surface is completely flat. For example, a rectangular shape 50 × 40 millimetres may be plotted as (50, 40). To create three-dimensional shapes the *z* axis is used.

Using datum points and coordinates

Paper and board products can be made using computer-aided design and manufacture. In this case, datum points and coordinates are used both in the CAD and CAM software in order to position the drawing effectively. This enables you to copy and tessellate shapes, using materials most effectively.

The datum point is where your coordinates are at 0,0,0. This is where each of the axes (*x*, *y*, and *z*) are at zero. This reference point is used so that the machine always knows where the start point is, and so that you can match your drawing with the cutting of the material and the bed of the machine.

Check your knowledge and understanding

1 Explain how costs can be reduced using materials management.
2 Explain how coordinates are used in marking out.

Timber-based materials

Cut materials efficiently and to minimise waste

Timber can be processed and shaped quickly and is generally easier to work with than metals and polymers, but this doesn't mean that it is any less important to consider measures that minimise waste.

As we have already discussed, timber is available in various stock sizes, and where possible, manufacturers and designers should plan to make use of these available sizes. If the finished dimension of a piece of timber is critical then the closest stock size should be chosen to minimise the amount of processing needed.

In commercial manufacture, waste timber can often contribute to the production of manufactured boards, but in school and college workshops and small factories, timber waste tends to be thrown away.

Use appropriate marking-out methods

Timber can be marked out using equipment including a pencil, rule, try square and marking knife. Care should be taken to mark out material lightly, as any marks will need to be removed before applying a surface finish.

If marking out a length of timber, you should always start by planing a surface flat, and ensuring that one edge is square and at 90 degrees to the surface. These surfaces are usually

KEY POINTS

- Reduction in material use can reduce environmental impact and helps to reduce cost.
- Nesting is when products are grouped together closely to reduce waste material.
- Marking out can be done by hand or using computer-aided design.

referred to as 'face side' and 'face edge'. These will give you an accurate datum or reference point from which to measure and mark out your timber.

When marking out a shape on a piece of board, consider the placement of the design on the material available. There is little point in marking out a shape in the centre of a large board when it will fit comfortably in the corner.

Where possible, try to choose the smallest piece of material that is suitable for the relevant design, and if marking out several items try to tessellate your designs as efficiently as possible.

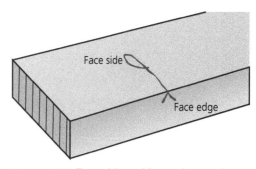

Figure 3.9.2 **Face side and face edge marks**

Tessellation involves marking out your shape to make best use of the material available to you. This may mean that you have to rotate your design in order to fit more on to your available material.

Use of datum points and coordinates

If a CNC router is being used to cut out a design it is easy to move and tessellate shapes on screen before manufacturing.

The data point on a CNC machine is the point at which the *x*, *y* and *z* coordinates are at zero. The datum point is important as the machine uses this fixed reference point when calculating the paths to take to cut out your design. The datum point differs between CNC machines, but generally it is either the top left- or bottom right-hand corner of the work bed. You must ensure that your material is correctly fitted in the machine to correspond to this datum.

If a design is to be manually marked out on a piece of manufactured board, then templates should be used where possible to increase accuracy and allow for efficient use of material.

Metal-based materials
Cut materials efficiently and to minimise waste

Metals are generally harder to shape than timbers and polymers, and also more expensive, which makes it is even more important to minimise any additional shaping or machining.

Metals are available in a wider selection of stock sizes and extrusions than woods and polymers, so there is more choice available to the designer when planning their projects.

If shaping or machining needs to take place, then a centre lathe or milling machine provide an efficient way to remove excess material. All of the waste material removed however contributes to the overall cost of the item.

Figure 3.9.3 **CNC machining an aluminium component**

Use appropriate marking-out methods

Metal can be marked out using equipment including a scribe, rule, engineer's square and dividers. Greater contrast between the surface of the material and the markings can be achieved by the use of layout fluid, sometimes referred to as engineers blue; this can be removed using thinners or methylated spirit.

Use of datum points and coordinates

In a similar way to the use of CNC machines when working with timber, metal CNC machines use comparable datum points for referencing the co-ordinates

Figure 3.9.4 **Layout fluid used when marking out metal**

used to machine the design. These are found in all CNC machines, including milling machines, centre lathes and plasma cutters.

Polymers

Cut materials efficiently and to minimise waste

Polymers can be cut and shaped using a variety of hand tools and workshop machinery, such as coping saws, files, scroll saws and band facers, but the most efficient method of working with polymers is through the use of a laser cutter.

Laser cutters are now commonplace in schools and colleges, with many machines having large bed sizes in excess of 500 square millimetres. This capability to cut large pieces of material means that materials management is an important consideration to minimise and reduce waste and unnecessary cost. If small cutting jobs are necessary, then it is good practice to source an appropriate sized piece of material, but one-off jobs do tend to be fairly wasteful. It is much more efficient to use a larger piece of material and layout as many designs as possible on the material.

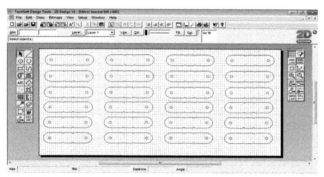

Figure 3.9.5 Wasteful arrangement of designs

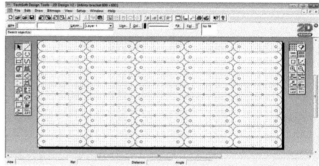

Figure 3.9.6 Efficient tessellated designs

Figures 3.9.5 and Figure 3.9.6 are screenshots of two files that have been set up for manufacture on a laser cutter. The material in both examples is the same dimension, but Figure 3.9.5 has 24 nets arranged on the work piece, whereas Figure 3.9.6 has 50. This illustrates the importance of tessellation and of planning the layout of your work to make the most efficient use of the material.

Use appropriate marking-out methods

Polymers can be marked out using equipment including a chinagraph pencil, rule and an engineer's square. If marking out is taking place while the polymer still has its protective film in place, a biro or permanent pen is a reliable marking-out technique.

Much like marking out on timber, where possible the use of a template is recommended when marking out several of the same designs. Tessellation should be considered to reduce waste and use the material efficiently.

Use of datum points and coordinates

As with other CNC machines, a laser cutter uses a datum point from which to calculate the movements of the head that holds the lens. Most laser cutters have a raised lip along the x and y axis that can be used to ensure that your polymer work piece is square and corresponds to the datum point on your CAD drawing.

Check your knowledge and understanding

1 Why is a datum point important?
2 What is tessellation?
3 Why is using a material in a stock size more cost efficient?
4 On what material would you find 'face side' and 'face edge'?
5 Why does a template help when marking out?

Textile-based materials

There are many factors to consider when planning fabric layouts so that fabric is used economically. Fabrics come in a number of different widths and many have patterns and directional naps that must be taken into account when planning layouts.

Cut materials efficiently and to minimise waste

If you are using a commercial pattern for your product the layout will have been done for you, and it is important to lay the pattern pieces as shown for the style and size you are making so that you use the fabric economically and include all the pieces needed to make the product.

If you have developed your own templates, you will need to plan the layout yourself. Make sure that you have put a seam allowance on all the pieces or that you allow for it when planning the layout, or your product will not be the right size when sewn. A standard seam allowance is 15 millimetres on domestic patterns, but will be less on commercial patterns, so that the fabric is used as economically as possible. You may also need to include a hem allowance – this will vary from about 15 to 70 millimetres depending on the product.

Wider fabrics may allow pattern pieces to be laid side by side, but when using narrower fabric the pieces may need to be laid end to end. Sometimes it is possible to dovetail the pattern pieces to use the fabric economically, but this will not work if the fabric has a directional nap or pattern. Place the largest pattern pieces on the fabric first making sure that they are placed on the correct grain line. Do not try to squeeze the templates into a smaller length or the product will not hang well and patterns may be distorted. Place smaller pattern pieces in the spaces left but take care not to overlap any.

Fabrics that have checks and prominent patterns need special consideration to make sure that the pattern is used to best effect on the product. The pattern needs to be symmetrical and checks matched across seams wherever possible. This is likely to require extra fabric and will add to costs.

The pattern templates should be laid out on a table or other large surface with the fabric width marked out. This enables you to try the pieces in different positions to find the most economical and to calculate the length of fabric needed. Fabric is usually sold

Figure 3.9.7 Remember to add a seam allowance to pattern pieces that you have made yourself.

Figure 3.9.8 A layout for this large busy pattern will need to be planned carefully so that it is balanced on the product.

in multiples of 10 centimetres so you will need to work out the length and round it up to the nearest 10 centimetres.

Using appropriate marking-out methods

Seam allowances can be marked on to the fabric using tailor's chalk and pencils or a special pen used to draw marks on fabrics. The marks from these pens disappear when wetted or after a certain length of time. A tracing wheel and paper or tailor's tacks can be used to transfer markings from the pattern to the fabric. Look back at Topic 2.8 Specialist tools and equipment for more details.

Using datum points and coordinates

The fabric selvedge is an important datum point when cutting fabric, and the straight grain line marked on the pattern template must always be parallel to the selvedge. This should be measured accurately using a tape measure and not just guessed.

When working with checked fabrics, it is usually desirable to have the centre line of the check pattern run along the centre line of the product, and care must be taken to ensure that the pattern template is placed accurately with the centres matched, otherwise the checks may appear unbalanced or out of line on the finished product.

> **KEY POINTS**
> - Fabric layouts need to use fabric as economically as possible to reduce costs.
> - Layouts need to be planned carefully to make sure that any pattern is used effectively.
> - Seam and hem allowances need to be included on pattern templates.
> - It is important to place the straight grain line of the pattern template parallel to the fabric selvedge.

Check your knowledge and understanding

1 Why is it important to consider the fabric pattern when planning layouts?
2 Why is it important to follow the grain when laying patterns on fabric?

Electronic and mechanical systems

When it comes to new electronic products, smaller is generally better. Tablet computers are among a multitude of electronic products that are designed in such a way that circuits take up as little space as possible. One of the keys to a smaller product is, of course, a smaller printed circuit board (PCB). Smaller PCBs use less material and also reduce costs. (Smartphones are generally bigger than previous generations of mobile phones, but this is because they are now small computers, rather than just telephones. The screen size has increased but the circuits inside have become smaller.)

Cut materials efficiently and to minimise waste

There are many software packages that allow circuit modelling, testing and PCB routing. The question that can often be asked after the PCB routing has been done is, 'Can the PCB be made any smaller?'. In Figure 3.9.9, a PCB has been designed by a software package for an electronic timer device. If you look at this design and ask the question, 'Could the PCB be made smaller?', the answer is yes. The size of the components that are mounted on the board cannot be altered so a starting point can often be to move them as close together as you feel is practical.

Care must be taken when moving components to ensure that tracks connecting components do not cross or touch one another.

The final PCB design, in Figure 3.9.10, has all the same components with the same connection, but is using considerably less material. If this design were to be created by PCB milling, the machining time would be less, saving energy and material in production. Also, the wear on the cutters would be less, as they do not need to do as many cuts. If the design were to be produced by etching, as well as saving on the amount of board required, the chemicals needed to develop and etch the image would last longer because of the smaller board size.

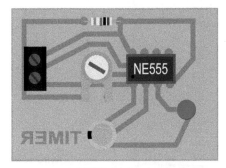

Figure 3.9.9 **A PCB design produced by a software package**

Using datum points and coordinates

If you manufacture your PCB using a CNC milling machine, the computer software changes the design into a code that operates the machine. The code then guides the cutting tool in three axes, the *x* and *y* axes move the cutter to various points across the bed of the machine and the *z* axis moves the cutter up and down. The code created by the software guides the cutter using the coordinates in these three axes. The cutting tool will start its movement from a point set by the machine manufacturer, called its home position. Once you have attached the copper-clad board you are going to cut to make your PCB onto the bed of the machine, you have to set a datum point. This is the starting coordinate for the code that will cut the PCB and could be the bottom right corner of your PCB design. When the code is run the cutter will move from the home position to points relative to the datum point that was set.

Figure 3.9.10 **The same circuit as shown in Figure 3.9.9, using less board material**

Using appropriate marking out methods

Manually making a PCB

If you are going to manufacture your PCB using the etching method you will have to manually mark out the size and shape of the board. The tools used for this are the same as you would use when marking out metals or polymers. If you were marking out a rectangular board you would use an engineer's try square. This tool allows you to draw lines at 90 degrees to the edge of the board you are going to cut. To draw the lines for the edge of your board you would use a scriber or a marking pen, and to measure you would use a steel ruler.

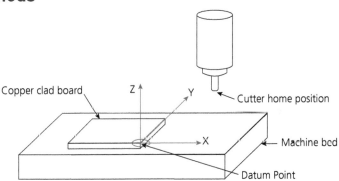

Figure 3.9.11 **Reference points and coordinates on a milling machine**

KEY POINTS
- Printed circuit boards should be designed to be as small as possible to reduce costs.
- The datum point is the point in a coordinate system from which measurements and movements are taken.

Check your knowledge and understanding

1 Explain why costs are lower when etching or machining a printed circuit board.
2 Explain why small is generally best when designing printed circuit boards.

3.10

Specialist tools and equipment

What will I learn?

In this topic you will learn about:
- → how to use a laser cutter
- → how to use a 3D router to machine timber-based materials
- → how to use a laser cutter to machine timber-based materials
- → how to use a vacuum bag with timber-based materials
- → how to use the CNC lathe
- → how to use the CNC milling machine
- → how to use a plasma cutter
- → how to blow mould polymers
- → specialist machines and hand tools
- → safe use of tools and equipment
- → how to use laser cutters and 3D printers to produce mechanical systems.

Having developed your design and decided what material you are going to use, you must now decide how you are going to make your prototype. You should use this chapter together with Topic 2.5 Using and working with materials and Topic 2.8 Specialist techniques to help you produce a production plan.

Papers and boards

Laser cutters

Figure 3.10.1 Laser cutter

Laser cutters are often used to cut paper and board in schools and colleges, but also in industry, particularly in the prototyping process. Laser cutters can be accurately set so that the exact speed and power is used in order to cut or engrave the surface of the material without charring. For example, a thin flammable material such as paper will use a much faster speed and lower power in order to cut through it effectively than a thicker material like acrylic. The laser cutter receives its instruction by reading a CAD (computer-aided design) file. Often, colours are used in the controlling software as codes to determine the settings needed for cutting and vector and raster engraving.

The laser cutter uses a powerful laser beam which is directed through mirrors on to a lens. This focuses the light onto the material underneath, which can be lifted or lowered to the optimum height. This piece of machinery is very safe to use. A laser cutter is an enclosed machine that will not work if the lid is open. This means very little risk to the user.

When creating paper and board products, a clean edge is sometimes required, and detailed shapes are needed in delicate materials. This is difficult to create by hand and very difficult to replicate if more than one of the same product is needed.

KEY POINTS

- Tools and equipment should be chosen due to a range of factors, including availability, cost and accuracy.
- A laser cutter is often used to create prototypes in paper and board products.
- The laser cutter uses a variety of speed and power settings, which makes it suitable for a wide range of materials.

Check your knowledge and understanding

1 Explain how a laser cutter works.
2 Discuss the advantages of using digital design and manufacture.

Timber-based materials

Hand tools

A wide range of hand tools are used when working with timber-based materials. Refer to Topic 2.5 Using and working with materials for more information on these.

The 3D router

The 3D router is capable of machining three-dimensional products from timber-based and polymer-based materials. Most types of wood can be machined, but close-grained hardwood produces the best results. Many manufactured boards can be used, but chipboard is often avoided due to its open, brittle nature. Modelling polymers such as PVC foam sheet work extremely well with the 3D router, producing high-quality models with a very good surface finish.

The 3D router requires a CAD drawing to be produced. At a basic level this can be a two-dimensional image using relatively simple two-dimensional CAD software. Areas of the drawing are filled with colour two-dimensional and the 3D router can then be programmed to machine each colour to a specific depth. The 3D router really shows its capability when used with sophisticated three-dimensional CAD software.

The CAD image shown in Figure 3.10.2 is for a portable speaker system to be used with a mobile phone. Having drawn the image, it is then processed using the 3D router's dedicated software, and a computer simulation checks to see if the drawing can be manufactured.

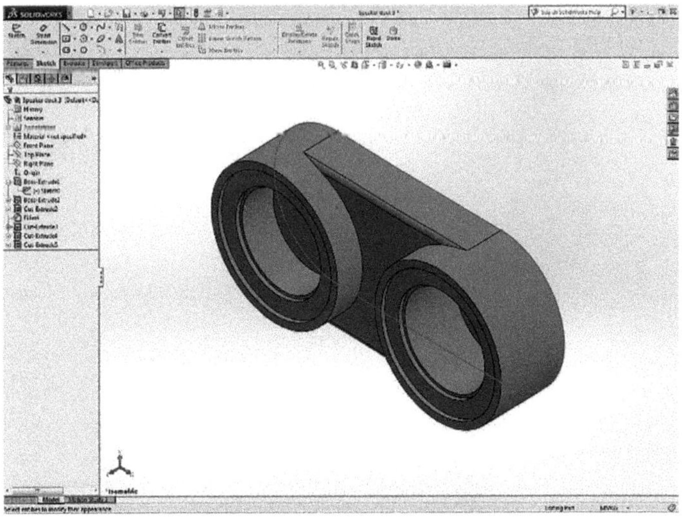

Figure 3.10.2 **A CAD image of a portable speaker system**

A suitably sized piece of PVC foam sheet is then clamped to the bed of the 3D router and the correct cutting tool is fitted. As the cutting tool spins, it is moved along the *x*, *y* and *z* axes (side to side, forwards and backwards, and up and down), powered by three stepper motors that are controlled by a computer. The combination of the three directional movement means that virtually any three-dimensional image can be produced.

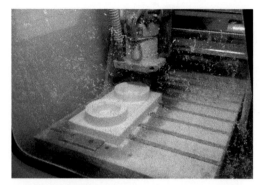

Figure 3.10.3 Machining a PVC foam model

Figure 3.10.4 Finished portable speaker system

If the designer or client is happy with the model then the process can be repeated with a timber-based material. Figure 3.10.4 shows the finished design machined from mahogany.

As with all CNC machinery, this process can then be repeated many times to produce accurate, consistent outcomes. As more products are produced the unit cost of each product is reduced because there are no additional set up costs each time.

Laser cutting

The laser cutter is normally used for cutting polymers but can also cut timber-based materials in thin sections of up to three millimetres. There are a number of manufactured boards that are specially produced to work with the laser, and will cut with minimal discolouration due to burning. The laser is very effective at etching onto the surface of wood and produces an accurately burned image similar to that achieved by pyrography. When etching, the laser cutter can work from a CAD-drawn image but can also work from an imported image. Monochrome line drawings can be downloaded from the internet and work very well.

How to cut and etch using the laser

A 2D CAD drawing is needed before the material can be cut or etched using the laser. If any etching is required then the image should be imported or drawn directly onto the shape. Different coloured lines/areas should be used for the cut lines and etched lines/areas. The image of the key fob shows the cut lines in black and the etched areas in blue.

The laser's power settings can then be set to cut black lines and etch the blue areas. This is done by altering the power and speed of the laser. A high power setting and a slow speed will cut through the material. A high power setting and a high cutting speed will etch the surface of the material. Power settings vary between different materials and models of laser and you should make sure that you know the correct settings before operating the laser.

Industrial lasers can cut far larger sizes of material and can also cut steel.

Vacuum bag

How to use a vacuum bag

- An accurate mould/former of the curved surface is produced.
- PVA or a resin based wood glue is then applied to the wood veneers or layers of flexible plywood
- Each layer is then carefully positioned over the mould/former
- This assembly is then placed into the vacuum bag.
- The bag is then sealed and the vacuum switched on, removing air from the bag.
- The vacuum will hold the veneer firmly in place until the glue dries.

A vacuum bag is a very useful piece of equipment that can be used when veneering or laminating wood. The bag consists of a heavy-duty polymer bag that is attached to a vacuum pump. The pump sucks the air out of the bag, which then puts even pressure onto the work piece. This process is particularly useful when veneering onto curved surfaces, where it would be very difficult to evenly clamp by any other clamping method.

Straps to initial hold the glued layers of flexible plywood

Former made from expanded polystyrene

Vacuum bag

Pipe leading to vacuum pump

Layers of flexible plywood glued together

Figure 3.10.5 How to use a vacuum bag

When veneering, a decorative veneer (a thin sheet of decorative wood) has to be glued onto a wooden surface. This is usually a sheet of manufactured board such as plywood. The plywood is coated in an adhesive such as PVA. The veneer is then located onto the plywood and placed into the vacuum bag. The bag is then sealed and the vacuum switched on. The vacuum will hold the veneer firmly in place until the glue dries. For more complex shapes, PVA may start to bond too quickly before everything is assembled and placed in the bag. In these cases a slower setting adhesive such as 'Cascamite' works better. When laminating, several sheets of veneer are glued together to form a solid, usually curved, wooden shape.

Figure 3.10.6 **Laser-cut wooden coasters**

ACTIVITY

Your work desk at home is no doubt cluttered with pens, pencils, rulers and other items of stationery that you use every day.

Use two-dimensional or three-dimensional CAD software to produce yourself a desk tidy that consists of a block of wood that has several holes, slots and recesses machined into it to hold specific items of stationery.

You should then machine this out on the 3D router.

You can even develop your design to hold other items, such as your smartphone, tablet, loose change and so on.

KEY POINTS

- A 3D router can convert two-dimensional and three-dimensional CAD drawings into three-dimensional products made from timber- or polymer-based materials.
- A laser cutter can cut thin sections of timber-based materials and can etch images onto the surface.
- A vacuum bag can be used as a clamping aid when veneering or laminating.

Check your knowledge and understanding

1 Which materials can be used with a 3D router?
2 Describe a process where would you use a vacuum bag.

Metal-based materials

CNC lathe

A CNC lathe can perform all the operations of a 'traditional' centre lathe, such as parallel turning, facing off, taper turning and screw cutting, but the processes are controlled by a computer. Sophisticated CNC lathes can even perform their own tool changes, and this significantly speeds up the manufacturing process. The CNC lathe can be used to manufacture components from most metals and polymers.

The advantages of the CNC lathe are that it can work to a high degree of tolerance; it can produce components far quicker than by traditional methods; it consistently produces identical components; and because it is totally enclosed it is safer to operate.

Figure 3.10.7 **A CNC lathe**

The CNC lathe works from a CAD drawing. Once the drawing is completed it is run through the machine's own dedicated software program, which turns the drawing into machine code

Figure 3.10.8 **A CNC milling machine**

(information that controls the lathe), and then it runs a simulation to see if the component can be made. A blank piece of metal/polymer is then fastened into the lathe, a datum point is set by moving the tool to the front edge of the blank piece of metal/polymer and the manufacturing process can begin.

CNC milling machine

As with the CNC lathe, a CNC milling machine will carry out all the operations of a 'traditional' milling machine, such as cutting slots, grooves, machining edges and smoothing large surface areas, but once again these processes are controlled by a computer. It has the same advantages as the CNC lathe; it machines to a high level of accuracy, is faster and is consistent. It works from a CAD drawing that is processed into machine code and runs this through a simulation before actual machining takes place.

Plasma cutter

A plasma cutter works in a similar way to a laser cutter, but it has the ability to cut through sheet metal and heavy metal plate. First, a CAD drawing is produced of the profile of the component to be manufactured. The dedicated machine software converts the CAD drawing into machine code and sets the speed and strength of the cutting power once the material and its thickness have been entered. The cutting medium is a high-velocity ionised gas that burns its way through the metal. This process is significantly quicker than other machining processes and it has the advantage of being able to cut very intricate shapes.

Industrial laser cutters can also be used to cut metals but are not normally found in the school workshop.

KEY POINTS
- A CNC machine works from CAD drawings.
- CNC machines are fast, accurate and consistent.
- A plasma cutter can cut thick sections of metal plate.

STRETCH AND CHALLENGE

Most CNC machines in schools and colleges operate on 3 axis. In industry many machines have 5 axis capability. Investigate 5 axis CNC machining and explore the manufacturing opportunities that it offers.

Check your knowledge and understanding

1 What are the advantages of using CNC machinery?
2 When would you use a plasma cutter?

Polymers

Blow moulding

You may have will have learned about blow moulding as an industrial process, but blow moulding can also be done in the school workshop to form a dome shape.

The process is a relatively simple one that is done on a vacuum-forming machine (vacuum forming is covered in Topic 2.8). Firstly, a clamping ring needs to be produced. This consists of a sheet of material, such as 3-millimetre plywood, that has had a hole cut into it that is the same size as the diameter of the dome. It is clamped onto the vacuum former together with a sheet of thermoplastic, such as high-impact polystyrene (HIPS). The HIPS is heated until soft and then blown into a dome shape. If your design requires a flat base on your dome then a restrictor can used. A restrictor is a flat piece of material that the blown dome will come into contact with while it is still warm. Once cool the dome can be removed from the vacuum former and trimmed.

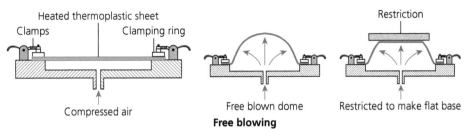

Figure 3.10.9 **Blow moulding**

KEY POINTS
- A vacuum former can be used to blow a dome. The polymer sheet is heated in the same way as when vacuum forming, but instead of sucking the sheet over a mould the sheet is blown.

Check your knowledge and understanding

1 Use notes and sketches to describe the process of blowing a dome.

Textile-based materials

Many different items of equipment and tools are used when making textile products and it is important to use them efficiently and safely in order to achieve high-quality results.

Sewing machines

Sewing machines are used to join fabrics together and produce decorative effects. There are many different models available, but all thread up and work in the same basic way. There are three main types.

A **basic sewing machine** can perform many sewing operations, such as straight and zigzag stitching, which allows some decorative stitches and buttonholes to be made.

An **automatic sewing machine** can sew many functional and decorative stitches that are preprogrammed into the machine, especially buttonholes.

Computerised sewing machines have a very wide range of preprogrammed stitches and can also stitch motifs that have been designed on a computer or downloaded from a website and scanned into the machine.

Figure 3.10.10 **A basic sewing machine and a computerised sewing machine**

Figure 3.10.11 **A zip foot has only one 'toe' and allows the machine to stitch close to the zip for a neat finish. It is also used when sewing in piping.**

Figure 3.10.12 **An overlocker trims and neatens the seam in one operation.**

It is important to read the instruction book that comes with the machine in order to use it correctly and safely. Most sewing machines come with a range of interchangeable feet that are used for different tasks, such as sewing in zips and piping, and making buttonholes. A walking foot helps to stop the top layer of fabric slipping over the lower one when machine sewing and is useful when sewing very slippery fabrics or many layers such as used in quilting.

Overlockers

Overlockers can perform a number of sewing proceses, although they are mostly used for neatening seams and other edges. An overlocker uses three or four large cones of thread on the top, but does not have a lower bobbin thread. Overlockers have a sharp blade that trims the edge of the fabric at the same time as it covers the cut edge with stitches to give a neat edge that does not fray.

Overlocker machines are commonly used in industrial manufacture as sewing and neatening only need one operation.

Ironing and pressing equipment

Irons have many uses when making textile products and help to achieve a good finish. Steam irons are most useful as they have a reservoir of water that makes steam when the iron is hot. Irons can be set to work at a range of temperatures between 60 °C and 200 °C, depending on the fabric being used.

An ironing board can be adjusted in height and provides a stable surface on which to use the iron. It should have a well-padded heat-resistant surface with a removable cover. A sleeve board is like a mini ironing board and is used when pressing small and narrow areas, such as sleeves.

It is important to keep the base of the iron and ironing board covers clean, especially when using dyes, batik wax and iron-on interfacings or other work could be spoiled by staining.

Cutting tools

Making textile products involves lots of cutting and a number of cutting tools are used for different purposes.

You will already be familiar with shears and rotary cutters. The table below shows some other cutting tools used in textile work.

Tool	Use
Pinking shears 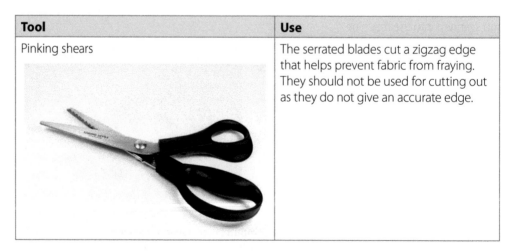	The serrated blades cut a zigzag edge that helps prevent fabric from fraying. They should not be used for cutting out as they do not give an accurate edge.

Tool	Use
Embroidery scissors 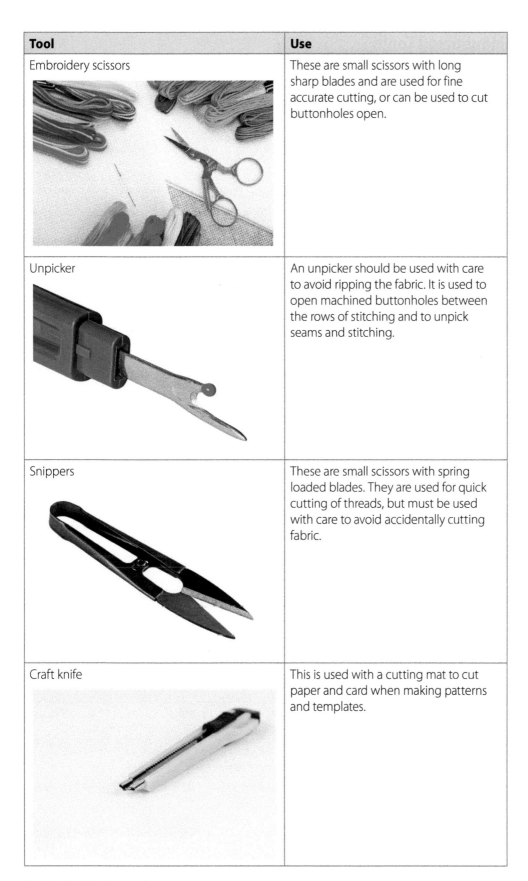	These are small scissors with long sharp blades and are used for fine accurate cutting, or can be used to cut buttonholes open.
Unpicker	An unpicker should be used with care to avoid ripping the fabric. It is used to open machined buttonholes between the rows of stitching and to unpick seams and stitching.
Snippers	These are small scissors with spring loaded blades. They are used for quick cutting of threads, but must be used with care to avoid accidentally cutting fabric.
Craft knife	This is used with a cutting mat to cut paper and card when making patterns and templates.

Table 3.10.1 **Cutting tools**

Textile-based materials

Measuring tools

Accurate measuring and marking out is essential if good-quality products are to be made and a range of measuring tools is available.

Pins and needles

There is a huge variety of pins and needles, and the type used will depend on the fabric and thread used and the task to be done.

Dressmaker's pins are 26 millimetres long and are usually made from steel or nickel-plated steel. Pins should always be sharp or they will snag fabric.

Different needles are used for hand and machine sewing.

The chart below shows the main types of hand sewing needles. The higher the number, the finer the needle. Using the correct needle makes the task easier; a thick needle may be easier to thread but it is difficult to push through fabric.

Hand needle type	Use
Sharps	Long sharp needles available in sizes 3–10 and used for general sewing such as tacking and gathering.
Crewel	Used for embroidery, these sharp needles have a long eye so the thicker embroidery thread will easily pass through. Available in sizes 1–12.
Betweens	Short sharp needles available in sizes 4–10 and used for fine work such as hemming and sewing fine fabrics.
Bodkins	Short blunt needles with a long eye and used to thread elastic and cords through casings.
Ball point	These have rounded tips that can pass through knitted fabrics more easily than sharp pointed needles.

The chart below shows the main types of machine needles. The higher the number, the thicker the needle.

Machine needle type	Use
General purpose	Available in sizes 60–110, these needles are used for most machine sewing.
Ball point	Available in sizes 60–100, they have rounded tips and are used when sewing stretch and knitted fabrics.
Denim	Used for sewing very tough fabrics as they have an extra sharp tip.
Leather	These have a wedge-shaped tip that slits a hole in the leather as it is stitched. This helps prevent the leather from tearing.
Twin and triple	Two or three needles are joined together in a row to create narrow tucks and decorative stitching.

KEY POINTS

- Sewing machines can be used to perform many different functions in the making and decorating of textile products.
- It is important to use the most appropriate tool for the task in order to achieve a high-quality finish.
- Care should be taken to use tools and equipment efficiently and correctly.

Check your knowledge and understanding

1 Why is an overlocker used to neaten edges?
2 When might a sleeve board be used?
3 Give two uses for an unpicker.
4 What is the difference between a sharps needle and a crewel needle?
5 Which sewing machine needle is finer – a size 100 or a size 70?

Electronic and mechanical systems

If you intend to make your own mechanical systems that use gears, then the use of more specialist equipment will make it much easier to achieve success. (Other equipment, including soldering icons & etch tasks are discussed in Topic 2.8.)

3D printers

The 3D printing process works by representing a whole object in thousands of tiny little slices, then making it from the bottom up, slice by slice. The tiny layers stick together to form a solid object. Each layer can be very complex, meaning some 3D printers can create moving parts, such as hinges and wheels, as part of the same object. You start by designing the three-dimensional object using a three-dimensional modelling software package (there are many modelling packages on the market, some of them free to download) that will enable you to draw out gears of different sizes with different numbers of teeth that will mesh together. Once you have drawn out the size of gears you require, the gears can be manufactured to high tolerances using a 3D printing machine.

Figure 3.10.13 A 3D printer in action

Depending on the 3D printer you are using, the gears should be of an acceptable quality and precision for use in your prototype. **Compound gears** can be constructed using 3D printers. You can also customise the items you make by adjusting the file if the gear you manufacture isn't to your liking, and you can make as many as you like at a relatively low cost, as the materials used in most 3D printers are generally inexpensive.

After you have generated a model in a CAD program when you are working in 3D, the model needs to be transformed into a machine code that the CNC machine can understand and follow. Some CAD software programs produce codes that will operate the milling machine, or 3D printer you have available, while with others you will need another program to convert your 3D drawing to a code for your 3D printer or CNC mill.

Some CAD/CAM software programs will give you the option to model virtually the manufacture of your design. This can save you time and save material costs. The computer simulation software checks to see if your design can be manufactured successfully, this gives you the opportunity to alter your design before it is manufactured by a CNC machine.

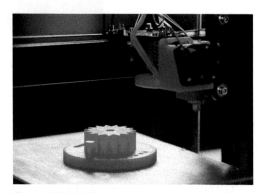

Figure 3.10.14 A compound gear constructed on a 3D printer

Laser cutters

Laser cutters work by directing a very powerful laser beam through mirrors or fibre optics onto a lens that that focuses the light onto a material, which they either cut or etch depending on how the laser cutter has been set up. Laser cutters cut materials in a similar way to other computer-controlled machines, the difference being that they use a beam of light rather than blades or cutters. Laser cutters can etch designs onto the surface of materials–when they do this they act like a printer, where they use the beam to etch the image onto the surface of the material. Users are protected from the laser beam by guards that cover the cutting area: the machines do not operate without the guards in place and if opened the machine and the laser stop.

To cut gears using a laser printer, it is best to draw the gears using a two-dimensional vector graphics drawing package. Again, there are a number of two-dimensional drawing packages on the market, some of them free to download. Once the gears are drawn using the software the laser cutter can be set up to cut them.

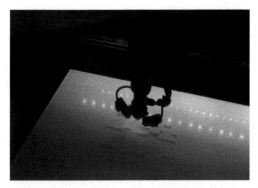

Figure 3.10.15 **Laser cutting machine**

The speed and power of the laser beam needs to be set using the setup software for the laser cutter you are using before cutting. Different materials, and different thicknesses of material, will need different power and speed settings. It is also possible to engrave onto the surface of a material when laser cutting. Different line colours can be used to control the feed (the speed the laser travels over the surface of the material), and the power (the intensity of the laser). One colour could be used for cutting the outline of your gear wheel, and another could be used for engraving on to the surface the number of teeth the gear wheel has.

Another advantage of using the laser cutter to cut gears is that the distance between the shafts that the gears are fixed to can be easily copied and used to produce the mounting for the gear assembly.

KEY POINTS 🎯
- Laser cutters and 3D printers can be used as precision tools to manufacture gears and gearboxes.

Figure 3.10.16 **Gears drawn out and used to produce drawing for the gear mounting assembly**

Check your knowledge and understanding

1 Give an example of a material that can be laser cut.
2 Discuss the advantages of making gears using 3D printers and laser cutters.

Specialist techniques and processes

What will I learn?

In this topic you will learn about:

→ how to select and use specialist techniques and processes appropriate for the material and/or task and use them to the required level of accuracy in order to complete quality outcomes

→ how to use them safely to shape, fabricate and construct a high-quality prototype, including techniques such as wastage, addition, deforming and reforming

→ how surface treatments are applied for functional and aesthetic purposes

→ how to prepare a material for a treatment or finish

→ how to apply an appropriate surface treatment or finish

→ how to make PCBs safely and accurately

→ how to apply a spray lacquer to protect a PCB.

We have covered a wide variety of manufacturing processes throughout this book and have learned about the correct use of tools and machinery, but what decisions need considering when choosing the most appropriate manufacturing method for a prototype?

Papers and boards

Using specialist techniques and processes to the required level of accuracy

When manufacturing any projects throughout your Design and Technology education, you will need to select the most appropriate techniques and processes to ensure quality results. Standard tools and equipment available to you in a school or college environment might be craft knives, steel rules, scissors and so on. These are useful tools when marking and cutting out, but you may be able to create a higher level of accuracy with alternative tools, such as a laser cutter (if one is available). In industry, die cutters can be used, but dies are often expensive to produce and so would sometimes not be suitable for small-scale production.

Surface treatments and finishes

In school or college, the most common finish for paper and board products is encapsulation. This provides a waterproof, glossy appearance to the final product and requires machinery commonly found in these environments. When encapsulating paper and board, you need to allow the bottom and top plastic layers to touch or the plastic may peel off at a later date.

Encapsulation is different to lamination. Encapsulation leaves a rim of plastic around the outside edge of the paper or board and so is watertight. Lamination does not leave this rim and uses plastic that is directly adhered to the surface of the paper or board. Water can then get in to the product through the edges of the material, but it leaves a glossy surface and gives the material additional strength.

Figure 3.11.1 Self-adhesive film or sticky-backed plastic

Sometimes it is not possible to finish the product using encapsulation as the material is too thick. This is true of some boards, such as foam core board or solid white board. A similar finish can be achieved using self-adhesive plastic sheet or sticky-backed plastic, although this is more difficult to apply. When applying this material as a surface finish, you have to be careful not to trap bubbles of air underneath the plastic as this can reduce the quality of your final outcome. The benefits of sticky-backed plastic are that you do not need to allow the plastic to reach past the paper or board material as it adheres to the paper or board rather than to itself.

Fixative spray or varnish can be used in small-scale production to increase water resistance and the effects of UV light. This provides a less glossy finish compared with laminating and the use of sticky-backed plastic. Fixative spray can be used to help reduce the risk of ink smudging.

Little preparation is needed in the case of finishing paper and board products, other than ensuring surfaces are clean and free from dust.

KEY POINTS
- Paper and board products in school or college are generally cut using hand tools or a laser cutter.
- Paper and board can be finished in a number of ways, but in school or college this is usually done by laminating.

Check your knowledge and understanding

1 Why might you want to apply a finish to paper or board?
2 What are the most common processes used in small-scale production for cutting paper and board?

Timber-based materials

Using specialist techniques and processes to the required level of accuracy

When working with softwoods and hardwoods it is important to consider the intended end use of the product; this will help determine the variety of material chosen, the strength and type of the joints or fixings used and the manufacturing process chosen.

Where a traditional timber outcome is required, such as a dining room table or chair, you may choose to use traditional cutting, shaping and joining techniques. There is nothing to say that the outcome has to be traditional, but many wood joints such as dowel joints and mitres are excellent ways to create strong timber products.

Where more organic shapes may be required, then laminating and steam bending may provide you with a reliable method to create curves and bends in timber. Traditional two-part laminating moulds may have limitations due to the size of object to be made, but vacuum-bag systems and polyurethane foam moulds allow for greater creativity. When these are used in commercial products, the mould being cut using a CNC hot wire system achieves accuracy.

In school and college workshops you are more likely to shape moulds using a bandsaw and sander, or in some cases a CNC router.

Manufactured boards, such as MDF and plywood, lend themselves to being ideal materials to machine accurately using a CNC router. Their large, flat, smooth surface can be easily clamped down using a vacuum bed, which speeds up the manufacturing process by removing the need for mechanical clamps. These manufactured boards are also easy to join and assemble using knock-down fittings, as seen in flat-pack furniture.

Surface treatments and finishes

Protective finishes for timber are applied to increase the durability of the material, to protect it from insect attack and decay, and to improve its resistance to weathering. The application of a protective finish is especially important in timber products that are going to be used outside, as they are more likely to be exposed to the elements. The cycle of a timber product becoming wet and then drying out, along with changes in temperatures and exposure to sunlight, quickly damages the surface, leading to cracking and shrinking. The application of preservatives or specific external varieties of paint will help to reduce this degradation.

It is rare that a finish for timber is applied to perform a purely protective role. Most surface finishes are also used to improve or alter the aesthetic appearance of a product. This could involve changing the colour of a timber by painting or staining, or by applying a clear gloss varnish or oil to help accentuate the timber's natural grain.

Preparation

With all surface finishes, the success and quality of the final finish is directly related to the preparation that has taken place on the material beforehand. It is important to sand down the wood to provide a clean, smooth surface. Glass paper should be used to rub down the surface of the timber – always going in the direction of the grain. The coarseness of the abrasive paper used can be selected depending on the application, but generally the higher the grit rating, the less coarse the glass paper.

Knots are found in timber, particularly in softwoods, as branches are located along the length of the tree trunk. Knots can be an attractive feature, but can also cause issues when preparing and finishing a timber. Knots are naturally resinous and sometimes need sealing before a finish is applied.

Sanding the surface of a timber is a sufficient level of preparation for finishes that are designed to soak into the surface of the timber, such as stains and oils. Where a surface finish such as paint is to be used, it is good practice to apply a primer to the timber to help seal the surface and help the paint adhere to the timber properly.

Application

Application to small products or pieces of furniture is most commonly undertaken by hand, using a brush or roller if painting. On larger items or products manufactured on a commercial scale, varnishes and paints can be applied by spraying. Some preservatives used in fencing and fence posts can be applied by dipping the pieces or by pressure treating, where a preservative is forced under pressure into the outside cell layer of timber.

Some preservatives can cause problems with the timber at the end of its useful life, as they can prevent the material from naturally degrading.

Figure 3.11.2 A cellulose varnish being applied to a softwood floor by hand. The varnish both accentuates the natural grain pattern and protects the timber from wear, increasing its durability.

KEY POINTS
- Timber can be shaped accurately into complex shapes using CNC machinery.
- Laminating can be used to create curved products in timber.
- End-grain is more absorbent than the surface of a piece of timber.
- Finishes for timber can be both protective and aesthetic.
- Thorough surface preparation is key to a high-quality surface finish.

Check your knowledge and understanding

1 Why are manufactured boards suitable for CNC machining?
2 In what two ways can a laminated shape be achieved?
3 What are the three reasons you apply a finish to a timber?
4 Why can knots cause problems when applying a finish?
5 Why would you use a primer?

Metal-based materials

Using specialist techniques and processes to the required level of accuracy

Metals are generally harder to work with than timbers and polymers due to their hardness. This is especially true if shaping and cutting metal by hand. The advantage of using metal in the manufacture of products is that more accurate tolerances can be achieved and there are a wide variety of manufacturing processes that can be used, particularly in commercial manufacture.

In a school or college workshop you are probably going to shape and cut metal using a range of hand tools such as files and hacksaws; achieving accuracy with these kinds of tools is challenging. If you have access to metal-specific bandsaws, centre lathes and milling machines, both the quality and accuracy of your work can increase.

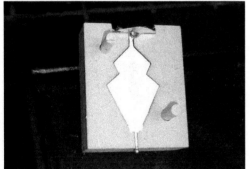

Figure 3.11.3 **Pewter casting using MDF moulds**

Sheet-metal manufacture can be undertaken in school workshops where shapes and developments can be cut out using tin snips and saws, but accuracy and quality will increase with the use of guillotines and sheet-metal folding equipment.

Reforming processes are a very effective way of shaping metals. Some schools and colleges have access to sand casting equipment, but pewter casting is a more common reforming process for low-melting-point alloys. Accurate moulds can be created from MDF or plywood using a laser cutter, which can then allow for small-scale batch production of accurate cast products.

Surface treatments and finishes

As you have already been introduced to the physical properties of metals you should remember that corrosion and oxidisation are two major factors to consider when thinking about an appropriate surface finish for metal. Corrosion and oxidisation occur when the material reacts with oxygen, so the application of a surface finish that provides an impermeable coating on the metal will prevent this oxidisation occurring.

As with timbers, the application of a finish to most metals also influences the aesthetic appearance of the product. Most ferrous metals can be painted or polymer-coated in a wide range of colours. Metals such as aluminium can also be painted, but often aluminium products are anodised to protect the surface from corrosion.

Ferrous metals can also be chrome-plated, which is an electroplating process. The chrome plating provides a hard protective coating with excellent aesthetic qualities. Where protection is more important than the aesthetic appearance, ferrous metals can be galvanised. Galvanising protects the metal by coating it in a zinc layer that is more reactive than the base metal. Galvanising is used in road barriers and lamp posts.

Preparation

Metal must be completely clean and grease-free before applying any surface finish, otherwise the finish being applied may not adhere correctly. Abrasive papers can be used to prepare metal surfaces for the application of a finish, and in large objects or when a large volume is being produced, more industrial methods such as sand blasting or bead blasting can be used. Here, small abrasive particles of sand or glass are fired under pressure at the surface of metal to remove any dirt or grease. This is an effective way of efficiently preparing large surface areas.

When applying a paint finish, most metals will benefit from a primer being applied beforehand. This primer will help the topcoat adhere to the metal.

Application

The application of a surface finish to small metal objects or products can usually be achieved with cellulose aerosol sprays. As the scale and volume increase, processes such as powder coating are more commonly used.

Figure 3.11.4 **This aluminium torch has been anodised to help protect the surface from oxidising, while at the same time providing an attractive aesthetic finish.**

KEY POINTS
- It is hard to achieve high levels of accuracy with hand tools.
- Metals can be shaped into intricate shapes by reforming by casting.
- Corrosion in ferrous metals is known as rust.
- Anodising can be used to add an aesthetic protective finish to aluminium.
- Surfaces must be clean and free of grease for a finish to adhere properly.

Check your knowledge and understanding

1 How can you accurately shape metal?
2 What can moulds be made out of for pewter casting?
3 Why does corrosion occur?
4 Why does galvanising help protect steel from corrosion?
5 When would sand blasting be used?

Polymers

Using specialist techniques and processes to the required level of accuracy

Although there are two categories of polymer, thermoplastic polymers such as acrylic and high-impact polystyrene are the most common varieties used in schools and colleges.

They can be cut and shaped using similar hand tools to timber, and are slightly easier to finish to a high standard than metals.

The most accurate way to shape and cut polymer in a school or college environment is by the use of a laser cutter. Intricate designs can be cut out and engraved in a range of thermoplastic polymers.

The use of CNC machinery in manufacture is a reliable way to achieve accurate, high-quality results, and it is no different for polymers. Three-dimensional printing is now a common method of producing high-quality intricate parts. Most 3D printers use PLA filament, but some can also use ABS to produce a more durable print, due to ABS being tougher than PLA.

Thermoplastic polymers can be moulded and formed with the addition of heat. In a school or college workshop, vacuum forming and plug-and-yoke press forming are two frequently used methods of creating accurate polymer products. As with all moulding processes, the outcome can only be as good as the original mould used to produce it; any blemish or defect found on the surface of the mould will be transferred to each and every moulding.

Surface treatments and finishes

Polymers are an interesting group of materials when discussing surface finishes, as most polymers require no surface finish and are often referred to a self-finishing.

Polymers can however be coloured by the addition of a pigment. The pigment is usually added to the polymer before it is moulded, where it combines with the polymer to create a material that is coloured from its core through to its outer surface. This means that the colour will never wear off and small scratches are masked and hard to notice.

Figure 3.11.5 Polymer car bumpers are painted to match the colour of the main metal panels of the car.

Polymers can also be painted. They would go through the familiar preparation process of being rubbed down with abrasive papers in order to provide a smooth, dirt-free surface and also have a polymer-specific primer applied. Care needs to be taken when applying a spray finish to a polymer, as the chemicals in some paints can react with the surface of the polymer.

Other aesthetic surface finishes can be applied to polymers through a series of printing processes or through the application of transfers. These tend to use the colour of the polymer as the background, and then logos or decals can be added in a contrasting colour. This process is commonly found in packaging such as carrier bags and carbonated drinks lids.

KEY POINTS
- Thermoplastic polymers are most commonly in schools and colleges.
- 3D printing and vacuum forming both use heat to shape and form thermoplastic polymers.
- Many polymers are said to be self-finished.
- Pigment can be added before the moulding process takes place.
- Polymers are the same colour from their core through to their surface.

Check your knowledge and understanding

1 How can you form polymers with heat?
2 How can you accurately cut polymers?
3 What is the benefit of a polymer having a pigment added before moulding?
4 Why are most polymers not painted?
5 Why should some paints not be used on polymers?

Textile-based materials

Using specialist techniques and processes to the required level of accuracy

It is important to cut textile materials on the correct grain for the product. In most cases, this will mean that the grainline on the pattern template runs parallel to the fabric selvedge or warp grain. Sometimes, the pattern template needs to be placed on the bias grain of the fabric so that the product has greater fluidity and drape. If the grain is not placed accurately the end product will not look and perform as intended.

Construction techniques used must be selected carefully with regard to the function and appearance of the product. For example, a seam can be inconspicuous or decorative, a badly made edge finish such as a binding or hem can reduce the strength of the product as well as being unsightly. Most schools and colleges will have an overlocker and this can be used to neaten seams and edges efficiently.

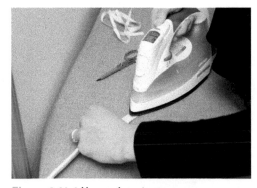

Figure 3.11.6 Use an iron to prepare materials.

Pressing work after each stage in its construction is important to achieve a high-quality finish, as it ensures that the fabric is flat and ready for the next stage. You will not get a good finish if you leave all the pressing until the product is completed.

The use of the iron can also save time in manufacture by preparing fabric for use in certain processes, for example, pressing a bias strip in half so that it is ready to apply to an edge.

Many parts of a product need to be interfaced to give them strength and shape. The correct choice and use of interfacing is important in making a high-quality product. The weight and colour of the interfacing needs to be chosen with care – if it is too heavy it will make the product too stiff and unable to hang correctly; if it is too light it will not support the fabric adequately. A black or grey interfacing will show through light-coloured fabrics and spoil the appearance. Interfacing needs to be applied before the construction or decorative process is started in order to be effective.

Surface treatments and finishes

The design for your product may include some decorative processes, such as printing or embroidery. You may be able to use a computerised embroidery machine to sew the pattern, or choose to hand embroider your product, which will give a less accurate design and finish.

If your school or college has fabric-printing facilities you may choose to screen print a design on your fabric. Many schools or colleges have a dye sublimation printer that will allow you to transfer your own designs or designs downloaded from a website onto fabric. When checking

Figure 3.11.7 Hand embroidery can be used to decorate a textiles product.

Figure 3.11.8 Each repeat of a fabric print must be accurate to ensure a high-quality product.

the accuracy of a print it is important to compare it with the original to make sure that the colour, position and size is the same as the original, especially when the print is repeated.

KEY POINTS
- Cutting fabric accurately on the correct grain is important for the appearance and performance of the product.
- Pressing work as you go along can help achieve a good finish.
- The choice of construction and decorative techniques affects the appearance and functioning of the product.
- Checking prints for accuracy will ensure that the fabric is aesthetically pleasing.

Check your knowledge and understanding

1 How can the choice of seam used on a product affect its appearance?
2 Why is pressing work at each stage of construction important?
3 Give two points to consider when choosing interfacing for a product.
4 Why does a repeat fabric print need to be checked?

Electronic and mechanical systems

Using specialist techniques and processes to the required level of accuracy

If you decide to make your own PCB, you will need to choose a method for creating it. Your choice will usually be based on the availability of the materials and equipment. Here is a brief summary of the different methods and their main features that will help you decide:
- The acid etching method requires a high level of safety measures. The chemicals used pose a risk to health, and the safety measures to protect yourself must be fully understood before beginning this process. It requires protective gloves and eye protection, and extreme care. The use of some specialist equipment and materials are also necessary, such as bubble or spray-etch tanks and chemicals, and it is quite a slow process. The quality of PCBs produced can vary, depending on the materials you use, but it is a good method for simple to intermediate complexity circuits.

To ensure accuracy using this method you need a good quality mask for your PCB, and all timings of exposure to the ultra violet light and to the etching chemicals must be accurate measured and adhered to.

- The mechanical machining method requires a special CNC machine that will mechanically remove unwanted copper from the board. If a machine is available to you, once you have had training regarding set up, it is a safe method of production, as it will not operate without the appropriate safety guards in place. This method is also good if you need to produce fine, relatively complex PCBs.

To ensure accuracy using this method, care is needed when setting up and positioning the copper clad board on the bed of CNC machine.

Surface treatments and finishes

If the PCB you produce for a product is to be used outdoors where it may be subject to damp environments you may wish to consider lacquering. You can purchase acrylic protective lacquers in aerosol cans that can be sprayed onto your finished PCB that will give protection from corrosion in damp environments. When spraying, you would need to follow the normal safety procedures for spraying: use a spray booth if available; if not use a very well-ventilated area, avoid contact with your skin and wear eye and respiratory protection.

KEY POINTS
- Etching PCBs requires high levels of safety measures.
- PCBs can be protected from corrosion by applying an acrylic lacquer from an aerosol can.

Check your knowledge and understanding

1 List the safety equipment you would use if you were to manufacture a PCB using the etching method.
2 Explain why you would lacquer a PCB.
3 Discuss the advantages in terms of safety of machining a PCB rather than using the etching method of manufacture.

PRACTICE QUESTIONS: designing and making principles

1 The product in Figure 3.P.1 is a famous light from the Memphis design group.
   a What do you think the intended target market might be for this product and why? **[2 marks]**
   b Designers from the past are often used to inspire new ideas. Name a designer who has inspired you and explain how this designer has influenced your work. **[10 marks]**

Figure 3.P.1 **Memphis lamp**

2 Explain how ICT can be used to increase the accuracy of manufacture. **[6 marks]**

3 Explain the term user-centred design. **[2 marks]**

4 Name one environmental concern and explain the issues surrounding this concern for designers, manufacturers and retailers. **[6 marks]**

5 Explain how you effectively evaluate your work during the making process and explain why evaluation of your own designs is important. **[6 marks]**

6 The product in Figure 3.P.2 is a chair that can be adjusted by the user. Analyse and evaluate this product, considering how effective the design might be for use in a hospital waiting room. You should discuss positive and negative aspects of the design in this situation. **[10 marks]**

Figure 3.P.2 **Adjustable chair**

7 The product in Figure 3.P.2 has an impact on the environment. Evaluate the environmental impact, discussing both positive and negative aspects of the design. **[4 marks]**

8 Where might ergonomics have been considered in the design of
the hairdryer in Figure 3.P.3?   *[4 marks]*

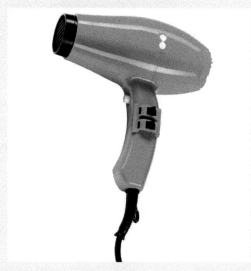

Figure 3.P.3 **Hairdryer**

9 The following data shows the height of 100 people. Calculate the average
in order to gain necessary anthropometric data for the design of a product.
Remember to only use the 5th to the 95th percentile.   *[2 marks]*

130–132 cm	1
133–135 cm	0
136–138 cm	0
139–140 cm	0
141–143 cm	2
144–146 cm	2
147–150 cm	5
151–153 cm	7
154–156 cm	3
157–159 cm	4
160–162 cm	5
163–165 cm	13
166–168 cm	27
169–171 cm	20
172–174 cm	6
175–177 cm	2
178–180 cm	2
181–183 cm	0
184–186 cm	0
187–189 cm	1
Total	**100**

10 Research is an important part of the design process. Name three different methods of research you are familiar with and explain how they might impact the success of your project. *[6 marks]*

Research method	How will they impact the success of your product?

11 You have been asked to design some new packaging for a glass bowl. Write four specification points for this product. *[8 marks]*

12 The Fairtrade symbol is often displayed on a range of different products. Explain what this symbol means to the consumer. *[2 marks]*

13 How can the purchase of fair trade products improve the life of the communities involved in making the product or supplying the materials? *[4 marks]*

14 Explain what orthographic drawings are and why they are useful for communication between designer and maker. *[4 marks]*

15 Explain the term 'tolerances' and why tolerances are important in the manufacture of products. *[5 marks]*

16 When mass producing products, how would a designer or manufacturer ensure that as little material as possible is wasted? *[4 marks]*

17 Explain the processes you would use to accurately mark out and create shapes. *[4 marks]*

18 Name a material and possible finish for this material. Explain how this finish may be applied to your chosen material. *[4 marks]*

19 You are cutting out a piece of material that should be 200 × 210.5 mm with a tolerance of ±2mm. Which of the following measurements are within tolerance? Show your working out.

201 × 213 mm

198 × 199 mm

201 × 212.5 mm

197.5 × 199 mm

4 Non-exam assessment

4.1

Non-exam assessment (NEA)

The non-examination assessment (NEA) accounts for 50 per cent of the marks for AQA Design and Technology GCSE. To complete the NEA, you will need to design and make a prototype that fulfils its purpose and meets the needs of the intended users.

When will the NEA be completed?

Each June the exam board will release a small number of themes or **contextual challenges** to schools and colleges. Your teachers will decide which theme(s) your school or college will be using that year.

For the following few months you will work on your project. Although your teacher will give you deadlines and target dates to help you manage your workload, the work will need to be finished well before your teacher needs to send your mark to the exam board. Getting the work completed on time is your responsibility. As you are unlikely to have experienced such a large project working mainly on your own, it is very important to plan your time so that everything is completed on time.

How long should I spend on the NEA?

The total project should take about 30 to 35 hours. You are allowed to finish off some work at home, but the majority of work and all practical work must be completed under your teacher's supervision. If they cannot be absolutely sure that it is your work, they cannot give you any marks for that aspect of your work.

What format will the NEA take?

You will produce a portfolio of design work that records any investigations, planning, design ideas as sketches, photographs of models, development drawings, evaluations and photographs of the prototype during making and in its finished state.

Your portfolio can be paper-based or produced in digital or electronic form. If your work includes a lot of CAD or photographs then it might be easier to present it as a PowerPoint or PDF file. It must be one file and not contain links to websites or other associated files.

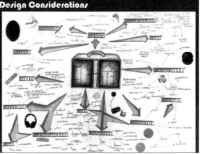
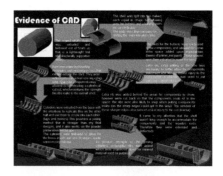

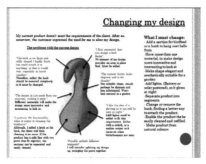
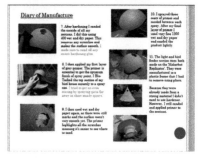

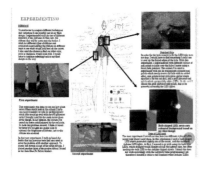
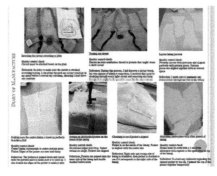
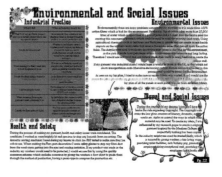

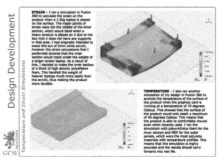
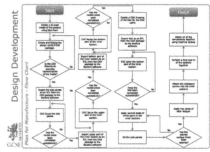

Figure 4.1 Portfolio examples

How will I be assessed?

You will be assessed against the six criteria shown in the table below. There will be a total of 100 marks available for your NEA project.

	Section	Criteria	Maximum marks
AO1 Identify, investigate and outline design possibilities	A	Identifying and investigating design possibilities	10
	B	Producing a design brief and specification	10
AO2 Design and make prototypes that are fit for purpose	C	Generating design ideas	20
	D	Developing design ideas	20
	E	Realising design ideas	20
AO3 Analyse and evaluate	F	Analysing and evaluating	20
	Total		**100**

*For the latest mark schemes, please also refer to the AQA website.

Sections A and B: Identifying and investigating design possibilities (20 marks)

Analysis and research into the contextual challenge

Your teacher will be given a small number of contextual challenges to choose from by the exam board; they will tell you the contextual challenge you will need to use for your NEA. Some examples of challenges include:

- a high-profile event
- addressing the needs of the elderly
- the contemporary home
- children's learning and development
- the world of travel and tourism.

These examples are sample challenges issued by the exam board. They will not be used in this form in your real NEA, but are provided in this book to help you to understand more about the approach that needs to be taken in GCSE Design and Technology.

One of your first tasks is to try to identify a **design opportunity** that relates to the challenge. (More information about identifying a design opportunity can be found in Topic 3.1 Investigation, primary and secondary data in this book.)

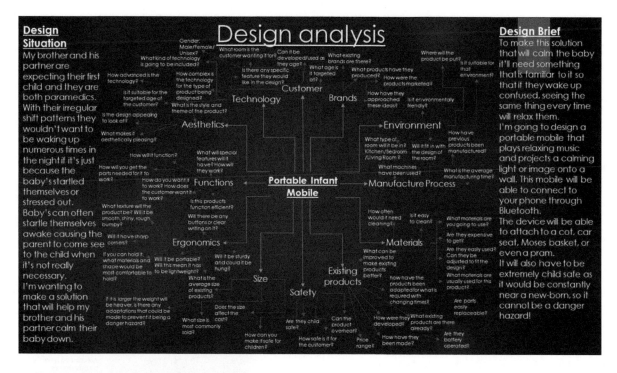

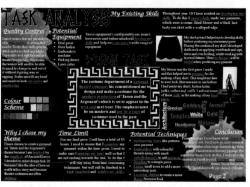

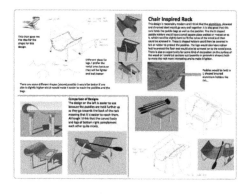

Initial Mechanical Ideas; Evaluation and Modelling of the gripper

The motors I am using spin at 13300 rpm at 5V DC with 2.55 N cm of torque. On the first Gripper the gears that the output from the motor goes through creates a ratio of 60:1. This means it takes 1/14 of a second to close the gripper from half way open. This is far too quick.

On the second gripper, all of the gears that it goes through create these ratios; 3.75:1 ; 16:1 ; 16:1 ; 16:1. Overall this creates a 15360:1 ratio. This means that in order for the gripper to close from being half open it takes 8 seconds. This is slow enough to allow people to respond to when the should stop allowing the motor to turn.

Both grippers use worm gears which means the gripper only moves when the motor turns. Both grippers also meet my specification of being able to pick something up. But because the second gripper operates within a sensible range I have decided to use this design, although it uses more materials and is less sustainable and environmentally friendly, it is better for the product otherwise the gripper would be hard to operate.

To ensure that my gripper design works I decided to model it using Lego. I have used Lego because I have got worm gears and spur gears already in some of my sets. I have created a simplified version of the second design, I have not included all of the gears, just the last worm gear and the spur gears that drive the claw. This enables me to prove that my design works. Once I had built this simple Lego model I know that my idea works this allows me to develop my idea further.

Lego gripper, showing spur gears and worm gear.

Lego Gripper Open

Lego Gripper Closed

Figure 4.2 Design opportunities and analysis

How will I be assessed?

Primary existing solutions

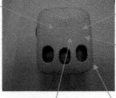

The small indents appear red, this tells you which fragrance is being diffused.

The three holes are where the scents are inserted. This is different to other products on the market because some only have one scent.

The product is small, this means it can easily fit in a room with being a big piece of furniture.

As the product is going to be used mainly by adults of both gender, white is a universal colour appealing to both men and women.

This is an view to the inside of the product. Here the fragrances are inserted.

There is a layer of plastic which protects the glass jar of the scented liquids from smashes and breaks. However this piece of plastic is very thin, meaning it can easily snap off, potentially causing damage to the product.

The gap between the back and front of the product is fairly small, this might make it difficult to change the fragrances and fingers could get stuck or hurt in the process.

The materials of this product are plastic. This makes the finish smooth to touch. However environmentally, plastic is made from crude oil which is a resource that is running out.

The outside shape of the product is rectangular with rounded edges. This is good because it means the product is safe, having no sharp sides that you could injure yourself on. However the rectangular outline makes it look boring and too simple.

The colour of the product is white and pale grey. I think that these colours are dull and boring, it would make it more interesting if their was a bright colour or design.

The edges of the product are fairly sharp, this could cause minor injury and scratches.

The brand of the product Air Wick can be seen on the front. This is good because customers sometimes are more likely to buy a product if they know the brand has a good reputation and quality for their products.

The holes in the product are where the different fragrances fit in. I like the idea of being able to see the liquid because you then know what different fragrances are being added.

Health and safety precaution are engraved into the product. This makes it look less appealing, however it means you are less likely to lose or throw away if they are on packaging or loose paper.

The product is held together by metal screws. This makes it more strong and less likely to fall apart. However they are not disguised very well, making the product look less appealing.

Being a plug in product it does mean there are limited choices of where you can put the product.

This product is fairly big, so if plugged in to a socket it may stop another appliance being used.

On the side of the product there is a push up or down switch. This enables you to change how strong the fragrance diffusing out is. Going from minimum to maximum strength.

An extra feature of this product is that when the scent oils runs out, they can be replaced and swapped for ones more suited to the customer. However this can be quite expensive, but long lasting.

This product cost around 15 pounds, but I think that this is quite over priced. Even though it can be reused the extra pieces you have to buy makes the price get higher and higher. As well as not being very interesting or exciting to look at.

The sides of this product are dull and plain. If a patter or different material was added then it may look more interesting, as well as standing out better on a shelf in a shop.

The product is not very wide meaning you are less likely to catch it and cause a break.

The product is po[...] by plug and socket. This is better than batteries as sometime[...] product can be on all the time and batteries can easily run[...]

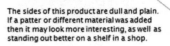
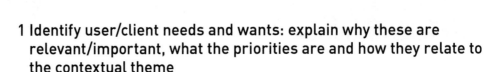

Figure 4.3

1 Identify user/client needs and wants: explain why these are relevant/important, what the priorities are and how they relate to the contextual theme

Figure 4.4

Once you have identified a design opportunity, then the user or client needs to be identified. These will make a useful focus for your designing activities, as what is being designed must meet their expectations.

2 Look at the work of others and investigate how they were influenced

It is unlikely that you will be the first person ever to try to design a solution to the design situation you have identified through your analysis, so it is worthwhile investigating what other designers and organisations have already developed. This will prevent you from spending time on reproducing something that already exists. This is an activity professional designers need to carry out to avoid patent and copyright issues and costly legal proceedings.

(More information about the work of other designers can be found in Topic 3.3 The work of others, in this book.)

3 Carry out some analysis and investigation on possible impacts of the product or system

All designers need to be aware of any possible impact the product or system they are developing might have: it could be social or economic, such as causing unemployment; or environmental, such as being excessively noisy. It might be moral, perhaps causing offence to others, maybe due to religious beliefs. Again, only investigation can ensure this does not happen unintentionally, so have a look at things that have already been developed and see if you can discover how they were influenced.

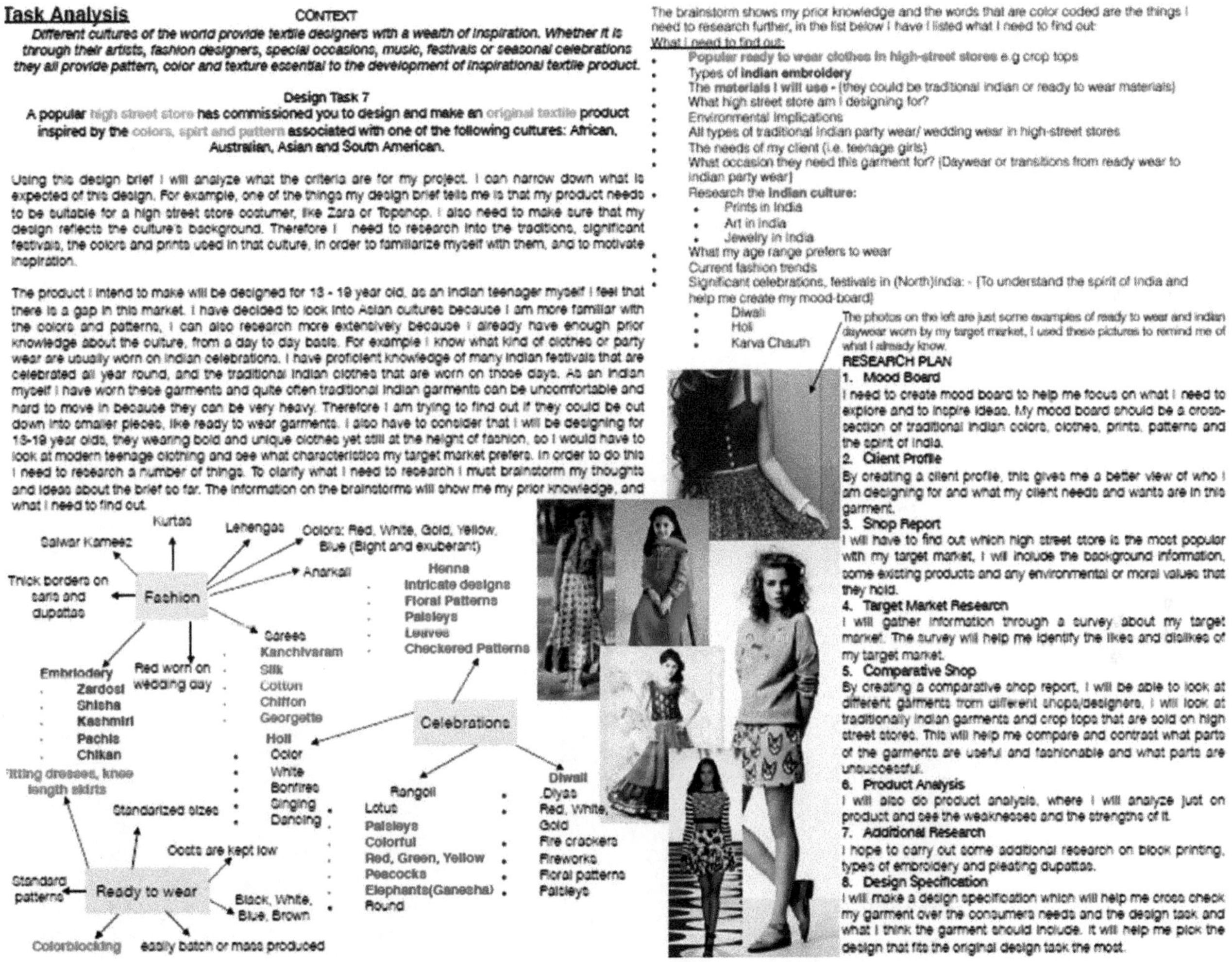

Task Analysis

CONTEXT

Different cultures of the world provide textile designers with a wealth of inspiration. Whether it is through their artists, fashion designers, special occasions, music, festivals or seasonal celebrations they all provide pattern, color and texture essential to the development of inspirational textile product.

Design Task 7

A popular high street store has commissioned you to design and make an original textile product inspired by the colors, spirit and pattern associated with one of the following cultures: African, Australian, Asian and South American.

Using this design brief I will analyze what the criteria are for my project. I can narrow down what is expected of this design. For example, one of the things my design brief tells me is that my product needs to be suitable for a high street store costumer, like Zara or Topshop. I also need to make sure that my design reflects the culture's background. Therefore I need to research into the traditions, significant festivals, the colors and prints used in that culture, in order to familiarize myself with them, and to motivate inspiration.

The product I intend to make will be designed for 13 - 19 year old, as an Indian teenager myself I feel that there is a gap in this market. I have decided to look into Asian cultures because I am more familiar with the colors and patterns, I can also research more extensively because I already have enough prior knowledge about the culture, from a day to day basis. For example I know what kind of clothes or party wear are usually worn on Indian celebrations. I have proficient knowledge of many Indian festivals that are celebrated all year round, and the traditional Indian clothes that are worn on those days. As an Indian myself I have worn these garments and quite often traditional Indian garments can be uncomfortable and hard to move in because they can be very heavy. Therefore I am trying to find out if they could be cut down into smaller pieces, like ready to wear garments. I also have to consider that I will be designing for 13-19 year olds, they wearing bold and unique clothes yet still at the height of fashion, so I would have to look at modern teenage clothing and see what characteristics my target market prefers. In order to do this I need to research a number of things. To clarify what I need to research I must brainstorm my thoughts and ideas about the brief so far. The information on the brainstorm will show me my prior knowledge, and what I need to find out.

The brainstorm shows my prior knowledge and the words that are color coded are the things I need to research further, in the list below I have I listed what I need to find out.

What I need to find out:

- Popular ready to wear clothes in high-street stores e.g crop tops
- Types of **Indian embroidery**
- The materials I will use - (they could be traditional Indian or ready to wear materials)
- What high street store am I designing for?
- Environmental implications
- All types of traditional Indian party wear/ wedding wear in high-street stores
- The needs of my client (i.e. teenage girls)
- What occasion they need this garment for? (Daywear or transitions from ready wear to Indian party wear)

Research the **Indian culture:**
- Prints in India
- Art in India
- Jewelry in India
- What my age range prefers to wear
- Current fashion trends
- Significant celebrations, festivals in (North)India: - (To understand the spirit of India and help me create my mood-board)
  - Diwali
  - Holi
  - Karva Chauth

The photos on the left are just some examples of ready to wear and Indian daywear worn by my target market, I used these pictures to remind me of what I already know.

RESEARCH PLAN

1. **Mood Board**
I need to create mood board to help me focus on what I need to explore and to inspire ideas. My mood board should be a cross-section of traditional Indian colors, clothes, prints, patterns and the spirit of India.

2. **Client Profile**
By creating a client profile, this gives me a better view of who I am designing for and what my client needs and wants are in this garment.

3. **Shop Report**
I will have to find out which high street store is the most popular with my target market, I will include the background information, some existing products and any environmental or moral values that they hold.

4. **Target Market Research**
I will gather information through a survey about my target market. The survey will help me identify the likes and dislikes of my target market.

5. **Comparative Shop**
By creating a comparative shop report, I will be able to look at different garments from different shops/designers, I will look at traditionally Indian garments and crop tops that are sold on high street stores. This will help me compare and contrast what parts of the garments are useful and fashionable and what parts are unsuccessful.

6. **Product Analysis**
I will also do product analysis, where I will analyze just on product and see the weaknesses and the strengths of it.

7. **Additional Research**
I hope to carry out some additional research on block printing, types of embroidery and pleating dupattas.

8. **Design Specification**
I will make a design specification which will help me cross check my garment over the consumers needs and the design task and what I think the garment should include. It will help me pick the design that fits the original design task the most.

Brainstorm (Figure 4.5):

- Kurtas
- Salwar Kameez
- Lehengas
- Colors: Red, White, Gold, Yellow, Blue (Bright and exuberant)
- Anarkali
- Thick borders on saris and dupattas
- **Fashion**
- Henna: Intricate designs, Floral Patterns, Paisleys, Leaves, Checkered Patterns
- Sarees: Kanchivaram Silk, Cotton, Chiffon, Georgette
- Embroidery: Zardosi, Shisha, Kashmiri, Pachis, Chikan
- Red worn on wedding day
- **Celebrations**
- Holi: Color, White, Bonfires, Singing, Dancing
- Fitting dresses, knee length skirts
- Standarized sizes
- Diwali: Diyas, Red, White, Gold, Fire crackers, Fireworks, Floral patterns, Paisleys
- Rangoli: Lotus, Paisleys, Colorful, Red, Green, Yellow, Peacocks, Elephants (Ganesha), Round
- Costs are kept low
- **Ready to wear**
- Standard patterns
- Black, White, Blue, Brown
- Colorblocking
- easily batch or mass produced

Figure 4.5

4 Produce a design brief and specification

In Section B you will be expected to produce a **design brief** and **specification**. (For more information on producing a design brief and specification see Topic 3.1 Investigation, primary and secondary data, in this book.) Statements need to be justified, so you will need evidence that they are required and meet the needs or wants of the user/client. Eventually, at some stage in the development of the product or system, you will need to use these statements to see if the prototype is either likely to or has achieved the required standard. You will need to evaluate your work at every stage of the process, not just when it is finished, so that you can improve it during the development process.

The design brief and specification will provide detailed and testable information (criteria) that will influence the next stages of designing.

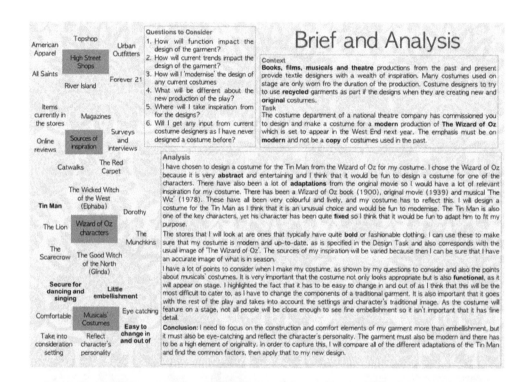

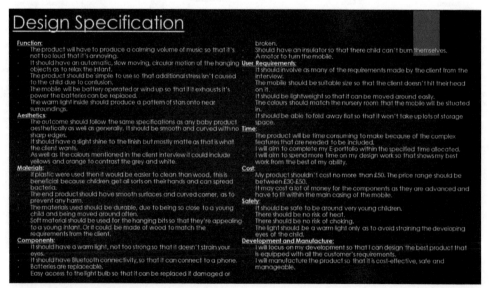

Figure 4.6 Design briefs and specifications

Design brief

Modern day hockey was founded in Scotland in 1900 and since then has disseminated globally, especially in the UK, with field hockey being the 3rd most watched sport in the 2012 London Olympics. But also it is evident as to what extent it has gained increasing stature across the globe with over 3,000,000 players spread over 100 countries and 5 continents and this doesn't include those participating at social levels either.

Due to this advancement in the sport there are currently many major field hockey tournaments including The FIH Hockey World Cup, The Olympics, The Commonwealth Games, The Asian Games, Champions Trophy and Sultan Azlanshah Trophy. Contributing to this cycle are hockey boards who work in coherence with these tournaments to organise teams and generally increase awareness and popularity in different countries, at expert levels but also club and school levels to promote hockey through society.

EHB (England Hockey Board) has approached me in cooperation with Slazenger (a major sports brand) and asked me to design for them a new product. They highlighted the fact that there are more field hockey players in college than in any other sports in the world. Hence forth they wanted me to design a device for inside sports luggage which is affordable for the students participating in hockey in order to achieve a consistent interest in hockey especially into the future. Slazenger are currently setting up a campaign to increase participation of sports and maintaining good health in teenagers in order to help them stay fit, healthy and adopt good habits for future years where it can be more important. They are planning to incorporate this new product within the range they are promoting alongside this campaign and so I have to remember to make this product attractive and helpful for participants in order to maintain interest in sports.

EHB and Slazenger highlighted the target market to be students and teenagers and so I have to focus on attractive designs which are suited for this age group. However, this also means I should make it relatively affordable as students and teenagers generally have low disposable income.

Therefore, considering low costs, I should make a comparatively small product in order to keep the cost of material down. However, this will also depend on the material I use. I need to calculate the cost of material in relation to the size so it isn't too expensive, this means I should research into different materials which are firstly suitable for manufacture of a small product and secondly will keep costs low in relation to amount used. For example, it may be a more expensive material but if this means the size can be kept small the overall cost may actually be less then alternate solutions.

I also need to find the opinion of the target market into what they find attractive and are willing to invest their money in. I could do this via interviews or focus groups for example. I should ask questions which I will identify in my problem analysis, this may include questions about power source or colour. As well, to ensure my product is easy to use, I should return to the target market for testing and prototyping, but also if I do find any problems (including safety problems) I can alter the product and incorporate the improvements into the final product. This will also ensure quality is high and that I stay focused on my goal and purpose throughout the project. On the other hand, within the designing process I should keep it simple in terms of options, buttons and shape in order for the design to be understood and utilized with ease in the depths of the bag.

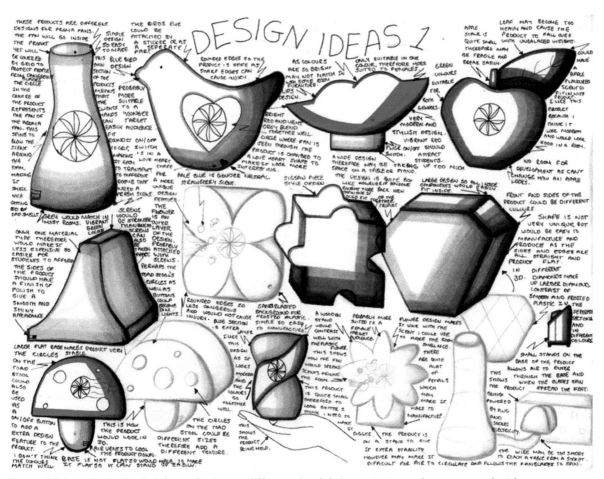

Figure 4.7 Sample design briefs can look very different, but it is important you show your product is required and meets the needs or wants of the user/client

Sections A and B: Identifying and investigating design possibilities (20 marks)

Sections C to E: Designing and making (30 marks)

Designing and making is divided into three parts: generating design ideas, developing design ideas and realising design ideas. These follow a normal sequence of designing, but you can go backwards and forwards as much as you need, as often better ideas will crop up, or you might find a difficulty that you had not foreseen and need to work round it.

5 Having identified what the important aspects of your design are meant to achieve, produce a range of ideas that might satisfy the design specification

Marks are available for devising different ways of solving the design problem. The process we use for this is called the generation of ideas. People will approach this in different ways, and that is perfectly acceptable. It will be the variety and quality of ideas that counts, as these will demonstrate your imagination, creativity and how capable you are of producing new and novel solutions, called innovating.

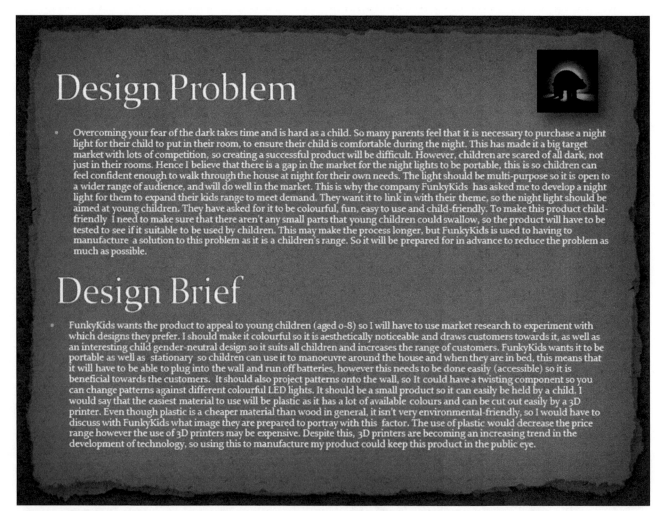

Design Problem

- Overcoming your fear of the dark takes time and is hard as a child. So many parents feel that it is necessary to purchase a night light for their child to put in their room, to ensure their child is comfortable during the night. This has made it a big target market with lots of competition, so creating a successful product will be difficult. However, children are scared of all dark, not just in their rooms. Hence I believe that there is a gap in the market for the night lights to be portable, this is so children can feel confident enough to walk through the house at night for their own needs. The light should be multi-purpose so it is open to a wider range of audience, and will do well in the market. This is why the company FunkyKids has asked me to develop a night light for them to expand their kids range to meet demand. They want it to link in with their theme, so the night light should be aimed at young children. They have asked for it to be colourful, fun, easy to use and child-friendly. To make this product child-friendly I need to make sure that there aren't any small parts that young children could swallow, so the product will have to be tested to see if it suitable to be used by children. This may make the process longer, but FunkyKids is used to having to manufacture a solution to this problem as it is a children's range. So it will be prepared for in advance to reduce the problem as much as possible.

Design Brief

- FunkyKids wants the product to appeal to young children (aged 0-8) so I will have to use market research to experiment with which designs they prefer. I should make it colourful so it is aesthetically noticeable and draws customers towards it, as well as an interesting child gender-neutral design so it suits all children and increases the range of customers. FunkyKids wants it to be portable as well as stationary so children can use it to manoeuvre around the house and when they are in bed, this means that it will have to be able to plug into the wall and run off batteries, however this needs to be done easily (accessible) so it is beneficial towards the customers. It should also project patterns onto the wall, so It could have a twisting component so you can change patterns against different colourful LED lights. It should be a small product so it can easily be held by a child. I would say that the easiest material to use will be plastic as it has a lot of available colours and can be cut out easily by a 3D printer. Even though plastic is a cheaper material than wood in general, it isn't very environmental-friendly, so I would have to discuss with FunkyKids what image they are prepared to portray with this factor. The use of plastic would decrease the price range however the use of 3D printers may be expensive. Despite this, 3D printers are becoming an increasing trend in the development of technology, so using this to manufacture my product could keep this product in the public eye.

Figure 4.8 Student sample - You need to match your brief to the problems you are trying to address

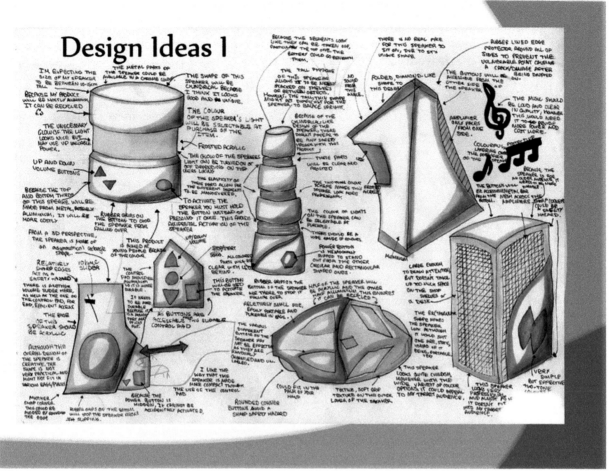

Figure 4.9

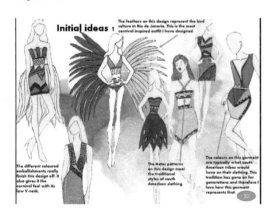

Figure 4.10

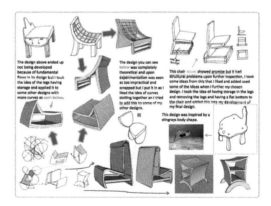

Figure 4.11

6 Evaluate your ideas using the design specification as a guide

It is possible at this stage that you come across problems or reach conclusions that are different to your first ideas or intentions. Check with your client/user to see if this makes a difference. You may need to change the design specification to avoid being too restricted. Try not to go in too many directions as this could slow you down as you refine the design.

During the development phase of the design, you will be involved in detailed thinking and recording of your thoughts as you continue to experiment with colour, shape, mechanical movement, control system design or decoration, depending on the function of your product or system.

7 Produce a series of detailed design sheets

These may be decorative or technically functional, or a combination, depending on the object you are designing and the stage you are at. They must convey to others why you have made those decisions and show all the important aspects of the design. Do not leave anything out (although small parts such as how buttons are to be fastened or handles fixed will not need more than a few words).

If you are using a CAD package for this stage, make sure the development is still clear.

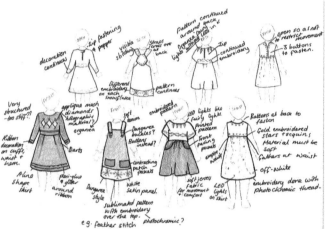

Figure 4.12

Designing The Turbine

- I decided that the best turbine to make would be a savonius turbine, as it is the simplest in construction, cheapest in construction materials, less complex in regards to balancing and is comprised of much less parts. Before making the physical model, I first planned it out on cardboard by measuring dimensions and cutting them out, then attaching them together to form a make-shift 'rough draft' for my turbine. The measurements were made onto cardboard with a ruler and compasses in order to retain accuracy and were then fixed together with masking tape and the finished 'rough draft' can be seen to the right. The dimensions worked fine, and the turbine fitted together well, with only a small amount of bending in the card, as a result of the flexibility of the material used and not because of the design. Satisfied that I could proceed with my design I then moved on to the next stage.

- I also would need something upon which to mount my wind turbine upon completion. I considered a range of mounting methods; from attaching it to a wall via a bracket, erecting it on a pole or simply attaching it to a plate and clamping it to a wall. I decided upon the last option because it allowed me to keep my turbine as compact as possible, taking up less space, making less of a visual image, costing less in production and also allowing for ease of erection as a commercial product. The mounting plate is a simple construct of wood, upon which the circuit board will be bolted to the side along with the motor, and the turbine affixed to the top, with the band running to the motor. For display purposes I will mount this plate to a base of wood, but the commercial product itself is intended to be bolted directly to a wall to take up the minimum space possible and reduce cost.

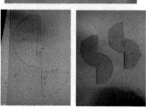

Figure 4.13

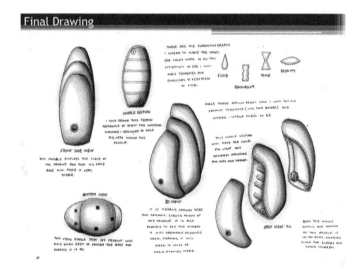

Figure 4.14

8 Try out ideas by modelling them; incorporate evidence of modelling in your portfolio

Try not to present just the finished item. You may want to try several alternatives. A good method is to make a model; this can be by building physical models with card or MDF, using cheap materials to replace expensive ones as you experiment. In systems, it is likely to involve computer simulations or breadboarding circuits; working with textiles might mean making a toile.

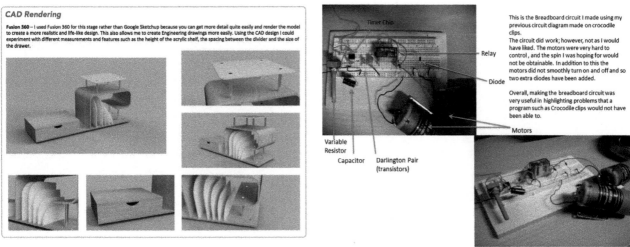

Figure 4.15

Figure 4.16

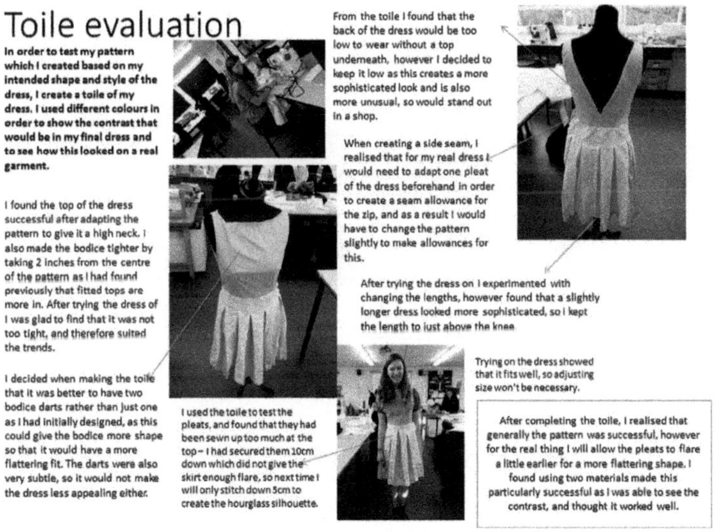

Figure 4.17

9 Produce a manufacturing specification

You will need to produce a manufacturing specification. This contains all the information somebody else would need if they were to make your item, such as lists of materials and components, scale drawings, patterns and any other useful information.

The next thing to do is to plan how it will be made. It could be a list, flow chart or a table, but the object is to make sure you can use your time efficiently and not hit any problems as you construct the prototype. Make sure you identify any quality control points, as these should ensure your work turns out as intended.

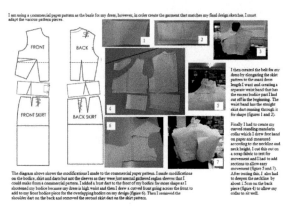

Figure 4.18

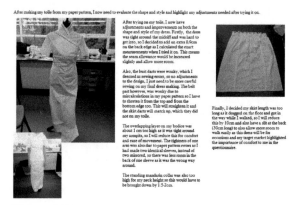

Figure 4.19

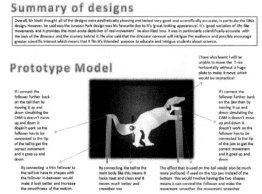

Figure 4.20

10 Obtain and prepare materials, construct the prototype and finish

One you have reached this stage it is time to make the final outcome. This is known as realising the design. Remember you are making a prototype, so in order to be capable of being tested to see if it is fit for purpose it needs to be made as well as you are able.

Making can be divided into three distinct stages: preparation – getting all the materials and components ready for use; construction – to shape and assemble; and finishing. Occasionally it works better if finishing happens before assembly, such as adding decorative motifs or varnishing the inside of a container.

Section F: Analysing and evaluating (20 marks)

Although the product or system is now complete, the design task continues. The prototype needs to be tested against the original design criteria. How well it meets those criteria needs to be analysed and the overall project evaluated. The opinions of clients and users needs to be sought and fed back into the evaluation. Potential modifications should be identified and there needs to be some consideration of how the product/system could be improved or put into commercial production.

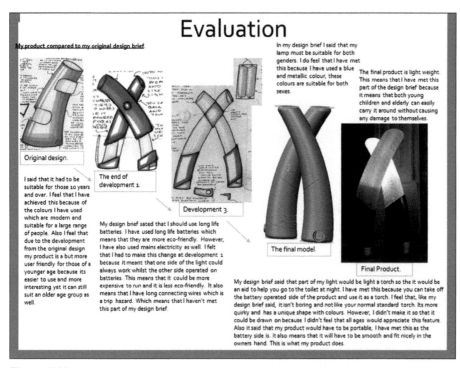

Figure 4.21

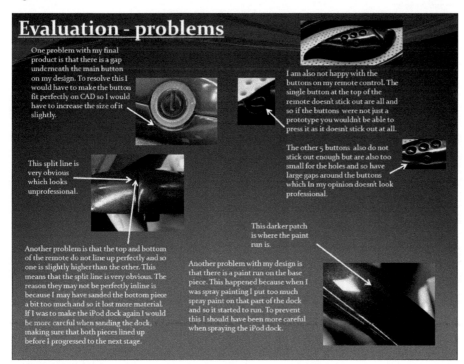

Figure 4.22

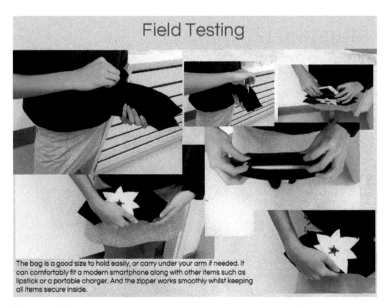

Field Testing

The bag is a good size to hold easily, or carry under your arm if needed. It can comfortably fit a modern smartphone along with other items such as lipstick or a portable charger. And the zipper works smoothly whilst keeping all items secure inside.

Figure 4.23

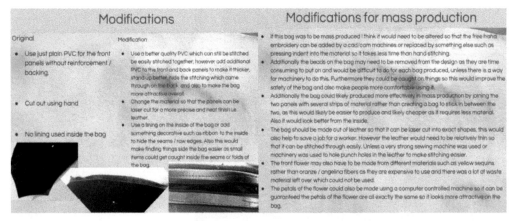

Modifications

Original

- Use just plain PVC for the front panels without reinforcement / backing.

- Cut out using hand

- No lining used inside the bag

Modification

- Use a better quality PVC which can still be stitched together, however add additional PVC to the front and back panels to make it thicker, stand up better, hide the stitching which came through on the back, and also to make the bag more attractive overall.

- Change the material so that the panels can be laser cut for a more precise and neat finish i.e. leather.

- Use a lining on the inside of the bag or add something decorative such as ribbon to the inside to hide the seams / raw edges. Also this would make finding things side the bag easier as small items could get caught inside the seams or folds of the bag.

Modifications for mass production

- If this bag was to be mass produced I think it would need to be altered so that the free hand embroidery can be added by a cad/cam machines or replaced by something else such as pressing indent into the material so it takes less time than hand stitching.
- Additionally the beads on the bag may need to be removed from the design as they are time consuming to put on and would be difficult to do for each bag produced, unless there is a way for machinery to do this. Furthermore they could be caught on things so this would improve the safety of the bag and also make people more comfortable using it.
- Additionally the bag could likely produced more effectively in mass production by joining the two panels with several strips of material rather than creating a bag to stick in between the two, as this would likely be easier to produce and likely cheaper as it requires less material. Also it would look better from the inside.
- The bag should be made out of leather so that it can be laser cut into exact shapes, this would also help to save a job for a worker. However the leather would need to be relatively thin so that it can be stitched through easily. Unless a very strong sewing machine was used or machinery was used to hole punch holes in the leather to make stitching easier.
- The front flower may also have to be made from different materials such as yellow sequins rather than orange / angelina fibers as they are expensive to use and there was a lot of waste material left over which could not be used.
- The petals of the flower could also be made using a computer controlled machine so it can be guaranteed the petals of the flower are all exactly the same so it looks more attractive on the bag.

Figure 4.24

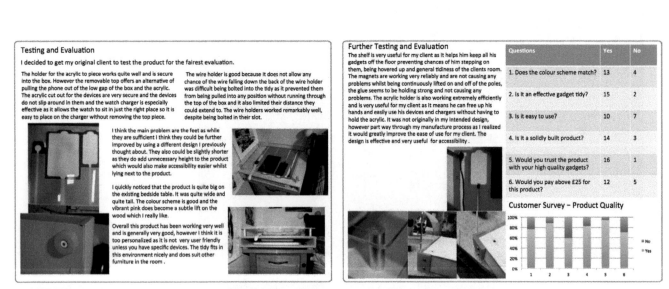

Testing and Evaluation

I decided to get my original client to test the product for the fairest evaluation.

The holder for the acrylic to piece works quite well and is secure into the box. However the removable top offers an alternative of pulling the phone out of the low gap of the box and the acrylic. The acrylic cut out for the devices are very secure and the devices do not slip around in them and the watch charger is especially effective as it allows the watch to sit in just the right place so it is easy to place on the charger without removing the top piece.

The wire holder is good because it does not allow any chance of the wire falling down the back of the wire holder was difficult being bolted into the tidy as it prevented them from being pulled into any position without running through the top of the box and it also limited their distance they could extend to. The wire holders worked remarkably well, despite being bolted in their slot.

I think the main problem are the feet as while they are sufficient I think they could be further improved by using a different design I previously thought about. They also could be slightly shorter as they do add unnecessary height to the product which would also make accessibility easier whilst lying next to the product.

I quickly noticed that the product is quite big on the existing bedside table. It was quite wide and quite tall. The colour scheme is good and the vibrant pink does become a subtle lift on the wood which I really like.

Overall this product has been working very well and is generally very good, however I think it is too personalized as it is not very user friendly unless you have specific devices. The tidy fits in this environment nicely and does suit other furniture in the room .

Further Testing and Evaluation

The shelf is very useful for my client as it helps him keep all his gadgets off the floor preventing chances of him stepping on them, being hovered up and general tidiness of the clients room. The magnets are working very reliably and are not causing any problems whilst being continuously lifted on and off of the poles, the glue seems to be holding strong and not causing any problems. The acrylic holder is also working extremely efficiently and is very useful for my client as it means he can free up his hands and easily use his devices and chargers without having to hold the acrylic. It was not originally in my intended design, however part way through my manufacture process as I realized it would greatly improve the ease of use for my client. The design is effective and very useful for accessibility .

Questions	Yes	No
1. Does the colour scheme match?	13	4
2. Is it an effective gadget tidy?	15	2
3. Is it easy to use?	10	7
4. Is it a solidly built product?	14	3
5. Would you trust the product with your high quality gadgets?	16	1
6. Would you pay above £25 for this product?	12	5

Customer Survey – Product Quality

Figure 4.25

5 The written paper

5.1

The written paper

The written paper accounts for 50 per cent of the total marks for the AQA Design and Technology GCSE, so you will need to ensure that you are well prepared for the types of questions you will be asked. There is just one paper to complete, but that paper contains three distinct parts: core technical principles, specialist technical principles and designing and making principles.

When will the written paper be taken?

The written paper will take place in the summer exam period around June. The date of the exam will vary each year, but information about this will be given to you by your school or college. You can also look at the AQA website to find out for yourself.

How long will I have?

You will be given two hours in which to complete the examination. This will be the total allocated time for all three sections to be completed. Once you have practised some examples of question papers you will have a better idea about how long each section might take you. You can use the number of marks as a guide (for example, section C is worth 50 marks so spend approximately 50 minutes on this section), but this is not always an accurate method. You may feel that the multiple-choice and short-answer questions at the beginning of the paper, despite being worth 20 marks, actually only take you on average 15 minutes. Try to practise questions enough to know which aspects you find most difficult, and revise and practise how to answer these sections.

What format will the written paper take?

The written paper will consist of a number of different question types. You will be credited for your knowledge, understanding and application, depending on the question.

There are 20 marks for Section A, 30 marks for Section B and 50 marks for Section C.

Multiple-choice questions are a feature of the first section of the examination. In these questions you will be given several choices and you have to select the correct answer from those choices. This is not necessarily an easy thing to do, so make sure your knowledge of a wide range of materials is adequate in order to answer these questions correctly. Below is an example of a multiple-choice question. You should fill in the space in the box (it is known as a lozenge) of the correct answer.

---

3   Which of the following is a smart material?

   A   Gore-Tex

   B   Kevlar

   C   MDF

   D   Shape memory alloy   *[1 mark]*

Short-answer questions also tend to be in Section A of the examination, so will be based on the core technical principles across the broad material areas. Short-answer questions will look similar to that shown below and will require simple-statement or one-word answers. They are worth between one and four marks.

> 12 Explain the potential advantages tidal power might have over conventional hydroelectric schemes using a blocked valley. **[4 marks]**

Most other questions throughout Sections B and C of the paper will be questions that require more explanation and examine higher level skills such as **analysing** and **evaluating**. They are normally worth between three and eight marks, and you will be expected to give an answer that covers a number of different points, demonstrating your knowledge of the subject in depth.

> 5 Explain why stock forms are used in the manufacture of products. **[4 marks]**

Comments

You must ensure that your answer relates directly to the question and illustrates clearly that you understand the question and can apply your knowledge to the scenario.

There will also be an **extended answer question** in the paper which will be worth between eight and ten marks. This will really test your depth of knowledge and will require you to be able to present a broad response which covers a number of points in depth. You may be required to analyse and evaluate in order to gain the highest marks. The below is an example of such a question.

> 1 Explain how the use of automation and the use of robotics has changed the way products are made. **[8 marks]**

Comments

In order to access the highest possible marks in this question, you need to present a really well-reasoned argument. You need to show analysis and evaluation in your response.

Maths is assessed throughout the examination in different forms, but all will be Design-and-Technology-specific questions that are relevant to your studies. **Maths questions** can occur in all stages of the paper.

COMMENTS

This question asks you to work out the percentage of 160 teenagers who like different styles. You can be awarded only one mark for two correct percentages.

24/160 × 100 = 15 per cent

40/160 × 100 = 10 per cent

It is also possible to find this answer by addition of the other percentages and subtracting this total from 100. If you do it this way you would also gain one mark.

1   You have been asked to design a product for teenagers and have collected some data to help you make judgements about the style of the product. Complete the table by calculating the missing percentage of teenagers who like different styles.

Style	Number of teenagers	Percentage of total
Modern	80	50
Traditional	24	
Retro	40	
Futuristic	16	10
Total	160	

[2 mark]

How will I be assessed?

There are four assessment objectives (AOs) in this qualification, and AO3 and AO4 are assessed in the written paper. The assessment objectives show you what percentage of the paper is assessing different skills.

AO3 is the analysis and evaluation questions in the paper, where the following may be examined:
● Design decisions and outcomes, including prototypes made by themselves and others.
● Wider issues in design and technology.

AO4 assesses how well you can demonstrate and apply knowledge and understanding of the technical principles and the designing and making principles.

Analyse and evaluate (AO3)	20 marks
Demonstrate and apply knowledge and understanding (AO4)	80 marks
Total	100 marks
Application of mathematics	15 marks from above

*For the latest mark schemes, please also refer to the AQA website.

General advice on answering exam questions

The best way to prepare for the exam is by practising the exam questions available to you in this book. It is also important to use specimen assessment material provided by AQA, as this will give you a clear idea of what is expected of you in the allotted time.

Time is really crucial. Make sure you practice exam questions under exam conditions and give yourself a set amount of time so that you are prepared in your real examination for how long questions will take you.

You will need to become familiar with the way questions are worded in order to help you to respond in a certain way. The following are commonly used command words in the GCSE Design and Technology exam:
● 'Evaluate' means that you need to make a 'judgement from available evidence'.
● 'Discuss' means that you should 'present key points'.
● 'State' means that you should 'express in clear terms'.

- 'Illustrate' means that you need to 'present clarifying examples'.
- 'Explain' means that you need to 'set out purposes and reasons'

You need to practise this so that you are clear as to what these mean in real terms.

Read the question carefully and use your prior experiences of practice questions to understand what the exam question is asking from you. It is good practice to read the question twice in order to ensure your understanding. In the extended writing question it might be useful to plan your answer out if you have time. You can use extra sheets to create a mind map of points you wish to cover. This will give you more structure to your answer. It is often preferable, and sometimes necessary, for you to give both sides of an argument. In your planning, try having two sections of the mind map for both sides of an argument as this will encourage you to remember to look at the issue objectively.

When you have finished the exam, make sure you spend any extra time looking through your answers. Remember you cannot have marks deducted so add any additional information you can think of to add to your answers can only improve your grade.

Section A: core technical principles

This section accounts for 20 marks out of 100 and consists of multiple-choice and short-answer questions examining the core technical principles.

The majority of questions are likely to be multiple-choice questions such as below.

These questions should help you to get started in the exam as they are only one mark each and don't require you to write a long answer.

---

3 Which of the following is a smart material?

A Gore-Tex

B Kevlar

C MDF

D Shape memory alloy   *[1 mark]*

In Section A you will also be asked short-answer questions, worth one or two marks each. In this question you are given one mark for each way stated.

---

5 State two ways in which you can prolong the life of a product.   *[2 marks]*

COMMENTS

The correct answer from the options above is D – Shape memory alloy. You should have learnt about this in the 'Developments in new materials' section of the specification. Remember, shape memory alloys change their shape when the temperature changes. They are used in products such as glasses frames.

COMMENTS

Typical answers could include reference to:
- **Adding finishes that protect the materials from damage (for example, varnish to protect timber from moisture).**
- **Creating easily replaced parts so that if one part of a product breaks, the consumer is encouraged to buy a new part rather than purchase a completely new product.**
- **Creating a design that does not date or go out of fashion quickly.**
- **Making a product simple in its design so that there is less chance of features not working or becoming obsolete.**
- **Using high-quality materials.**
- **Using standard components that can be replaced.**
- **Any two such points would warrant two marks.**

Section B: specialist technical principles

This section accounts for 30 marks out of 100 and consists of longer response questions that assess the specialist technical principles.

1 Give two reasons why standard components are used in the manufacture of commercial products. *[2 × 2 marks]*

COMMENTS

You can be awarded two marks for each answer so you must make sure you fill in both spaces with two separate points. The quality of your response will determine whether you gain one or two marks for each correct answer. The mark scheme gives guidance of this in the chart below.

2 marks	Complete reason demonstrating both knowledge and understanding of the use of standard components and how they can be of benefit.
1 mark	Simple description with little explanation or some misunderstanding of why standard components are used.
0 marks	Nothing worthy of credit.

The kind of response you will be expected to make might be:
- **Standard components can be supplied by another manufacturer who specialises in their production. This means that they make the component in large quantities so the product can be purchased for less.**
- **Standard components are made from specialist companies to ensure a consistent quality.**
- **Standard components can be made by other companies and bought in by the manufacturer of a product. This reduces the need for expensive and bulky machinery set up to produce only small parts of the product.**
- **Different products can be joined together if standard components are used. This allows for greater flexibility for the user (for example, Ikea furniture uses standard components so that the products can be customised and used alongside other products).**
- **Standard components are used to allow the consumer to replace parts easily if they break or are lost.**

Section C: designing and making principles

This section accounts for 50 marks out of 100 and consists of questions that assess the designing and making principles.

---

1   The product in Figure 5.1 is a chair that can be adjusted by the user.

Analyse and evaluate this product, considering how effective this design might be for use in a hospital waiting room. You should discuss positive and negative aspects of the design in this situation.   **[8 marks]**

Figure 5.1 **Adjustable chair**

COMMENTS

Many of the questions in this final section require some discussion and you will be expected to write in full sentences rather than using one-word answers or bullet points.

Typical points you may include in your answer are:
- comfortable seating for users who may experience discomfort
- back rest offers support
- arm rests are comfortable and padded, offering extra support
- design is old-fashioned and may be more suitable for an older target market – this may be appealing to one group of users and not another
- tilting motion of the chair helps user get up from the chair – this may be particularly beneficial to elderly or injured users
- unattractive colour scheme that is uninteresting
- takes up a lot of space, so will mean fewer people can sit down in that room
- not wipe clean, so may harbour germs and may begin to look dirty quickly
- children may trap their fingers in moving parts
- people in wheelchairs may need assistance to get on it
- the chair needs power points in order to work, which will be inconvenient in terms of layout and will require electricity which is costly
- people can put things down the side of the cushions or items can fall out of their pockets and become lost between the cushions
- they are much more expensive than the usual polypropylene chairs
- they don't stack away, which may be necessary to increase the space in an area
- users can trip over the wires, so they may need to be embedded in the floor.

In order to gain the highest marks in this question you would need to ensure that you write both positive and negative points and make a judgement about the chair's suitability for a hospital environment.

GLOSSARY

Air seasoning a natural method of drying out green timber.

Batch production when a limited number of the same product is made during a particular period of time.

Bulk buying when materials or products are bought in large quantities they usually cost less per unit than buying just a few. This is because the costs of setting up the manufacturing are the same no matter how many are made.

Casting a method of heating metal into a molten state and pouring it into a pre-prepared mould.

Cellulose fibres fibres found in trees and plants which can be used to make paper and card.

Composite a material that combines the properties of the materials that were used to make it.

Compound gear a number of gears on the same shaft that rotate at the same speed.

Continuous production runs constantly and is highly automated.

Conversion the process of sawing a tree trunk into planks.

Cross filing a method of shaping metal.

Deforestation large areas of trees cut down, often due to mining, drilling, farming or logging.

Depth-stop a mechanical means of setting how deep a drill bit will cut.

Design movement a style of design particularly popular with a group of people or within a period of time.

Dimensional tolerance the difference between the maximum and minimum acceptable size.

Down time when a machine has stopped working and no products are being made. This could be for maintenance, because the machine has developed a fault, or the time taken to set the machine up for a new operation.

Draw filing a method of smoothing the edge of metal.

Drilling the process of making a hole in the Earth's surface, usually to extract liquids or gas.

Ecological footprint the impact of a person or community on the environment; the amount of land needed to supply the natural resources they use.

Ergonomics the relationship between people and products, and how they use and interact with them.

Extrusion a reforming method of production that involves forcing a molten polymer through a die to produce a regular cross-section polymer product.

Face edge the surface of a piece of wood that is known to be straight and true.

Face side the surface of a piece of wood that is known to be straight and true.

Fairtrade mark the mark that identifies products for which a fair price has been paid to the producers

Farming the use of land for growing crops or keeping animals for food.

Felling the process of cutting a tree down.

Flow soldering a method of soldering where solder paste is melted by heating the printed circuit board.

Freehand sketching drawing done without the use of rulers or drawing aids. It is a way that a designer can quickly express thoughts and ideas.

Friction the resistance to movement when surfaces rub together.

FSC the Forestry Stewardship Council.

Go–no-go gauge a special tool to check the acceptable tolerance.

Green timber wood that has not been seasoned.

Human factors considerations that are concerned with people.

Injection moulding a reforming method of production that involves forcing a molten polymer through a die and into a mould.

Integrated circuit a miniature electronic circuit on a semiconductor material.

Isometric drawing means 'equal measure'. It is a technique of presenting a design sketch in three dimensions.

Jig a three-dimensional device that aids a production process.

Kiln seasoning a relatively quick method of drying out green timber using steam.

Lacquering the process of applying a coating material to another material.

Laminating a method of bending wood by slicing into thin veneers and gluing back together in a curve.

Life-cycle assessment a method of assessing the environmental impact of the manufacture and use of products, at all stages from raw materials through to disposal or recycling of the product.

Marking out the process of applying a drawing on to a material.

Mass production manufacturing in large quantities over a long period of time. This typically uses a production line.

Milling a method of cutting metal to produce slots, grooves and flat surfaces.

Mining the extraction of minerals and metals from the ground.

Modern material a material that has recently been developed due to its use for certain applications.

Moisture content the amount of moisture in a timber.

Ore a rock which contains metal.

PIC a programmable interface controller is a device that can be programmed using a computer to control complex systems.

Pick and place machine a machine used to place surface-mounted components onto printed circuit boards.

Plug the bottom part of a press-forming mould.

Press forming a process that forms pre-heated thicker thermoplastic sheet between two moulds.

Primary research any type of research where you collect new information yourself (for example, through interviews, surveys or observations).

Printed circuit board (PCB) a board that physically supports and electrically connects electronic components using copper tracks and pads on a glass-reinforced polymer board.

Prototype an early sample, model, or release of a product built to test a concept or process.

Quality control a process that ensures every product is manufactured to the same standard.

Rendering the addition of colour, or texture, to enhance a sketch to better communicate design intent.

Secondary research gathering existing data that has already been produced (for example, using books, newspapers, magazines or the internet).

Smart material a material that changes its properties in response to changes in its environment.

Smelting the process of extracting metal from ore.

Standard components components produced in a set size for a range of uses.

Steam bending a method of bending wood by steaming, bending and cooling.

Stock forms set sizes of commercially produced materials.

Strip heater a machine used to produce bends in thermoplastic sheet.

System diagram a diagram that breaks down a system into its three main component parts. More complex systems may have more than one input, one process, and one output.

Template a two-dimensional shape that aids cutting out a shape.

Tolerance the acceptable difference between the upper and lower given size.

Torque the turning force that causes rotation.

Turning a method of making a wood blank round, or producing cylindrical items in metal.

Vacuum forming a process that uses heat to soften a thermoplastic sheet, then sucks it around a mould.

Yoke the top part of a press-forming mould.

INDEX

Index

fibre-reinforced 88
and forces/stresses 88
forming 131–2, 196–8
fractional distillation 108–9
functionality 75
joining 132
marking-out 131, 196, 306
materials management 306–7
modification for specific purpose 130
pigments 57, 130, 155, 217, 326
and pollution 99
printing 217
production aids 196
quality control 200
recycling 98, 108
scales of production 168
selection 75–8, 297
shaping 131–2, 326
and social factors 77
sources/origins 108–9
and specialist technical principles 75–8
and specialist techniques and
    processes 196–201, 314–15, 326–7
and specialist tools and equipment
    314–15
stock forms, types and sizes 155–7
surface treatments and finishes 217–9,
    326
textiles 111
thermoforming 56–7, **57**, 131
thermoplastic 130, 196, 326
thermosetting 57, **57**, 125, 130–1, 196, 297
tolerances 198, 301
using and working with 130–2
working properties 56–8, 130
see also specific polymers
polypropylene (PP) 56, 109, 177
polystyrene, expanded 64
polyurethane foam 322
polyvinyl acetate (PVA) 123
polyvinyl chloride (PVC) 56
see also PVC foam sheet
poppers 159
positive temperature coefficient (PTC) 33
postmodernism 240, 246, 248
powder coatings 216
powders, polymer 156
power stations, fossil fuel 13, 14
PP *see* polypropylene
pre-assembled programmable project
    boards 299
precious metals 296
preservatives, timber 213, 323
press forming 3, 198, **201**
press studs 159
pressing tools 202–3, 316, 327
pressure sensors 33
primary data 228–9
primary research 226–9, **226**

primers 214, 215, 325
printed circuit boards (PCBs) 83
    accuracy 328–9
    cleaning 206
    coatings 221
    drilling 206
    etching 142, 143, 205–6, 208–9, 309,
        328–9
    manufacture 140, 141–2, 205–9, 309,
        328–9
    materials management 308–9
    populating 206
    quality control 208–9
    surface treatments/finishes 329
    surface-mount technology 207–8
printing
    3D printing 10, 200, 292, 319
    digital printing 219
    dye sublimation printers 327–8
    offset lithography 173
    paper/board 210–11
    polymers 217
    relief printing 211
    roller printing 219
    screen printing 211, 217, 219, 327
    stereo lithography printers 200
    sublimation printing 219
    textiles 219
Proban 133
problem solving, through design 233,
    252–3, 254
process devices 33–4, **34**
product analysis/evaluation 229–30
product life cycle 5
product lifespan 11
production
    and new and emerging technologies 10
    see also scales of production
production aids
    electronic/mechanical systems 205
    metals 184–6
    paper/boards 170–2
    polymers 196
    textiles 202
    timber 174–6
productivity 4
programming, microcontrollers 34
propylene 109
prosthetics 24
prototypes 164–5, **164**, 225, 284
    aesthetics 291
    and clients wants/needs 290–1
    demonstrating innovation through 291
    design 289–92
    development 289–92
    evaluation 292
    and functionality 291
    and marketability 292
    and materials 289–90

for the non-exam assessment 346–7
    rapid prototyping 200, 292
    virtual prototypes 290, 291
protractors 171
PSE *see* planed square edges
psychological factors 228
PTC *see* positive temperature coefficient
pulleys 44
pulp, paper 100–1, 102
pulping 100
punk 8, 246
PVA (polyvinyl acetate) 123
PVA glue 172
PVC (polyvinyl chloride) 56
PVC foam sheet 311
Pyrovatex 133
quality control **184**, **208**
    electronic/mechanical systems 208–9
    metals 194
    paper/boards 173–4
    polymers 200
    textiles 204
    timbers 183
Quant, Mary 244, 256
quarter cuts 103–4
questionnaires 227–8, 229
questions
    closed 227
    open 227
quilting 80, 139
rack and pinion mechanisms 42
radio 283
Rams, Dieter 247
rapid prototyping 200, 292
    see also 3D printing
raw edge 61
raw materials 9
    access to 3
    costs 71, 83
    and social factors 71
    and sustainability 5
reciprocating motion 36
recycling 5, 6, 12, 70, 72, 93
    mechanical/electronic systems 84, 99
    metals 97
    paper/boards 74, 96, 100, 102
    polymers 98, 108
    textiles 80, 98
    timber 76, 97
Red and Blue Chair 245
reducing 5, 93
reduction cells 106–7
reforestation 97
reforming 324
refusing 93
registration marks 173
relief printing 211
remote controls 82
rendering **263**, 264, 269–70, 276

Index